Lecture Notes in Energy

Volume 77

Lecture Notes in Energy (LNE) is a series that reports on new developments in the study of energy: from science and engineering to the analysis of energy policy. The series' scope includes but is not limited to, renewable and green energy, nuclear, fossil fuels and carbon capture, energy systems, energy storage and harvesting, batteries and fuel cells, power systems, energy efficiency, energy in buildings, energy policy, as well as energy-related topics in economics, management and transportation. Books published in LNE are original and timely and bridge between advanced textbooks and the forefront of research. Readers of LNE include postgraduate students and non-specialist researchers wishing to gain an accessible introduction to a field of research as well as professionals and researchers with a need for an up-to-date reference book on a well-defined topic. The series publishes single- and multi-authored volumes as well as advanced textbooks. **Indexed in Scopus and EI Compendex** The Springer Energy board welcomes your book proposal. Please get in touch with the series via Anthony Doyle, Executive Editor, Springer (anthony.doyle@springer.com).

More information about this series at http://www.springer.com/series/8874

Hassan Qudrat-Ullah • Muhammad Asif
Editors

Dynamics of Energy, Environment and Economy

A Sustainability Perspective

Springer

Editors
Hassan Qudrat-Ullah
College of Industrial Management
KFUPM
Dhahran, Saudi Arabia

Muhammad Asif
School of Computing, Engineering and
Built Environment
Glasgow Caledonian University
Glasgow, UK

ISSN 2195-1284 ISSN 2195-1292 (electronic)
Lecture Notes in Energy
ISBN 978-3-030-43577-6 ISBN 978-3-030-43578-3 (eBook)
https://doi.org/10.1007/978-3-030-43578-3

This Springer imprint is published by the registered company Springer Nature Switzerland AG
The registered company address is: Gewerbestrasse 11, 6330 Cham, Switzerland

To my children (Ali, Anam, Umer, and Umael) and grand-children (Abdur-Rahman and Khadija)

Hassan Qudrat-Ullah

To my parents

Muhammad Asif

Preface and Acknowledgements

Energy and environmental prosperity are imperative for sustainable development. At the same time, the economies of all countries, and particularly of the developed countries, are dependent on secure supplies of energy. The book addresses the vital and interwoven areas of energy, environment, and economy in the crucial sustainability perspective. It provides comprehensive coverage of the crucial topics covering fundamental theoretical and technical details, empirical data, and case studies taking into account local and international perspectives. The critical sustainability challenges and emerging response and trends especially depleting fossil fuel reserves, global warming and climate change, new energy trends and technologies, and international and global policies and protocols are covered.

The primary aim of this book is to disseminate the roles and applications of various theoretical, modeling, and empirical approaches in dealing with the dynamics of energy, environment, and economy in an integrated manner. The key focus is better energy policy decision making under the constraints of energy–environment–economy interactions and sustainability perspective. This edited book is a compilation of chapters contributed by experts in the broader fields of energy, environment, economy, and sustainable development. The book has drawn a fine representation balance in terms of the three key subject areas: energy, environment and economy. It has also attracted contributions from authors bringing together a great deal of experiential diversity – they have come from countries in Australia, Europe, Middle East, South Asia, North America, and South America. The contributions thus offer a reflection of the respective topics not only from the perspectives of developed and developing nations but also of the economies in transition. The edited book has involved a rigorous review process of the received chapters.

This book contains five parts. Part I, "An Introduction: Dynamics of Energy, Environment, and Economy; A Sustainability Perspective," has one chapter. It presents an overview of all the chapters of this book. Part II of the book, "Dynamics of Energy and Economy," consists of four chapters including (i) *The Framework of Renewable Energy Systems and Trends to Promote the Energy Security of the State of Baja California, (ii) Importance of Energy for Industries and Role of The Energy Sector in Turkey's Economy, (iii) Impacts of Electricity Subsidies Policy on Energy*

Transition, and *(iv) ZEMCH Strategic Framework for Low Carbon Solutions in Sustainable Housing Delivery*. Part III, "Dynamics of Energy and Environments," showcases four unique contributions addressing the challenges of energy–environment dynamics: *(i) Role of Hybrid Energy System in Reducing the Effects of Climate*, *(ii) Sustainable Solar Energy*, *(iii) Coupling Behavior-based Intervention with Pro-environmentalism? The Dynamics of Energy Usage, Crisis, and its Conservation*, and *(iv) Dynamics of the Wind-power Supply Chain to Reduce Emissions: How it Affects Transmission Congestion?* Part IV, "Dynamics of Energy Policy and Climate Change," of this book highlights the role of institutions, energy policies, energy use-behavior, and increasing demand for a better understanding of climate change and energy policy dynamics in a sustainability context. In this category, we have five chapters: *(i) Climate Change and Energy Policies: European Union-Scale Approach to a Global Problem*, *(ii) The Role of Institutions in Energy Policy and Environmental Protection*, *(iii) Access and Limitations to Clean Energy Use in Nigeria*, *(iv) The Nexus of Climate Change and Increasing Demand for Energy: A Policy Deliberation from the Canadian Context*, and *(v) Energy Policy and Climate Change*. The final part V, "Finale," overviews the key insights and learning points as well as the future research possibilities contained in this book in the chapter "Conclusions and Future Research."

We are grateful to the authors of the various chapters for their contributions. It had been a lengthy process from the initial long abstracts to developing the full chapters and then revising them in the light of reviewers' comments. We sincerely acknowledge the authors' willingness to go through this process. We also acknowledge the work and knowledge of the members of our review panel.

Thanks to all the people at Springer US, especially Christopher, HoYing, and Clement with whom we corresponded for advice and facilitation in the production of this book. We wish to record our thanks to Mrs. R. Santhamurthy (SPi Technologies India Private Ltd.) who prepared a camera-ready copy of the manuscript with her usual professionalism, we wish to record our thanks to her. We are grateful to our families for their sacrifice, support, and prayers all along.

Finally, the first author would like the acknowledge the financial support provided by the Deanship of Research at King Fahd University of Petroleum and Minerals (KFUPM) for funding this through project No. BW191001.

Dhahran, Saudi Arabia
Glasgow, United Kingdom
April 2020

Hassan Qudrat-Ullah
Muhammad Asif

Contents

Part I
Introduction

Chapter 1
Introduction: Dynamics of Energy, Environment, and Economy; A Sustainability Perspective

Hassan Qudrat-Ullah

Abstract A better understanding of energy-environment-economy interactions and a genuine appreciation and recognition of the alarming effects of climate change are the essential pre-requisites for any effective energy policymaking in a country. In this context, this book contributes with 13 unique chapters covering three major themes (i) dynamics of energy and economy, (ii) dynamics of energy and environment, and (iii) dynamics of energy policy and climate change. Methodologically, a range of theoretical, mathematical, econometric, empirical, simulation, optimization models and perspectives are used by authors to investigate various issues related to energy-environment,-economy interactions. This chapter presents a systematic pre-view of the content of this book.

Keywords Energy policy · Energy-environment-economy interactions · Econometric models · Electricity supply and demand · Climate change · Environmental emissions · Sustainability · Decision making · Policymaking · Renewable energy · Electricity energy security

1 Introduction

Energy and environmental prosperity are imperative for sustainable development. At the same time, the economies of all countries, and particularly of the developed countries, are dependent on secure supplies of energy. On the other hand rising demand for energy especially with fossil-based supplies poses serious threats and dangers of environmental emissions. Climate change effects add to the complexity of energy policymakers. Unless renewables make a big impact on the supply side of energy, the noble goal of achieving sustainability and sustainable development will

H. Qudrat-Ullah (✉)
School of Administrative Studies, York University, Toronto, ON, Canada
e-mail: hassanq@yorku.ca

H. Qudrat-Ullah, M. Asif (eds.), *Dynamics of Energy, Environment and Economy*, Lecture Notes in Energy 77, https://doi.org/10.1007/978-3-030-43578-3_1

remain elusive. Therefore impetus this book is to address the vital and interwoven areas of energy, environment and economy in the crucial sustainability perspective. Figure 1.1 provides an integrated view of the critical interactions among key variables of the energy sector: energy policy influences the supply and usage of energy, the trajectory of sustainable development activities, and regulatory oversight for the environment. In return, the status of the environment (e.g., level of emissions), energy supply and demand gap, and nature of economic development activities provide critical input to energy policymaking. Likewise, energy policy. Energy usage, sustainable economic development, and environment interactively affect each other.

This book provides comprehensive coverage of these crucial topics covering fundamental technical details, empirical data, case studies taking into account local and international perspectives. The critical sustainability challenges and emerging response and trends especially depleting fossil fuel reserves, global warming and climate change and international and global policies and protocols are covered. The book is written in a reader-friendly manner presenting vital details in the form of text boxes, tables and illustrations.

Unified by the common goal of making better decisions in the sustainable production and consumption of energy, we invited potential authors to submit and showcase their work related to the major theme of this book, "Dynamics of Energy, Environment and Economy". In response to our call, after a double-blind review process, 13 contributions (chapters) authored by well-reputed scholars are presented in this book. Methodologically, a range of theoretical, mathematical, econometric, empirical, simulation, optimization models and perspectives are applied by the scholars in their chapters to investigate the issues relevant to the understanding of the dynamics of energy-environment- economy interactions.

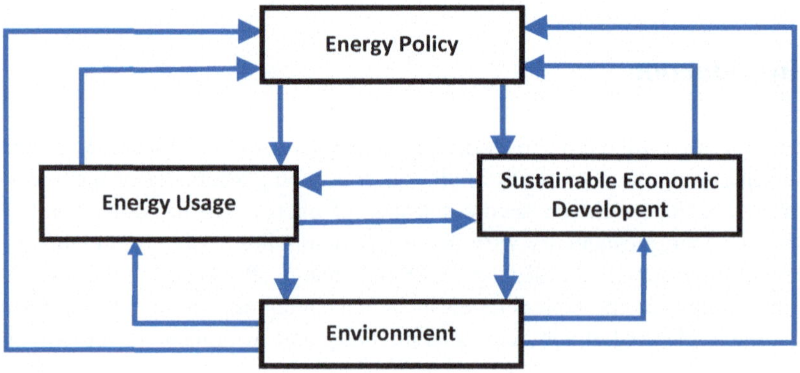

Fig. 1.1 A macro view of energy-environment-economy interactions

1.1 Methodology

In our call for contributions to this volume on "Dynamics of Energy, the Environment, and the Economy: A Sustainability Perspective," we went through various email lists of professional associations IEA, IRENA, IPCC. We also posted the call for chapters at the message boards of a few international conferences on the related topics. Personal invitations were also sent to target authors as well. We received a total of 21 "two-page" long abstracts as the expression of interests. Based on the initial screening by our review panel, the authors of 16 chapters were invited to submit the complete chapters. We received 16 chapters from the contributors that went through a double-blind review process. The reports from the independent reviewers were sent to the authors to address the issues and incorporate the suggestions made by the reviewers. Only 13 chapters made it to the final stage of acceptance. The final versions of these 13 chapters have been edited and included in this volume.

1.2 Research Categories

The chapters thus compiled are classified into five categories following the structure of the book. The first category, the current one, presents the introduction and preview of "Dynamics of Energy, the Environment, and the Economy: A Sustainability Perspective" The second category examines the dynamics of energy and the economy. Four state-of-the-art chapters in this theme include (i) *The Framework ofRenewable EnergySystems and Trends to Promote theEnergy Securityof the State of Baja California, (ii) Importance of Energy for Industries and Role of The Energy Sector in Turkey's Economy, (iii) Impacts of Electricity Subsidies Policy on Energy Transition, and (iv) ZEMCH Strategic Framework for Low Carbon Solutions in Sustainable Housing Delivery.* The third category showcases four unique contributions addressing the challenges of energy-environment dynamics: *(i) Role of Hybrid Energy System in Reducing the Effects of Climate, (ii) Sustainable Solar Energy, (iii) Coupling Behavior-based Intervention with Pro-environmentalism? The Dynamics of Energy Usage, Crisis and its Conservation, and (iv) Dynamics of the Wind-power Supply Chain to Reduce Emissions: How it Affects Transmission Congestion?* The fourth category in this book highlights the role of institutions, energy policies, energy use-behavior, and increasing demand for a better understanding of climate change and energy policy dynamics in a sustainability context. In this category, we have five chapters *(i) Climate Change and Energy Policies:European Union-Scale Approachto a Global Problem (ii) The Role of Institutions in Energy Policy and Environmental Protection, (iii) Access and Limitations to Clean Energy Use in Nigeria, (iv) The Nexus of Climate Change and Increasing Demand for Energy: A Policy Peliberation from the Canadian Context, (v) Energy Policy and Climate Change.*

The final category, Finally, overviews the key insights and learning points as well as the future research avenues contained in this book.

2 Dynamics of Energy and the Economy

2.1 *A Framework of Energy Security Through Renewables*

Energy security and economic development go hand in hand. Lopez-Leyva et al. (this volume) in Chap. 2, "The Framework of Renewable Energy Systems and Trends to Promote the Energy Security of the State of Baja California" presents an analysis and conclusion regarding the role of energy security in the development of Mexico. According to them,

> The concept of security is very important for the economic and social development of any region, state or country. In particular, the concept of energy security is now more important due to the negative effects of climate change and the high dependence on fossil fuels. Thus, the traditional energy security based on the use of fossil fuels is becoming highly questionable and renewable energies seem to be the best option to provide true energy security taking care of the environment and society in general. This chapter presents the current framework of renewable energy systems that impacts the energy performance and availability in the state of Baja California (which is part of the Cali-Baja region, formed by the hypothetical union of the states of California and Baja California) with the objective of promoting the energy security of the state. Finally, an analysis and conclusion regarding the energy security of the state are presented based on the energetic potentialities considering the autochthone renewable energy sources and the current energy regulatory framework of Mexico.

2.2 *Role of the Energy Sector in Turkey's Economy*

Through a comprehensive analysis utilizing the input-out modeling, in Chap. 3, "Importance of Energy for Industries and Role of The Energy Sector in Turkey's Economy", Topcuoglu and Orla (this volume) examined the importance, impact, structure and relationship of the energy industry with other industries for the case of Turkey. They used the input-output model to analyses 2002 and 2012 data. They obtained input coefficients matrix, Leontief matrix and Leontief inverse matrix via the Input-Output tables of 2002 and 2012. They analyzed key sectors in the economy, forward and backward linkage effects and income, output and employment multipliers of all industries with the input-output model. According to the results of their analysis, the energy industry was found to be the key sector in both years and the energy industry has an important place in the Turkish economy. Their key findings are (i) although the output multiplier value in the energy sector is high, employment and income multiplier values are found to be low, (ii) the low value of the income multiplier in the energy industry shows that Turkey is dependent on foreign energy and (iii) the high production multiplier value in the energy industry means that the degree of structural correlation is high.

2.3 Dynamics of Electricity Subsidies Policy on Energy Transition

In Chap.4, "Impacts of Electricity Subsidies Policy on Energy Transition" Dhakouani et al. (this volume) present a critical review and analysis of the impact of electricity subsidies policy energy transition in the case of Tunis. According to the authors of this chapter,

> Energy subsidy programs are socio-economically designed to offer modern and affordable energy, accessible for specific social groups, protect the domestic industry, stimulate economic development and protect the environment. However, in several countries, energy subsidies have deviated from their objectives and become an energy budget burden and a sustainable development barrier. Many questions arise: what are exactly energy subsidies? How are they implemented in a country mechanism? And what are their real effects?

In this chapter authors present a review of energy subsidies: definitions, typologies, measurement approaches and effects. They also examine the reforms to help decision-makers phase out energy transition barriers related to subsidies. They present the case of the Tunisian power system in deeper detail, characterized by a heavy burden of end-users electricity subsidies and an energy transition aiming 30% of renewable energies by 2030, against 3% in 2019. Using a holistic approach, based on hybrid energy systems modeling, they present insights on reforming electricity subsidies and achieving sustainable development. This approach, according to authors, links subsidies, pricing, emissions, demand and supply of the power system through the advanced version of OSeMOSYS. Finally, dynamics between energy, economics and environment are appealed within an integrated analysis of electricity subsidies policy.

2.4 Energy Efficiency as a Framework for Low Carbon Solutions

Low carbon solutions are at the heart of the UN's Sustainable Development Goals. Noguchi (this volume) in this chapter, Chap. 5 "ZEMCH Strategic Framework for Low Carbon Solutions in Sustainable Housing Delivery", presents a strategic framework for low carbon solutions in sustainable housing delivery. According to Noguchi,

> In the light of Sustainable Development Goals acknowledged by the United Nations, homes today need to be socially, economically, environmentally and humanly sustainable. This diverse sustainable housing production challenge necessitates multidisciplinary stakeholders' effective collaboration on the R&D actions. To form the global collaboration platform, the ZEMCH Network was established in 2010. ZEMCH is an acronym of Zero Energy Mass Custom Home which was conceptualized with the aim to function as a new domain for sustainable housing development in global contexts. For the purpose of ZEMCH design, production and marketing knowledge generation, collection and dissemination, the global network initiated and operated the academic conferences, industry-focused technical study

tours and the design training workshop. In consequence, several housebuilders and housing manufacturers, who participated in the knowledge transfer activities, embarked on ZEMCH delivery in their local contexts.

In this chapter Noguchi crystallized ZEMCH strategic framework for low carbon solutions in sustainable housing delivery through reviewing the design, production and marketing innovations applied to ZEMCH practices selected in Japan, Canada and Scotland.

3 Dynamics of Energy and Environment

3.1 On Dealing with the Effects of Climate Change Through Hybrid Energy Systems

Continuing with the theme of dealing with the effects of climate change, Bhattacharjee et al. (this volume) present in their chapter, "Role of Hybrid Energy System in Reducing the Effects of Climate," simulation-based analysis of the role of hybrid energy systems in reducing the effects of climate change. According to them,

> Climate change is a very rising topic nowadays since the climate of this world is changing rapidly day by day. In the technical field, it is seen that so many things or techniques used here, which have a very bad impact on our environment like the use of non-renewable energy sources, emission of greenhouse gases and so on. At present electric power generation is mainly dependent upon non-renewable sources. Due to the rapid uses of non-renewable energy sources, its storage reserves are decreasing rapidly. So an alternate source is required and that is the renewable energy source, nowadays renewable sources are utilized but in a small amount. Renewable sources are environmentally friendly, so using renewable energy sources are more preferable than non-renewable sources for the betterment of our environment. Due to rising environmental concerns day by day, the utilization of renewable energy need to be increased as much as possible. There are so many remote or island places in this world where huge numbers of renewable sources are available which can be used for power generation. And the most important thing is that they have no effect (or very less effect) in this environment. So our goal is to model and simulate a grid-connected solar-wind hybrid energy system which is used to solve the problems regarding power generation.

In this chapter, Chap. 6, the authors present a 24-h case study analysis by taking the real-time data of solar radiation and wind speed of a selected location. The results of this analysis indicate that the hybrid system is profitable and environmentally friendly. Their analysis includes a detailed discussion on climate change, the harmful effects of non-renewable energy sources on the environment and the need for a renewable energy-based hybrid energy system to combat climate change. They hop, by this explanation we will get to know more about how renewable energy sources mitigate two problems – climate change & power demand.

3.2 Solar Energy and Environmental Emissions Reductions

Advancing the theme on "renewable energy and environment", George (this volume) in his chapter, Chap. 7 "Sustainable Solar Energy" delineates in detail the solar technology and its impact on dealing with environmental emissions. According to Geroge,

> Solar energy is one of the most rapidly developing renewable energies in the world. The global solar PV market has been exceeding future market projections and giant untapped markets are only now waking up. For many years proposals have been made about using the Sahara desert as a place where a gigantic solar farm would be able to power the world. But is solar energy always the best solution out there? What can hold it back? What are some major drawbacks it has? Noτ all solar energy applications are created equal. Even amongst certain applications – for example PV- there are numerous technologies that are competing for greater efficiency, lower cost etc. In fact this can go even further, as apart from the more mainstream PV and solar thermal panels, other applications exist such as solar parabolic troughs, solar concentrators, solar desalination, solar ponds etc.

In this chapter Geroge provides an overview of various solar technologies, explains their characteristics and discusses case studies and scenarios for best practice in regards to maximizing energy yield, minimising environmental impacts and making sure that solar energy is used in the most sustainable way possible.

3.3 Energy Use Behavior as a Tool for Pro-environmentalism

Prabatha et al. (this volume) in their chapter, Chap. 8, "Coupling Behavior-based Intervention with Pro-environmentalism? The Dynamics of Energy Usage, Crisis and its Conservation," provides a comprehensive review and analysis of the dynamics of energy use, crises, and its conservation. According to the authors of this chapter,

> Higher cost and crisis of energy vis-à-vis increasing consumption have been a twin but a contested paradox. The rapidly growing energy demand has prompted many countries, including Canada to undertake manifold energy-saving initiatives. However, these are predominantly technology-driven and no apparent measures are taken yet to address and modify the end-users' behavior in the residential sector. In order to reduce the rate of growth of residential energy consumption, it is critical to engage the end-users through better education and awareness while using the inherent pro-environmental behavior (PEB). This is even more critical e.g. for the new immigrants and first nation Canadians. Given this background, this chapter essentially presents (i) the importance of behavior-based, non-technical interventions on end-users' perceptions of energy conservation; and (ii) its impact on the nature of consumption at the household level. Empirical findings from the East and current practices from the West are drawn to investigate these phenomena. The other part sheds light on the prospect and need for behavior-based interventions toward reduced energy consumption. While the time of use (TOU) is in effect, some forms of PEB exists among the residential users in Ontario. With this taken into account, the paper calls for a renewed policy insights on 'investment' and 'curtailment' behavior approaches to assess the 'longevity' effects on energy consumption. This, in turn, stems the foundational need for

'collaborative' think-tank for the multi-disciplinary professionals, including engineers, urban developers, environmentalists, planners, sociologists, economists, and psychologists.

3.4 On Dealing with Environmental Emissions Through Wind-power: A Case Study in Brazil

Promoting renewable wind-power as a mean to deal with the environmental emissions, Herra et al. (this volume) in this Chap. 9, "Dynamics of the wind-power supply chain to reduce emissions: How it affects transmission congestion?," present a case study in the context of Brazil. They posit that delays in new transmission line construction produce drawbacks that can lead to a precarious energy planning. According to them,

> As there is a delay between the licensing process and complete lines, it could create uncertainty as well as transmission congestion. In this context, it is essential to estimate the impact the transmission congestion might have on the reduction of emissions. To assess the effects of power grid congestion on emissions; first, it was necessary to design a simulation model, then some simulation scenarios were provided for analyzing the dynamic behavior of the wind-power supply chain.

In this chapter, they present the findings of the case study conducted in Brazil. Their analysis shows that options to reduce emissions through the current energy policy of Brazil might is affected by delays in transmission construction.

4 Dynamics of Energy Policy and Climate Change

4.1 On Dealing with the Effects of Climate Change through Energy Policies

Adding to the theme of "energy policy and climate change," in Chap. 10, "Climate Change and Energy Policies: European Union-Scale Approach to a Global Problem", Sahin and Ayyldiz (this volume) present an econometric-based analysis of energy resources and climate change. The goal of their research was to examine the relationship between climate change and energy resources within the scope of the European Union by using static panel data method and Root Mean Square Error methodologies for the period of 1990–2018. Specifically, they evaluated the effects of the consumption in nonrenewable and ecological energy resources on CO_2 emissions. Under the related purpose, the variables of CO_2 emissions, which have the greatest impact on the climate change parameter among the greenhouse gases; oil, natural gas, coal, nuclear energy, hydroelectric energy, wind energy, geothermal energy, solar energy and biomass energy-related to energy parameter were used.

According to their findings of the unit effective fixed-effects model; it was concluded that coal, natural gas and oil consumption increased CO_2 emissions while ecological energy consumption decreased CO_2 emissions. Oil consumption was the most influential variable in CO_2 emissions. Root Mean Square Error findings indicate that the variable which has the highest effect on CO_2 emissions is geothermal energy consumption; and the lowest effect variable is the consumption of oil energy.

4.2 *Energy Policy and Environmental Protection in Pakistan*

Asif and Majid (this volume), authors of this Chap. 11, "The Role of Institutions in Energy Policy and Environmental Protection,", makes the case for the role of institutions in shaping the energy use and environmental awareness in a country. They begin with the assertions that the sustainable development of a country largely depends on its energy production and efficient utilization of available natural resources. In this regard, the role of institutions is very much imperative to set the rules of the game. In their chapter, authors discuss (i) the nature of institutions and types of social institutions, (ii) the theories related to the institutions, and (iii) the concept of energy and environmental constraints. They contribute with a case study on the environmental protection act and energy policy of Pakistan.

4.3 *Access and Limitations to Clean Energy Use in Nigeria*

Popoola and Adeleye, in Chap. 12 "Access and Limitations to Clean Energy use in Nigeria" contribute with a review and analysis of various factors and experiences of clean energy usage in Nigeria. They begin by asserting that the energy situation in Nigeria has always been a paradox. According to the authors,

> despite having abundant energy resources in the country, widespread energy poverty is faced by the citizenry: about 60% (74 million) are not served with electricity, while another 94% (171 million) do not have access to clean energy. In a bid to cushion the effect of energy poverty, households and business enterprises in Nigerians relied on the constant use of generators, which is not eco-friendly, is costly and is harmful to human health.

In their study they adopted a think-through thematic methodological analysis, which involves the mapping of the country's potential clean energy sources. Then the thematic literature reviews were integrated to investigate the clean energy experience in the country. Taking into consideration the geopolitical classification of the country, they conducted interviews to examine the energy conditions in the country and the limitation to the maximization of clean energy within their locality, as well as the perception of its acceptability within the country. Their findings show that the main factors limiting the use of clean energy in Nigeria are exorbitant costs of installation and maintenance, inadequate investment in the energy sector;

non- involvement of the private sector, and the subsidies granted to generators of energy from fossils.

4.4 The Nexus of Climate Change and Increasing Demand for Energy

Energy use and its impact on climate change is the subject of this chapter, Chap. 13 "The Nexus of Climate Change and Increasing Demand for Energy: A Policy Deliberation from the Canadian Context." Karunathilake et al. (this volume) discuss this topic by conducting a comprehensive review. According to them,

> Canadian energy demand has been increasing due to population, industrial, and economic growth, and the effects of climate change have gained more visibility. Energy use is a major contributor to anthropogenic climate change. Therefore, global scale energy management strategies are paramount in climate change mitigation. However, the complicated 'marriage' between climate change, energy demand and consumption, and the policy instruments too. Therefore, this paper attempts to study the effect of policy instruments on energy demand and identify other causes behind the demand trends. A comprehensive review of governmental policies assessed the consistency and effectiveness of existing policy instruments. Communication models for the participatory involvement of stakeholders in mitigation initiatives as well as the financial benefits and offsets were critically evaluated. Often, the views of some stakeholder groups, including the individual households and citizens are not successfully captured in policies. There is an apparent gap between the regulatory instruments and policies at the territorial, provincial and local governments. Most stakeholders possess limited knowledge due to missing or partial information about energy demand and the outcomes of various policies.

Authors hope that this paper will trigger scholarly discussion focusing on the dynamics of energy demand and regulatory instruments and policies for climate change mitigation.

4.5 Energy Policy and Climate Change

In the final contribution to the theme, "Energy Policy and Climate Change". Dubey (this volume) presents in his Chap. 14 "Energy Policy and Climate Change" a comparative analysis of several countries' energy policies and their relationship with climate change. Dubey begins his analysis by posing some serious questions. He says,

> Is economic growth and development a zero-sum game vis-à-vis the environment; i.e. must economic welfare always come at the cost of environmental damage? This is one of the most fundamental questions facing humanity today – whether there is a way for both the economy and the environment to flourish, or if we must always make trade-offs between what's best for each. Avoiding knee-jerk reactions wholly in favor of either side is important for society, as evidenced by the disorder experienced in several countries around the world at the hands of self-declared guardians of the environment. The fact is that climate change

> is not limited to a national boundary but is tied to international actions and reactions. While the price an economy pays for a sustainable future is purely a local price, the social costs resulting from environmental damage is shared globally. Hence, even if a government completely capitulated to the environmental extremists' demands, there would still be environmental impacts to mitigate. In addition to developing energy policies to reduce carbon emissions, policymakers must act to ready their economies to deal with climate change. Can energy policies be designed to simultaneously provide social and environmental benefits? The answer is an emphatic yes.

In this chapter then he shares examples of a select subset of countries, reviewing their climate change agendas and targets, in conjunction with supporting environmental policies. Specifically, he provides a systematic assessment of the economic impacts of such measures, and the long-term impacts of such policies on energy supply and demand.

In the final chapter of this book, Chap. 15, Asif (this volume), conclude with key insights gleaned from the showcased contributions in this book. Both the researchers and the practitioner can get a quick glimpse of this book by ready this finally.

5 Concluding Remarks

For this book project, we started our journey in search of "theoretical and empirical underpinnings of the dynamics of energy, environment, and economy within the context of sustainability perspective". In this quest, we are successful in presenting 13 unique contributions in three major themes: (i) dynamics of energy and economy, (ii) dynamics of energy and environment, and (ii) dynamics of energy policy and climates change. With regards to the theme of "dynamics of energy and economy" we have four leading contributions addressing the key issues pertaining to energy security, renewable energy supply, low carbon economy, and sustainable economic development. Also four chapters from renowned scholars of the filed support the theme "dynamics of energy and environment". Finally, the "energy policy and climate change" theme have five equally unique contributions covering the critical aspects of dealing with the effects of climate change. These critical aspects include the effective design of energy policies, enactment of regulatory measures, and the initiation of clean energy initiatives. Consistent with the objective of this volume *"in the context of a global perspective,"* we have six state-of-the-art contributions from Brazil, Mexico, the US, Canada, the Middle East, Pakistan, Australia, and Europe.

It is worth noting that contributions in this volume have deployed a range of theoretical, mathematical, econometric, empirical, simulation, optimization perspectives and models to investigate the critical issues relevant to the integrative nature of sustainability through a better understanding of energy, environment, economy interactions. Also, by utilizing the identified research gaps summarized in the final chapter of this book, researchers in the energy systems domain can continue their research on important issues related to climate change and energy-environment-economy dynamics.

References

Asif M (this volume) Conclusions and future research. In: Qudrat-Ullah H, Asif M (eds) Dynamics of energy, the environment, and the economy: a sustainability perspective. Springer, New York

Asif M, Majid A (this volume) The role of institutions in energy policy and environmental protection. In: Qudrat-Ullah H, Asif M (eds) Dynamics of energy, the environment, and the economy: a sustainability perspective. Springer, New York

Bhattacharjee S, Das U, Chowdhury M, Nandi C (this volume) Role of hybrid energy system in reducing the effects of climate. In: Qudrat-Ullah H, Asif M (eds) Dynamics of energy, the environment, and the economy: a sustainability perspective. Springer, New York

Dhakouani A, Znouda E, Bouden C (this volume) Impacts of electricity subsidies policy on energy transition. In: Qudrat-Ullah H, Asif M (eds) Dynamics of energy, the environment, and the economy: a sustainability perspective. Springer, New York

Dubey K (this volume) Energy policy and climate change. In: Qudrat-Ullah H, Asif M (eds) Dynamics of energy, the environment, and the economy: a sustainability perspective. Springer, New York

George P (this volume) Sustainable solar energy. In: Qudrat-Ullah H, Asif M (eds) Dynamics of energy, the environment, and the economy: a sustainability perspective. Springer, New York

Herrera M, Dyner I, Maldonado M (this volume) ZEMCH strategic framework for low carbon solutions in sustainable housing delivery. In: Qudrat-Ullah H, Asif M (eds) Dynamics of energy, the environment, and the economy: a sustainability perspective. Springer, New York

Karunathilake H, Nahiduzzaman K, Prabatha T, Sadiq R, Alam S, Shah P (this volume) The nexus of climate change and increasing demand for energy: a policy deliberation from the Canadian context. In: Qudrat-Ullah H, Asif M (eds) Dynamics of energy, the environment, and the economy: a sustainability perspective. Springer, New York

López-Leyva J, Arreola-Sevilla R, Álvarez A, Valadez-García A, Ponce-Camacho M (this volume) A framework of renewable energy systems and trends to promote the energy security of the state of Baja California. In: Qudrat-Ullah H, Asif M (eds) Dynamics of energy, the environment, and the economy: a sustainability perspective. Springer, New York

Noguchi M (this volume) ZEMCH strategic framework for low carbon solutions in sustainable housing delivery. In: Qudrat-Ullah H, Asif M (eds) Dynamics of energy, the environment, and the economy: a sustainability perspective. Springer, New York

Popoola A, Adeleye B (this volume) Access and limitations to clean energy use in Nigeria. In: Qudrat-Ullah H, Asif M (eds) Dynamics of energy, the environment, and the economy: a sustainability perspective. Springer, New York

Prabatha T, Nahiduzzaman K, Karunathilake H, Hewage K, Sadiq R, Alam S, Shah P (this volume) Coupling behavior-based intervention with pro-environmentalism? The dynamics of energy usage, crisis and its conservation. In: Qudrat-Ullah H, Asif M (eds) Dynamics of energy, the environment, and the economy: a sustainability perspective. Springer, New York

Sahin G, Ayyldiz F (this volume) Climate change and energy policies: European union-scale approach to a global problem. In: Qudrat-Ullah H, Asif M (eds) Dynamics of energy, the environment, and the economy: a sustainability perspective. Springer, New York

Topcuoglu A, Orla IO (this volume) Importance of energy for industries and role of the energy sector in Turkey's economy. In: Qudrat-Ullah H, Asif M (eds) Dynamics of energy, the environment, and the economy: a sustainability perspective. Springer, New York

Part II
Dynamics of Energy and Economy

Chapter 2
Framework of Renewable Energy Systems and Trends to Promote the Energy Security of the State of Baja California

Josué A. López-Leyva, Jessica Estrada-Lechuga, Efraín A. Mejía-González, Ricardo X. Gómez-López, Roberto J. Arreola-Sevilla, Ariana Talamantes-Álvarez, Alfredo Valadez-García, and Miguel A. Ponce-Camacho

Abstract The concept of security is very important for the economic and social development of any region, state or country. In particular, the concept of energy security is now more important due to the negative effects of climate change and the high dependence on fossil fuels. Thus, the traditional energy security based on the use of fossil fuels is becoming highly questionable and renewable energies seem to be the best option to provide true energy security taking care of the environment and society in general. This chapter presents the current framework of renewable energy systems that impacts the energy performance and availability in the state of Baja California (which is part of the Cali-Baja region, formed by the hypothetical union of the states of California and Baja California) with the objective of promoting the energy security of the state. Finally, an analysis and conclusion regarding the energy security of the state are presented based on the energetic potentialities considering the autochthone renewable energy sources and the current energy regulatory framework of Mexico.

Keywords Energy security · Renewable energy · Energy regulatory framework · Energy industry · Sea energy · Baja California

J. A. López-Leyva (✉) · J. Estrada-Lechuga · E. A. Mejía-González · R. X. Gómez-López
R. J. Arreola-Sevilla · A. Talamantes-Álvarez · A. Valadez-García · M. A. Ponce-Camacho
Center for Innovation and Design (CEID), CETYS University, Ensenada, Mexico
e-mail: josue.lopez@cetys.mx; jessica.estrada@cetys.edu.mx; atenogenes.mejia@cetys.edu.mx; ricardoxavier.gomez@cetys.edu.mx; roberto.arreola@cetys.edu.mx; ariana.talamantes@cetys.edu.mx; alfredo.valadez@cetys.mx; miguel.ponce@cetys.mx

H. Qudrat-Ullah, M. Asif (eds.), *Dynamics of Energy, Environment and Economy*, Lecture Notes in Energy 77, https://doi.org/10.1007/978-3-030-43578-3_2

1 Introduction

Nowadays, energy security is a critical parameter for the social and economic development of any country, state, and region. Also, the use of renewable energy helps and directly supports the conservation of the environment, because the concentrations of diverse pollutants expelled to the environment are reduced. In fact, energy security and the use of renewable energies are factors that increase competitiveness, and not only for countries and regions but also for companies (Bilan et al. 2017; Tvaronavičienė et al. 2015; Nyga-Łukaszewska and Chilimoniuk-Przeździecka 2017; Perrels 2003). However, ideal or near-optimal scenario regarding energy security and renewable energies is not a trivial mission. In order to clarify, the elaboration of the framework for the two concepts mentioned is a challenge for any country, region, and company (Ang et al. 2015; Wang and Zhou 2017; deLlano-Paz et al. 2016). Thus, the California – Baja California Region, also called Cali-Baja region (which is formed of the hypothetical union of the states of California and Baja California) is considered very important for the economic, social and educational development of both sides. Because the state of Baja California is located on the side of California state of the United States of America, there are important challenges and trade-offs that have to be considered in order to increase competitiveness. In fact, the Baja California-California border is the busiest land border crossing around the world. The traffic is not only related to people but with a wide variety of materials, products and services. This means a strong competition between industries and, in some way, between people and governments (Powers 2005; Kozo et al. 2018). In this context, energy security is an important key in order to increase competitiveness based on the independence of resources, with energy being the principal resource used on a large scale by the whole society and industries. Thus, an adequate energy framework based on national policies with the permanent cooperation of society allows increasing regional competitiveness. Particularly, in the state of Baja California, agro-business, fish farmer business and industries highly promote energy security and sustainability to strengthen the regional economy based on the uses of renewable energy systems and good sustainability practices (Melgar 2010; Valenzuela and Qi 2012). In addition, the government also supports the energy framework and energy security using energy from different generation plants, e.g. high-enthalpy geothermal, wind and photovoltaic (Pinti et al. 2019). Considering the aforementioned, this chapter aims to share with the reader the energy framework of the state of Baja California and the current national policy related to sustainability in Mexico. In addition, important details will be shown about our case study based on real data on energy security that is supported by particular users, small, medium and large businesses using Renewable Energy Certificates.

The chapter is organized in the following way. Section 2 describes the general energy demand of the state of Baja California as a starting point of the energy security concept. Next, Sect. 3 presents the current energy policy of Mexico and of the state of Baja California. Section 4 describes the energy security of the state of Baja

California. Sections 5 and 6 show some important information related to the renewable energy systems used in Baja California and mention the state energy potentialities based on sea energy. Next, Sect. 7 presents the relation of the autochthone energy generation and energy security of Baja California in order to increase the competitiveness. Finally, Sect. 8 presents the conclusion.

2 Energy Demand of the State of Baja California

Currently, society has a high dependence on systems that consume electricity as part of daily activities. Thus, each system has particular energy consumption specifications, so that the sum of all the particular consumptions will be known as Energy Demand (ED). In addition, the energy demand of a system is not always the same, it presents certain variations related to the turn-on and turn-off, holiday periods, days off, weather, among others parameters and conditions. Generally, there is a greater energy demand in the summer season, when the use of cooling systems increases due to the intense heat that occurs. In terms of the energy supply used, it is provided by the National Electricity System (NES, or SEN for its acronym in Spanish), and this is composed of the public and private sectors; the public sector is integrated by the infrastructure of the Federal Electricity Commission (FEC, or CFE for its acronym in Spanish), and by the Independent Power Producers (IPP, or PIE for its acronym in Spanish), among which are the permittees of the private sector (Secretaria de Energía 2013). To determine the variation of energy consumption, the National Energy Centre (NEC, or CENACE for its acronym in Spanish) makes the forecasts of maximum energy demand have all available generation resources as a preventive measure. In addition, the Secretary of Energy (SENER) publishes an annual Energy Balance (EB, BNE for its acronym in Spanish), which presents information related to the supply and demand of energy, the energy produced, how much is exchanged with the outside (i.e. export) and how it is allocated to the different sectors and economic agents, this serves as a comparative basis for system variations (Secretaria de Energía 2016a). To elaborate on this type of reports, certain statistical information is considered, which is determined by means of indicators, associated with population growth and economic development, as well as environmental. On the other hand, for the generation of energy a primary source is needed, in 2008 Mexico was considered as the country that uses the highest percentage of oil products as the primary source of energy for the generation of electricity (secondary energy source). As mentioned, the generation of electricity is determined by the NES, which at the national level is organized by nine regions given its operational infrastructure. Its operation is under the responsibility of eight control centers (Secretaria de Energía 2013).

In addition, the country is divided into nine regions to determine a statistic of the electricity market. In 2012, the northwest region, which corresponds to the state of Baja California, had total sales of 28,817 GWh. In particular, the Baja California state had 9797.78 GWh, representing 34% of total sales. In order to clarify, the state of Baja California is categorized in another region, since it is not connected to the national network. Therefore, it can be said that the state of Baja California is an autonomous state since it generates its own electricity; and its operation is administered from the city of Mexicali. Also, the Baja California state has an energy base composed of petroleum derivatives and natural gas as primary energy sources. However, the state does not own these resources natively, for example, natural gas is imported since there is no connection with the National Gas Pipeline System (NGPS, or SNG for its acronym in Spanish). In addition to these sources for the production of electricity, the state of Baja California is one of the most important generators of electricity in the field of renewable energy, mainly using geothermal energy (Secretaria de Energía 2016b). Regarding the electricity generation of the state, there are private and public power plants. In 2008, the installed capacity of electricity generation was 2341 MW, equivalent to 4.6% of the installed capacity in the country. For 2009, it was increased to10 MW, because some wind farms were installed in Mexicali (Muñoz-Meléndez et al. 2012). Also, the power plants are distributed in two sectors, the Coast and Valley sector. The coastal sector is composed of Tijuana, Rosarito, Tecate, and Ensenada, and the valley sector is made up of Mexicali. Table 2.1 shows the infrastructure and generation capacity of these two sectors. By 2010, there were 11 electrical units for Tijuana, Rosarito and Tecate, with a total capacity of 1348 MW. It is worth mentioning that Turbo Gas (TG) units are used as a reserve for any contingency. Ensenada city has 2 TG units. In this sector, there is a total installed capacity of approximately 1400 MW, but considering the effective capacity (without the reserve) there is an amount of approximately 1190 MW. In the case of the Valley sector, there are 13 geothermal energy units in Mexicali, where 5 units are considered for reserve. There is a Combined Cycle Plan (CCC) unit and a TG unit which is for backup. So, the installed capacity of the state is 2700 MW approximately, where the effective capacity is 2416 MW approximately (Nieblas-Ortiz and Quintero-Núñez 2006).

On the other hand, there are also electric power plants located in the city of Mexicali with the main objective of exporting energy to the United States of America. There is a combined cycle unit with a capacity of 310 MW which is managed by the company, *Energía de Baja California*, and the thermoelectric plant of

Table 2.1 Information of coast and valley sectors

Coast Sector		
Locations	Units	Power generation (MW)
Tijuana-Rosarito-Tecate	6 TC	320
	2 CCC	818
	3 TG	210
Ensenada	2 TG	52.43
Valley Sector		
Cerro Prieto	13 GEO	720
Mexicali	1 CCC	506
Mexicali	1 TG	72.5

Mexicali operates a 650 MW combined-cycle unit. The company *Energía Azteca X S. de R.L de C.V* operates part of the La Rosita plant and has a capacity of 750 MW, of which 660 MW is for the FEC and 90 MW is for export. The La Rosita plant belongs to *InterGEn* (North American firm), this plant provides 600 MW that are for export by the private sector. Another company that is in the private sector is Sempra Energy which provides 600 MW, giving a total of 1200 MW between the two companies that represent 40% of exports, both plants are combined cycle and operate with natural gas. Considering the electrical public and private service of the state, there is an installed capacity of 3000 MW, approximately. With these facilities, the electricity sector produces 9156.6 GWh, which represents 5.7% of the total energy produced in Mexico. In addition, the Baja California state has an installed capacity of 166 MW based on wind power plants with an annual generation of 272.6 GWh (Muñoz-Meléndez and López-Vallejo 2018). In order to carry out energy trade in the border with the United States of America, a high power electrical interconnection is used between the states of California and Baja California. This electrical transfer is made through the NES (at the Mexico side) and the Western Electricity Coordinating Council (WECC) at the USA side. This generates a backup for contingencies and the possibility of having an operational reserve margin. However, there is a project to interconnect the Baja California area with the National Interconnected System (NIS, or SIN in Spanish), which consists of the construction, installation, operation, and maintenance of an electric power transmission system with the rest of Mexico. The capacity of this high power electrical interconnection will be 1500 MW, with a length of 700 km and a voltage level of approximately 500 kV. The aforementioned project will allow the sharing of national and state generation resources in the function of the demand, and to address power and energy activities in the NIS and the electric companies of California (WECC). In particular, not all of the aforementioned electrical plants operate 24 h a day. The plants work according to the demand of the population, for this, a cost-benefit analysis is carried out to satisfy the demand in the most efficient way possible. Next, Fig. 2.1 shows the percentage of population and energy consumption for various sectors and localities during the period 1990–2004. In particular, it can be observed that Ensenada has the smallest amount of population (only 15%), while the Tijuana-Rosarito-Tecate region (T-R-T) has the largest population (51%). However, the highest energy consumption is in Mexicali, due to the extreme weather conditions.

It is important to mention that Baja California represents 3.1% of energy users nationwide with a consumption of 4.5%. Specifically, users of the commercial and industrial sectors of the state represent 10%, which have an energy consumption of 56%. However, the domestic sector presents the highest number of users (8%) but has a consumption of 32%. Among the high demand consumers are the industrial and domestic sectors, and those that consume less energy are the commercial, agricultural and public lighting sectors. Regarding the industrial and service sectors, in the period 1996–2004, the following percentages were obtained: Mexicali consumed 46%, Tijuana 42% and Ensenada 12%. For the case of the industry sector was as the following: 44% Mexicali, 48% Tijuana and 8% Ensenada. The high percentage of consumption for Mexicali and Tijuana can be due to manufacturing and

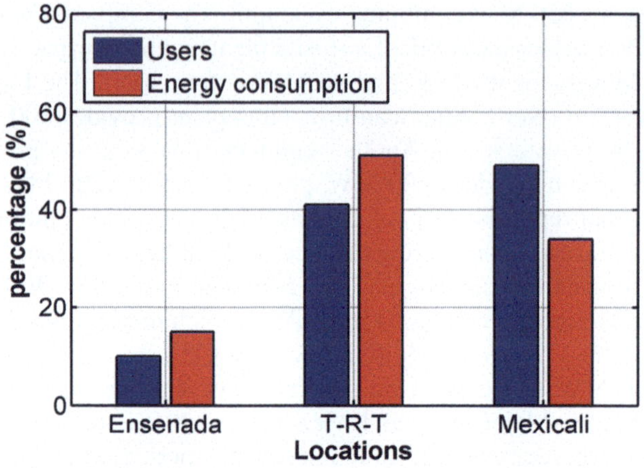

Fig. 2.1 Energy consumption and users for each location. Own figure elaborated using the energy profile of Baja California. (Muñoz-Meléndez and López-Vallejo 2018)

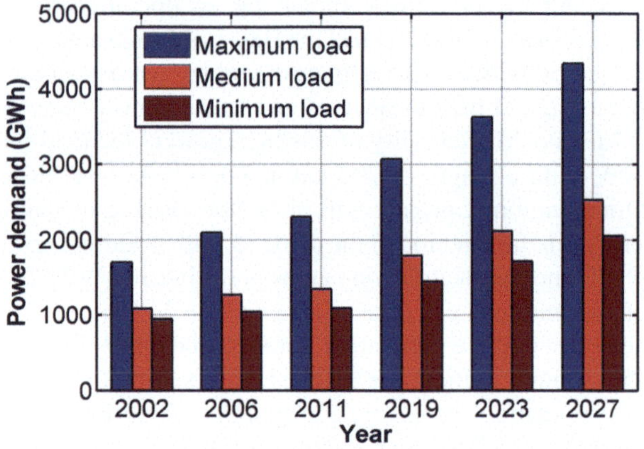

Fig. 2.2 Power consumption projection for Baja California state

assembly companies, production of glass and steel which are companies that require a high energy demand. Considering the aforementioned, Fig. 2.2 represents the power consumption projection for the Baja California state, considering the year 2002 as the starting analysis point, up to 2027 as the final projection point. The maximum, average and minimum electrical loads were calculated based on the information provided by the NEC. It is important to mention that all the electrical loads were increased due to the annual increment considering the economic and population behavior.

In particular, Fig. 2.3 shows the gross power consumption projection for Baja California state, where the gross projection is calculated considering the energy sales, the owner's use consumption, losses, among other parameters. Thus, the gross power projection for 2027 is 23,330 GWh, which imposes an important challenge because the actual capacity is 16,702 GWh, approximately. Considering the

Fig. 2.3 Gross power consumption projection of the Baja California state

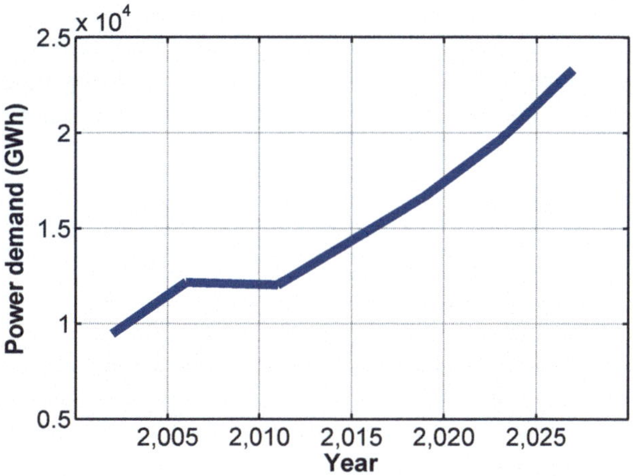

aforementioned, the energy demand of Baja California is well defined and analysed for various sectors and regions.

3 Actual Energy Policy of Mexico

Energy policies are aimed at establishing the commitments that exist to achieve a better energy performance based on energy management processes. Thus, energy policies are agreements, norms, laws, decrees or any parameter established for correct energy management. This means that the energy policy can be referred to the policies of a company to manage the use of energies that should be based on ISO50001, up to the proposed laws in a country with the same objective, work that corresponds to the federal government in the case of Mexico. In Mexico, the energy policy is related to various documents, such as the energy reform, the energy transition law and the new Mexican energy model. Several international agreements and treaties have also been signed since Mexico is always open to this kind of agreements. In total, Mexico collaborates in 68 international instruments, of which 8 are international treaties, 23 specific cooperation instruments and 37 energy cooperation agreements, all thanks to the negotiations that the Secretary of Energy (SENER) has had in recent years (Secretaria de Energía 2018). In addition, there are other agreements related to the mitigation of climate change, the implementation and use of alternative energies and other agreements that lead Mexico to use energy in a global way, that is, complying with international agreements that force to think and do globally and not only locally.

In particular, it is important to analyse the objective of the Mexican Energy Reform, modified for the last time in 2018. According to official information from the Mexican government, the energy reform aims to "detonate the potential of the Mexican economy to create quality jobs, reduce energy costs and increase government revenues to canalize them to social programs, and invest in the human resources required by the country and a long-term savings fund in favor of future generations" (Capetillo Aguirre 2016). To achieve these objectives, the reform proposes the following official and direct arguments and premises (ProMexico 2017; Garcia-Sanchez 2018):

1. Maintain the Nation's ownership over the hydrocarbons found in the subsoil.
2. Modernize and strengthen, without privatizing, the companies Mexican Petroleum (PEMEX, for its acronym Spanish) and the Federal Electricity Commission as State Productive Enterprises, being completely public and Mexican.
3. Reduce the country's exposure to financial, geological and environmental risks in petroleum and natural gas exploration and extraction activities.
4. Allow the Nation to exercise, exclusively, the planning and control of the National Electric System, for the benefit of a competitive system that allows reducing the prices of electric power.
5. Attract more investment into the Mexican energy sector to boost the development of the country and promising energy opportunity for foreign companies.
6. Offer a larger energy supply at better prices.
7. Ensure international energy supply efficiency, quality, and reliability standards, as well as transparency and accountability throughout the different activities of the energy sector.
8. Fight corruption in the energy sector effectively.
9. Strengthen the management of oil revenues and promote long-term saving for the benefit of future generations.
10. Foster socially and environmentally responsible development.

On the other hand, there is also the Energy Transition Law (ETL, or LTE for its acronym in Spanish) which aims to create greater competition in the energy sector to try to diversify the way to obtain energy and involve more companies, looking for conventional companies and users that are interested in generating clean energy. In particular, the LTE aims to be a law with a transversal policy that takes into account various sectors, such as environmental, social and economic, to obtain national energy from different sources, in the most efficient and environmentally friendly manner (Heselhaus 2018). Below, some important aspects are shown as part of the strategy followed for this energy transition to be carried out successfully:

1. Support for energy savers and clean energy producers.
2. Establish medium and long-term goals. Thus, it was established that by 2021, 30% of the electricity would be from clean energy sources, 35% by 2024, 37.7% by 2030, and finally, 50% by 2050.
3. Strengthen public institutions.
4. Promote investments with better offers in the market. Thus, Clean Energy Certificates (CEC, or CEL for its acronym in Spanish) began to be used, which are accreditations that the FEC sells to companies to produce clean energy.

5. Invest in better infrastructures, technologies, and research related to the energy sector.

In general, the ETL bases are: reduce the generation of polluting emissions, gradually increase the use of clean energies, take advantage of the concept of sustainability, and jointly use the various energy and environmental laws (Alfredo Anaya Orozco 2019).

In 2019, with the change of President, a new energy strategy was proposed at a national level based on producing, processing and generating, with the aim of achieving national energy self-sufficiency in the coming years and eliminate the importation of energy (Sordo 2019). Although this new strategy focuses initially on the production of national gasoline, then it will focus on the hydroelectric plants upgrade, which could help the energy sufficiency of the state of Baja California considering the project to interconnect with the National Interconnected System.

On the other hand, each state of Mexico has its own laws related to the generation of energy. For example, Baja California has the Law of Renewable Energy and the Law for Boosting Efficiency. These laws encourage the efficient use of energy, especially for the industrial sector (with support of the Clean Energy Certificates), and propose various sources of clean energy with potential use for the supply of electricity for the public and private sectors. It is important to mention that a CEC covers the generation of 1MWh of clean energy from a user in the industrial sector. However, the exact cost of the CEC is not fixed, since it will depend on the supply and demand that exists in the market. In fact, the purchase and sale related to the CEC are done through the NEC in a conventional manner, but the companies can make their own contracts and bilateral business related to CEC with other companies. In particular, the clean energy options that are allowed for the CEC are the following (Del Razo 2016; Ruiz-Mendoza and Sheinbaum-Pardo 2010):

1. Wind energy.
2. Solar radiation in any of its forms (thermal and photovoltaic).
3. Ocean energy in its different forms.
4. Geothermal systems.
5. Bioenergetics determined by the Law of Promotion and Development of Bioenergetics.
6. The energy generated by the use of the calorific value of the methane and other associated gases in waste disposal sites, livestock farms and in wastewater treatment plants, among others.
7. The energy generated by the use of hydrogen through its combustion or its use in fuel cells, while complying with the minimum efficiency established by the Energy Regulatory Commission (ERC, CRE for its acronym in Spanish) and the emission criteria established by the Ministry of Environment and Natural Resources.
8. Energy generated by hydroelectric plants.
9. Among other important and suitable clean energy options.

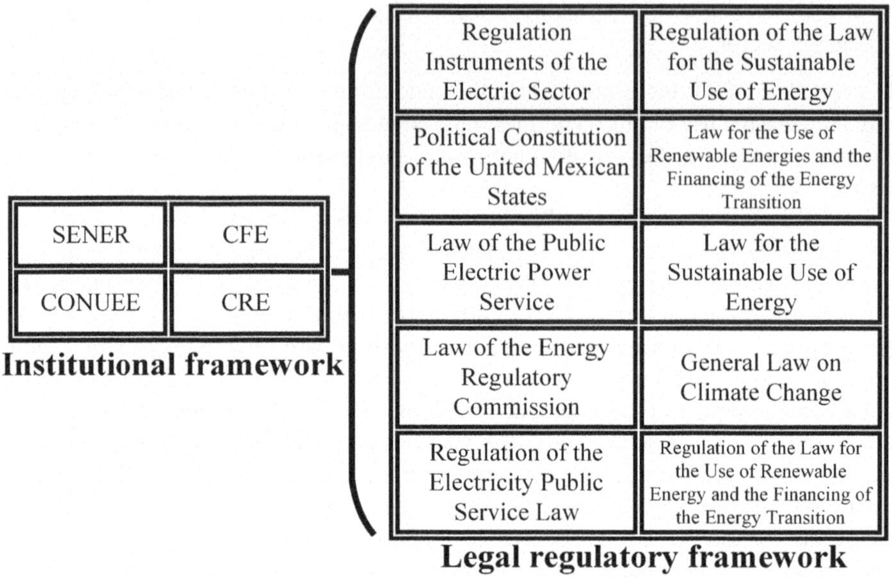

Fig. 2.4 Energy regulatory framework of Mexico. All acronyms are presented in Spanish to facilitate future research work, and some of them were defined in English in the chapter

Figure 2.4 shows the general energy regulatory framework of Mexico, which also has influence over the state of Baja California. In particular, this regulatory framework consists of institutional and legal regulatory frameworks. It is important to clarify that, due to the energy reform and energy transition law of Mexico, new institutions that support energy development have been established. However, Fig. 2.4 shows a general idea of the regulatory framework at the national and state levels.

4 Energy Security of the State of Baja California

The concept of energy security has changed over the years. In 1990, when most of the energy sources depended on oil, the main concern was having prices that could facilitate obtaining oil worldwide. However, there were factors that affected the concept of energy security, such as technological development, the effects on the economy due to globalization, the diversity of energy sources, the change suffered by society at a global level, the economic diversity, the instability of the resources required to produce energy, the influence of society on the decisions of governments that could affect the environment, the effect on the economy caused by terrorism, the availability of food in the world, among others (Winzer 2012). Therefore, the concept of energy security had to be based on certain characteristics based on the mentioned factors. Thus, in general, energy security defines that energy must be available, be accessible, people must be able to afford it and the quality of energy must be acceptable (Herrero 2016).

Therefore, the International Energy Agency (IEA) defines energetic safety as "The uninterrupted availability of energy sources at an affordable price", and this is the definition that is currently used. In addition, energy security has two perspectives: short and long term (Kester 2019; Rosas-Casals et al. 2014). With regard to the short term, there are some important arguments that define it, such as:

1. The acquisition and use of energy must be safe for all users.
2. The energy supply must be generated by different energy sources.
3. The energy supply must be based on a technological resource in order to prosper and promote the development of both economy and society.

To achieve the aforementioned, it is necessary that in the development plan and energy policies there are contracts that can regulate the energy supply and the necessary tools to improve the development of the energy sector, since the tools must react to sudden variations that may occur because of the supply and demand of energy resources, that is, the previous proposes a smart and bidirectional electricity network. Thus, the approach in which the short-term perspective is developed is strictly related to energy diplomacy and the promotion of stable international legal frameworks related to the generation, transmission, distribution and use of energy.

On the other hand, the long-term perspective must ensure that the economic investments that are developed, in order to face a future energy demand, can be efficient through an energy model that can generate a transition towards sustainable energy security to discover and use new techniques to generate energy. In particular, the priority of this perspective is the development of the necessary infrastructure for the generation of clean energy in the face of the challenges of climate change. The above must be achieved without incurring unnecessary expenses or uses. The investments in technological infrastructure, including investments in research, are other actions considered within energy security since it aims to reduce the inactivity of the development sector due to lack of investment and the imbalance related to the costs of energy research. In fact, research projects on energy issues developed in the country contribute enormously to knowing the current state of current energy demand and proposing technological solutions. In addition, Mexico has a broad and diverse potential to generate energy through renewable sources, and as a consequence, it provides part of the increase in energy demand, reducing dependence on hydrocarbons, which implies long-term energy security (Cox 2018; Mohammad et al. 2018).

In the case of the state of Baja California, there are abundant natural and energetic resources (which will be mentioned in the next sections), whose conservation is necessary for local and national development. In fact, in addition to natural and energy resources, and the regulatory framework of Mexico, there are some important laws and regulations in Baja California that support energy security. Thus, considering the aforementioned, Fig. 2.5 shows the basic framework at the state level to ensure energy security through the synergic work between the energy regulatory framework, the energy industry and the market sector.

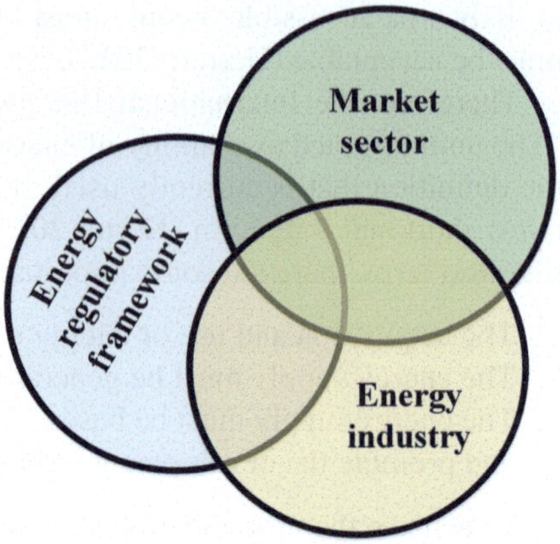

Fig. 2.5 Basic framework in order to ensure the energy security of the state of Baja California

5 Small and Large Scale Renewable Energy Systems in Baja California: Present and Future

According to Fig. 2.5, the energy industry, the energy regulatory framework and the market sector working together can ensure the energy security of Baja California. All of them are implied in the small, medium and large scale renewable energy system as support to energy security. Next, the current situation and analysis of the small, medium and large scale renewable energy systems in Baja California is presented in order to establish the trend in the state. Undoubtedly, the cheapening of energy production equipment, coupled with greater publicity of awareness and savings, has led to more and more people and businesses choose to produce clean energy, not only for their consumption but also for sale. Considering the above, renewable energy producers are classified in the so-called small-scale systems (small and medium scale) and large-scale. In both cases, there is a consumer-producer system also called distributed generation, in which, the public electric network has the capacity to provide and receive electric power (Deming 2012). With respect to small-scale systems, the Federal Electricity Commission in its document, "Model interconnection contract for renewable energy source or small-scale cogeneration system" describes that small-scale systems are those with generating capacity up to 30 kW, at voltages below 1 kV. On the other hand, medium-scale systems are usually formed by a large power generator block, either by wind farms, solar panels, hydroelectric plants, biomass plants. etc., with capacity less than or equal to 500 kW and voltage greater than 1 kV and less than 69 kV. In addition, these systems must be able to operate and remain connected despite fluctuations that do not exceed a range of +5% at −10% of the nominal voltage at the final interconnection point. A very important aspect regarding small and medium scale systems is that they do not require permissions from the Energy Regulatory Commission, so they are known as exempt generators. Finally, high-scale systems are those that have voltages higher than 69 kV and lower than 400 kV (Gómez-Hernández et al. 2019).

5.1 *Small and Medium Scale Energy Systems*

According to the reports of the Energy Regulatory Commission, in 2015 there were 16,979 small and medium-scale contracts for distributed generation, and in 2018, there were 94,893 contracts nationwide, which are divided into 34 contracts to generate electricity using biofuel, 15 contracts using wind energy, and 94,844 contracts for photovoltaic generation. In fact, by the end of 2016, in Mexico, there were a total of 172 companies dedicated to the generation of energy through solar radiation (Comisión Reguladora de Energía 2019). In addition, 75% of the total of small and medium-scale interconnection contracts were signed by users of the residential sector and 20% by users of the commercial sector (Comisión Reguladora de Energía 2017). Thus, from 2011 to 2016, there was a 121% growth in cumulative installed capacity (kW), which is why an increase in the number of contracts is predicted, estimating approximately 732,957 contracts for the year 2023 (Comisión Reguladora de Energía 2019). The growth trend is clear. At this point, society has a greater knowledge of the production of renewable energies, and service installers of solar panels and wind turbines, have greater ability and capabilities to offer the user credit options for the acquisition of equipment, coupled with a large increase in companies that commercialize products for the production of clean energy. In Baja California, at the end of 2018, there were a total of 4348 small and medium-scale contracts, with an installed capacity of 20.91 MW. In order to compare, the state of Jalisco (also in Mexico) has the largest number of contracts at a national level, 17,097 contracts with an installed capacity of 88.86 MW. Respect to the trend of small and medium-scale energy systems in Baja California, the installation of renewable energy systems in the commercial (i.e. micro, small and medium enterprises) and residential have been increased. On the other hand, the educational sector is also carrying out important actions regarding the generation of energy for its own use based on renewable resources. The main reason for that is the national and international regulations related to accreditations and sustainability indicators of educational centers, such as universities and research centers. In fact, apparently, if the small and medium energy systems are well distributed in a region (e.g. city, county, etc.), considering the maximized use at roofs and open areas, the potentiality of energy generation is extraordinary. Considering the mentioned, the energy policy at the state level has been promoting the installation of such systems through tax incentives to businessmen and domestic users.

5.2 *Large Scale Energy Systems*

As was mentioned in Table 2.1, the state of Baja California has an important generation capacity which is distributed in the Coast and Valley sectors. In addition, some important implemented and future projects are also present. For example, the eolic farms, called Rumorosa I and II provide a maximum of 60 MW. Also, the eolic farm, called Energia Sierra Juarez provide a maximum of 155 MW. In addition, the combined cycle power plant, called Baja California III, generates a maximum of 324 MW. Finally, the project Rumorosa Solar proposes a maximum power generation of 53 MW. Thus, the tendency of the state of Baja California regarding the generation of energy is to increase the installed capacity, which has as its first objective the energy supply to all the needs of the users at the state level, and second, to strengthen the market sector for the export of energy. However, neither at the state nor at the national level, the concept of generalized energy storage to facilitate the integration of large-scale heterogeneous flexible resources with energy storage capacity to participate in multiple markets is not a trend. In fact, this an issue that when resolved, will drastically increase the energy efficiency of the state of Baja California (Yao et al. 2019). Regarding the trend of large-scale energy systems, there is a clear difference with the small and medium-scale systems. Mainly, large-scale systems require large investments for the generation of energy in a specific place, regularly away from the main energy users. The above means that the private and public energy industry, the market sector and the energy regulatory frameworks at the national level have to works together in order to minimize the risk of the investment, and maximize the profit, which means a relationship with the energy security. For that, the development of large scale energy systems requires more care and usually are designed for long terms. Thus, although the state of Baja California has important advantages to produce energy with its own natural resources (solar energy, wind energy, sea energy and geothermal energy) and increase installed capacity, current on-working large-scale projects are considered sufficient in the short and medium terms.

6 Sea Energy Renewable Systems: Potentialities of Baja California

The water is a vital and abundant resource, that divides the continents of the planet. Thus, the oceans are the example of giant masses of water, in fact, each continent has the proximity of the sea that contains water bodies in smaller portions. The water is also found in areas within the mainland, and it can descend from the mountains through streams and form rivers. In general, the characteristic of the water is that it is generally in motion and, therefore, generates a force that can be harnessed to generate energy as a possible solution to the growing energy demand (Bankes and Trevisanut 2015). According to the report on the world situation of renewable energies in 2015, the technologies where water as a resource is involved are those that use waves and tides as a source of energy production. In fact, the use of waves and tides (waves and tides power systems) are the natural options that have generated most technological advances at the moment (Skeer and Leme 2018). In the case of Mexico, if the sea energy is used effectively, it can become one of the main clean energy-producing technologies, such as the case of Russia, which positioned itself in 2011 as the country with the greatest installed power using sea energy renewable systems, 87,400 MW, with which it is capable of an average annual production of 190,000 GWh (Quintero-González and Quintero-González 2016).

According to the Secretary of Energy in Mexico, the global supply of sea energy is estimated at 82,950 TWh, approximately, during the period of a year, of which the ocean thermal energy (based on the thermal gradient) contribute with 53%, Marine current power with 36%, tidal and wave energy with 9%, and osmotic power (based on salinity gradient) with 2% (Secretaria de Energía 2013; Anon 2003). During 2009, the Autonomous University of Mexico conducted a series of simulated studies to estimate the potential of sea energy in Baja California. The results were amazing, considering several simulated hydroelectric turbines, an installed capacity of 20,000 MW was achieved, which generated 5500 GWh/year. However, this research only considers a single energy generation option using water from the oceans, which implies that using more options, the power generation capacity would have to increase considerably (Hiriart Le Bert 2009).

Thus, the potential of the state of Baja California to produce sea energy has attracted foreign companies to carry out studies and analysis in order to develop renewable energy technologies, such as the Finnish company, AW-Energy Oy, which is the world leader in the development of the use of waves to generate energy. This interest has also existed to collaborate with the Mexican renewable energy company ENAL (Alternative Energies, Studies and Projects), which states that they aim to perform the steps and processes necessary to make a contract that gives way to an international transaction or business, even if it does not have a commitment effect in reality. In particular, these companies have shared the interest of designing and developing a wave park with a capacity of 10 MW on the Mexican coasts of the Pacific Ocean (Instituto Mexicano del Petróleo 2017).

The National Autonomous University of Mexico (UNAM for its acronyms in Spanish) in 2017, conducted a study in the Gulf of Mexico, the Caribbean Sea and the Gulf of California, in order to analyze the conditions and characteristics of the swell in these areas. In general, the upper part of the Gulf of California has the potential to produce energy using wave motion in the range of 38–41 kW/m, while in the Gulf of Mexico and the Caribbean Sea, the energy potential is 6 kW/m (see Fig. 2.6). These mentioned Mexican energy potentials are very different from the great potentials in other regions in the world, for example, Australia or New Zealand can reach an energy potential of 100 kW/m, and in the Atlantic Coast, the energy potential is 30–75 kW/m. In this way, the energy potential of sea energy world-wide is in the range of 2000–4000 GW, in the same way, with current technology only 2% of this potential can be exploited (Instituto Mexicano del Petróleo 2017; Calero and Viteri 2013). In addition, because the state of Baja California has a wide extension of coasts, approximately 1380 km, and based on the aforementioned, the energy potential of the sea energy is a highly suitable option to promote the energy security of the state. In fact, according to our research, there are not real sea energy projects with real energy and financial potential for the state of Baja California.

Fig. 2.6 Sea energy potential of the state of Baja California and Mexico. This map was adapted using the open-access map licensed under Creative Commons Attribution 4.0. (Source: National Institute of Statistics and Geography of Mexico)

7 Relation of the Autochthone Energy Generation and Energy Security of Baja California

As previously mentioned, the state of Baja California has important options for the energy generation based on autochthone renewable energy resources such as the geothermal energy, sea energy, among others. In fact, the actual and growing synergy between renewable energies and energy security without completely considering the traditional energy security based on fossil fuel is the trend throughout the world (Valentine 2011; Nie and Yang 2016). This means that the state of Baja California also has to propose a strong relationship between its autochthone renewable energy resources and its energy security. Regarding wind energy, Fig. 2.7 shows some relevant locations with high wind power in Baja California. In particular, most of these places are located in mountainous regions at high altitudes, i.e. the Juarez mountain range is at 1700 m over the sea level, San Pedro Martir mountain range is at 3096 m and San Miguel mountain range is at 2100 m. All the locations mentioned present a high wind speed, more than 8.8 m/s, which means a potential wind power generation more than 800 W/m^2. Also, Fig. 2.8 presents some important locations with high solar power in the state of Baja California. In particular, the minimum and maximum solar radiations are 2200 and 2300 kWh/m^2, respectively.

In fact, any project related to power generation using renewable energy located at these locations, firstly, will require deep analysis to propose countermeasure for many problems and difficult situation to generate and transport the energy from remote location to the final user. Otherwise, if the concept of energy security is not fully understood, or the correct application of the elements of the concept of energy

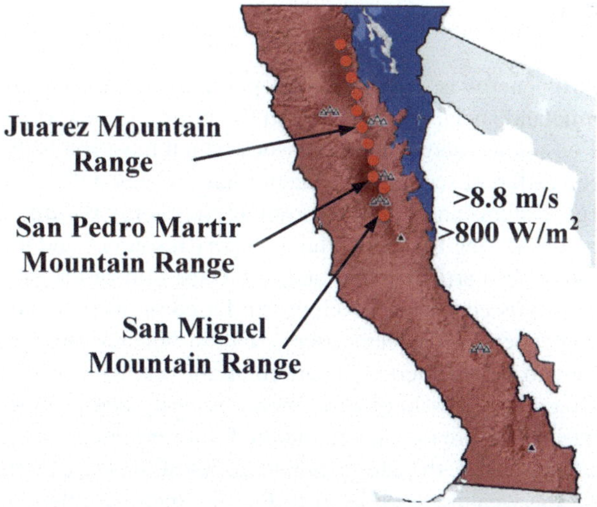

Fig. 2.7 Locations with high wind power in the state of Baja California. This map was adapted using the open-access map licensed under Creative Commons Attribution 4.0. (Source: National Institute of Statistics and Geography of Mexico)

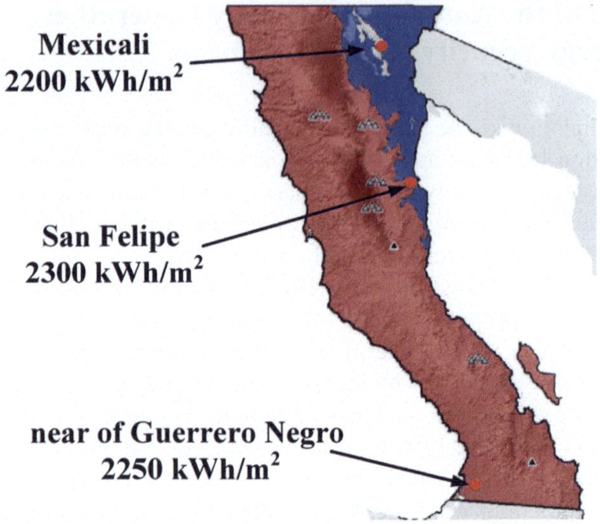

Fig. 2.8 Locations with high solar power in the state of Baja California. This map was adapted using the open-access map licensed under Creative Commons Attribution 4.0. (Source: National Institute of Statistics and Geography of Mexico)

security is not applied using the highest quality standards, the risk of failure to achieve energy security is very high.

8 Conclusion

Nowadays, security energy is a relevant concept for all countries in order to guarantee their economic and social development in the future. In fact, many social problems are based on many energy security issues which have to resolve through the synergic between different sector. This chapter has presented the framework of the renewable energy systems and trends to promote the energy security of the state of Baja California considering the autochthone energy resources and the national and state energy policy. In particular, the state of Baja California has an important potential to enhance its economic development based on its own natural resources to produce enough energy to achieve energy self-sufficiency and be a very important state company to export energy. In particular, the optimal use of energy options in Baja California guarantees long-term energy security and propose a social and economic scenario to increase and maintain future-oriented jobs, among other important advantages. Thus, the energy potentialities of the Baja California state are present, but also the weaknesses related to the international relation (i.e. geopolitics). In fact, a negative combination of energy security levels and geopolitics of Mexico (and Baja California state) can disrupt regional stability. But, under other conditions, it can promote sustainable competitiveness based on energy security. In this case, Mexican internal policy must strengthen the various processes that impact national energy security. Also, the results and information presented in this chapter should be used by different sectors (e.g. academic and scientific) in order to improve and propose academic and scientific programs with a direct impact on the social, public and private sectors. However, there is always a risk that does not allow achieving energy security, for example, distorted markets and non-clear regulatory frameworks. For these reasons, governance, governability and social participation are highly important and relevant concepts and activities, in order to broad political consensus related to the energy security of any nation, region or state. Finally, an energy project at the state and national level should not only consider energy security but also the sustainability of the project based on the dynamic cooperation among the different sectors, mainly the universities and academic centers to promote the developing of the energy clusters.

Terminologies

Energy security: it has been defined by the United Nations as "the continuous availability of energy in varied forms, in sufficient quantities, and at affordable prices".

Renewable energy: it is the energy from sources (e.g. biomass, hydropower, geothermal, wind, solar, sea, among others) that are naturally replenishing but flow-limited; renewable resources are virtually inexhaustible in duration but limited in the amount of energy that is available per unit of time.

Energy regulatory framework: related to the model of regulation, politics, laws, among other elements related to the energy sector of any country, state or region.

Sea energy: refers to the energy carried by ocean waves, tides, salinity, and ocean temperature differences. In particular, it is related to all forms of renewable energy derived from the sea.

Baja California: state localized at the northwest of Mexico and at the side of the state of California of United States of America.

Energy industry: it is a category of stocks that relate to producing or supplying energy. The energy industry includes companies involved in the exploration and development of conventional and renewable energies.

References

Alfredo Anaya Orozco (2019) Visiones Políticas para la Transición Energética en México rumbo al 2030. Comisión de Energía de la Cámara de Diputados. https://www.cemda.org.mx/visiones-politicas-para-la-transicion-energetica-en-mexico-rumbo-al-2030/. Accessed 7 June 2019

Ang BW, Choong WL, Ng TS (2015) A framework for evaluating Singapore's energy security. Appl Energy 148:314–325. https://doi.org/10.1016/j.apenergy.2015.03.088

Anon (2003) An energy overview of Mexico. https://doi.org/10.2172/821399. Accessed 10 June 2019

Bankes N, Trevisanut S (2015) Energy from the sea: introduction. In: Bankes N, Trevisanut S (eds) Energy from the sea. Brill, Nijhoff, Leiden, pp 1–12

Bilan Y, Strielkowski W, Karbach R et al (2017) Secure development of country and competitiveness issues: case of Germany's energy security. J Secur Sustain Issues 6:329–342. https://doi.org/10.9770/jssi.2017.6.3(1)
Calero R, Viteri D (2013) Energía Undimotriz, alternativa para la producción de Energía Eléctrica en la Provincia de Santa Elena. Rev Cient Tecn UPSE 1:1–12. https://doi.org/10.26423/rctu.v1i2.15
Capetillo Aguirre H (2016) Country nuclear power profiles. https://www-pub.iaea.org/MTCD/Publications/PDF/cnpp2016/countryprofiles/Mexico/Mexico.htm. Accessed 12 June 2019
Comisión Reguladora de Energía (2017) Potenciando la Generación Distribuida en México: Nuevos Instrumentos de Regulación. https://www.gob.mx/cms/uploads/attachment/file/181534/C_Marcelino_Madrigal_Paneles_10Enero_Vf.pdf. Accessed 2 July 2019
Comisión Reguladora de Energía (2019) Evolución de Contratos de Pequeña y Mediana Escala/Generación Distribuida. https://www.gob.mx/cms/uploads/attachment/file/483322/Estadisticas_GD_2019-1.pdf. Accessed 5 Dec 2019
Cox E (2018) Assessing long-term energy security: the case of electricity in the United Kingdom. Renew Sustain Energ Rev 82:2287–2299. https://doi.org/10.1016/j.rser.2017.08.084
Del Razo C (2016) A snapshot of the Mexican clean energy obligations system. Mex Law Rev 9:81–90. https://doi.org/10.1016/j.mexlaw.2016.09.004
deLlano-Paz F, Martínez Fernandez P, Soares I (2016) Addressing 2030 EU policy framework for energy and climate: cost, risk and energy security issues. Energy 115:1347–1360. https://doi.org/10.1016/j.energy.2016.01.068
Deming WE (2012) Integrating distributed generation, CHP and alternative energy. Distrib Gener Altern Energy J 27:5–6. https://doi.org/10.1080/21563306.2012.10531120
Garcia-Sanchez G (2018) The fine print of the Mexican energy reform. Soc Sci Res Netw Electron J 36. https://doi.org/10.2139/ssrn.3189024
Gómez-Hernández DF, Domenech B, Moreira J et al (2019) Comparative evaluation of rural electrification project plans: a case study in Mexico. Energy Policy 129:23–33. https://doi.org/10.1016/j.enpol.2019.02.004
Herrero R (2016) La seguridad energética y la estrategia global de seguridad de la unión europea. UNISCI J 42:83–98. https://doi.org/10.5209/runi.53788
Heselhaus S (2018) Energy transition law and economics. In: Mathis K, Huber B (eds) Energy law and economics. Economic analysis of law in European legal scholarship, vol 5. Springer, Cham, pp 19–41
Hiriart Le Bert G (2009) Potencial energético de las mareas para generar electricidad. Rev Dig Univ 10:1–10
Instituto Mexicano del Petróleo (2017) Reporte de inteligencia tecnológica energía del océano https://www.gob.mx/cms/uploads/attachment/file/279710/IT_OCEANO_VERSION_311017.pdf. Accessed 2 July 2019
Kester J (2019) The energy security paradox: rethinking energy (in) security in the United States and China. Int Aff 95:230–231. https://doi.org/10.1093/ia/iiy244
Kozo J, Zapata-Garibay R, Rangel-Gomez MG et al (2018) The border health consortium of the Californias—forming a binational (California–Baja California) entity to address the health of a border region: a case study. Front Public Health 5:1–6. https://doi.org/10.3389/fpubh.2017.00368
Melgar L (2010) Energy transition: a path toward sustainable development for Mexico. Latin Am Policy 1:98–113. https://doi.org/10.1111/j.2041-7373.2010.00006.x
Mohammad A, Reza H, Bilal E et al (2018) Long-term policy evaluation: application of a new robust decision framework for Iran's energy exports security. Energy 157:914–931. https://doi.org/10.1016/j.energy.2018.05.176
Muñoz-Meléndez G, López-Vallejo OM (2018) Building a transregional governance architecture for electricity integration in the Baja California–California border region: the cases of Tijuana and Mexicali. Latin Am Policy 9:349–372. https://doi.org/10.1111/lamp.12148

Muñoz-Meléndez G, Díaz-González E, Campbell Ramírez HE et al (2012) Baja California: Perfil Energético 2010–2020. http://www.investinbaja.gob.mx/docs/es/infraestructura/servicios/electricidad/perfil-energeticoBC2010-2020.pdf. Accessed 1 July 2019
Nie P, Yang Y (2016) Renewable energy strategies and energy security. J Renew Sustain Energy 8:065903. https://doi.org/10.1063/1.4967714
Nieblas-Ortiz EC, Quintero-Núñez M (2006) Gestión ambiental transfronteriza para la generación eléctrica en la región California, Estados Unidos-Baja California, México. Reg Soc 18:3–35
Nyga-Łukaszewska H, Chilimoniuk-Przeździecka E (2017) Modelling energy security and international competitiveness: the export perspective. Entrep Bus Econ Rev 5:71–84. https://doi.org/10.15678/eber.2017.050204
Perrels A (2003) Reconciling competitiveness and environmental objectives. Appl Energy 76:75–87. https://doi.org/10.1016/s0306-2619(03)00049-7
Pinti DL, Castro M, López-Hernández A et al (2019) Cerro Prieto geothermal field (Baja California, Mexico) – a fossil system? Insights from a noble gas study. J Volcanol Geotherm Res 371:32–45. https://doi.org/10.1016/j.jvolgeores.2018.12.010
Powers B (2005) Energy, the environment, and the California – Baja California border region. Electr J 18:77–84. https://doi.org/10.1016/j.tej.2005.06.001
ProMexico (2017) Mexico as an ally of the world leaders in industrial innovation. http://mim.promexico.gob.mx/work/models/mim/templates-new/Publicaciones/Estudios/Mexico-ally-world-leaders-industrial-innovation.pdf. Accessed 10 June 2019
Quintero-González JR, Quintero-González LE (2016) Sistemas de producción y potencial energético de la energía mareomotriz. Rev Ing Invest Des 16:39–45. https://doi.org/10.19053/1900771X.4078
Rosas-Casals M, Marzo M, Salas-Prat P (2014) Sovereignty, robustness, and short-term energy security levels. The Catalonia case study. Front Energy Res 2:1–10. https://doi.org/10.3389/fenrg.2014.00016
Ruiz-Mendoza BJ, Sheinbaum-Pardo C (2010) Mexican renewable electricity law. Renew Energy 35:674–678. https://doi.org/10.1016/j.renene.2009.08.014
Secretaria de Energía (2013) Prospectivas del sector eléctrico 2013–2027. https://www.gob.mx/cms/uploads/attachment/file/62949/Prospectiva_del_Sector_El_ctrico_2013-2027.pdf. Accessed 7 June 2019
Secretaria de Energía (2016a) Balance Nacional de Energía 2015. https://www.gob.mx/cms/uploads/attachment/file/248570/Balance_Nacional_de_Energ_a_2015__2_.pdf. Accessed 6 June 2019
Secretaria de Energía (2016b) Prospectivas de Energías Renovables 2016–2030. https://www.gob.mx/cms/uploads/attachment/file/177622/Prospectiva_de_Energ_as_Renovables_2016-2030.pdf. Accessed 10 June 2019
Secretaria de Energía (2018) Reporte de avances de energías limpias, Primer Semestre 2018, México 2018. https://www.gob.mx/cms/uploads/attachment/file/418391/RAEL_Primer_Semestre_2018.pdf. Accessed 2 July 2019
Skeer J, Leme R (2018) Renewable energy in the energy future. In: Rachid S, Prabhu P (eds) Agriculture & food systems to 2050. World Scientific, pp 473–502
Sordo AM (2019) Exploration and exploitation. In: Wionczek MS (ed) Energy policy in Mexico, 1st edn. Routledge, New York, pp 198–246
Tvaronavičienė M, Mačiulis A, Lankauskienė T et al (2015) Energy security and sustainable competitiveness of industry development. Econ Res 28:502–515. https://doi.org/10.1080/1331677x.2015.1082435
Valentine SV (2011) Emerging symbiosis: renewable energy and energy security. Renew Sust Energ Rev 15:4572–4578. https://doi.org/10.1016/j.rser.2011.07.095
Valenzuela JM, Qi Y (2012) Framing energy efficiency and renewable energy policies: an international comparison between Mexico and China. Energy Policy 51:128–137. https://doi.org/10.1016/j.enpol.2012.03.083

Wang Q, Zhou K (2017) A framework for evaluating global national energy security. Appl Energy 188:19–31. https://doi.org/10.1016/j.apenergy.2016.11.116

Winzer C (2012) Conceptualizing energy security. Energy Policy 46:36–48. https://doi.org/10.1016/j.enpol.2012.02.067

Yao Y, Zhang P, Chen S (2019) Aggregating large-scale generalized energy storages to participate in the energy and regulation market. Energies 12:1024–1046. https://doi.org/10.3390/en12061024

Chapter 3
Importance of Energy for Industries and Role of the Energy Sector in Turkey's Economy

Abdullah Topcuoğlu and I. Orkun Oral

Abstract It is aimed to examine importance, impact, structure and relationship of the energy industry with other industries by using input-output analysis for Turkey. In this context, input-output model is used to analyses 2002 and 2012 data. Input coefficients matrix, Leontief matrix and Leontief inverse matrix were obtained via the Input-Output tables of 2002 and 2012. Key sectors in the economy, forward and backward linkage effects and income, output and employment multipliers of all industries were analyzed with the input – output model. According to the results of the analysis, the energy industry was found to be the key sector in both years. This shows that the energy industry has an important place in Turkish economy. Although the output multiplier value in the energy sector is high, employment and income multiplier values are found to be low. The low value of the income multiplier in the energy industry shows that Turkey is dependent on foreign energy. In addition, the high production multiplier value in the energy industry means that the degree of structural correlation is high.

Keywords Input-output model · Key sectors · Employment multiplier analysis · Output multiplier analysis · Income multiplier analysis · Energy sector · Turkish economy

A. Topcuoğlu (✉)
Department of Economics, Ardahan University, Ardahan, Turkey
e-mail: abdullahtopcuoglu@ardahan.edu.tr

I. O. Oral
Bilecik Seyh Edebali University, Bilecik, Turkey
e-mail: ibrahimorkunoral@ardahan.edu.tr

H. Qudrat-Ullah, M. Asif (eds.), *Dynamics of Energy, Environment and Economy*, Lecture Notes in Energy 77, https://doi.org/10.1007/978-3-030-43578-3_3

1 Introduction

Energy is one of the main input of all production sectors. Energy has a substantial role on the economic and social development processes for all countries. The reason for this importance of energy comes from the structural relations with other sectors of the economy as an input in the production process. Energy is an important factor in determining the level of development for countries and in the decision-making process of their international policies.

From an economic perspective, it can be stated that energy is directly or indirectly related to all goods and services produced in an economy. In this respect, the relationship between production and energy leads to the relationship between growth and energy. Thus it is a known fact that countries need energy intensively to ensure the economic development.

The need for energy increases parallel to development level of countries. From this perspective, it can be said that the relationship between energy demand and economic growth in developing countries is weaker than in developed countries. However, investments in modern and renewable energy sources are generally low in developing countries due to low income and limited resources.

The energy sector is of great importance for Turkey as a developing country. There is an increasing trend in energy demand with parallel to growth process in Turkey and it increases the importance of energy supply. The Turkish economy is becoming more and more dependent on external energy sources each year due to the lack of domestic energy production. There are various reasons for foreign energy dependency as; insufficient local resources, insufficient production and intensive use of fossil-based fuels. Oil and natural gas resources are main exported energy supply from foreign countries (Yılmaz et al. 2016). Although Turkey is an externally dependent country in terms of non-renewable energy sources, it has been trying to meet the energy demand by focusing on renewable energy sources especially in recent years. Turkey also follow the global trends on increasing renewable energy sources. Thus, Turkey will reduce dependence on foreign energy through investments in renewable energy sources made by both government and private sector.

Since energy has a significant share in the production process of all goods and services in the economy, determining the impact of industrial energy consumption on the economy will be a key performance indicator for a country. In this context, input-output analysis is an important tool to determine the effect of energy consumption on economy and industry, since it enables to examine all industries in the economy.

The basic views and assumptions of the Input-Output model form the basis of many different econometric analyzes. Therefore, one of the most used econometric analysis methods is the Input-Output model (Baumol 2000; Grealis et al. 2017). Input-Output analysis provides insight into the understanding of economic relations between the Government, economic actors, industries and households and has an important role in this respect. It is also an important tool for economic decision-makers, both in the process of analyzing and analyzing the past and in preparing future plans. (Burrows and Botha 2013).

It is aimed to determine the importance and current status of energy sector in Turkey by using input-output model. Therefore, Input-Output tables for 2002 and 2012 will be included in the analysis. Results provide significant knowledge on importance of energy and current status of energy industry. In addition, key sectors, forward and backward linkage coefficients and income, output and employment multipliers of all industries are analyzed via the input – output model. Moreover, both the status of the energy sector and the relations of other sectors with the energy sector are determined. Results from the data analysis provide an important guidance for both private sector and public administration in case of energy planning.

The remain part of this study is organized as follows: Section 2 provides information on the energy sector and its position in the Turkey economy; Section 3 describes the research methodology, literature research and Input-Output model used in this study; Section 4 comprise of discussion on the empirical results and explain them; and finally, conclusion is presented in Sect. 5.

2 Energy Sector

People have always needed energy from the early ages to the present. Therefore, energy and demand for it have an important place in human life. However, as a result of increasing population and needs, countries' energy demand also increases. Since the industrial revolution, especially with the industrialization of European countries, the demand for energy has increased in these countries. The fact that energy demand started in non-European countries has increased the importance of energy at the global level and made it an indispensable element in the production process (Karadas et al. 2017).

Along with the progress in economic needs and increases in national income, the energy demand of the countries is also significantly affected by this progress. In each country, increases in domestic production cause more energy consumption and demand. Energy, which is one of the indispensable inputs of economic and social development, can be said to be on the top of the agenda of all the world countries with its accelerator feature. Energy is particularly critical for countries that are scarce in resources and have to meet country demand with imports.

The issue of energy, which is at the top of the agenda of the countries, is an important element of sustainable development together with sustainable energy, sustainable environment, and economy. Because of the relationship between energy needs and environmental conditions, countries develop sustainable development strategies to protect environment and sustain economic growth. The sustainable energy approach encompasses policies, technologies, and practices that enable the continuous supply of the demanded energy with minimum funding, and minimum environmental and social costs. Countries have started to invest in renewable energy sources based on their sustainable energy supply strategies.

The development of energy demand in Turkey has been similar to the progress in other countries. With the increasing population, economic development and investments in industry, the demanded energy has been increasing in recent years. With the exception of small European countries, Turkey has ranked first in Europe in terms of growth in demand for electricity, coal and natural gas over the last 10 years. Turkey's energy demand was 84.2 Mtep (million tons of equivalent oil) in 2005, up 54% in 2015 to 129.7 Mtep. Turkey's production from domestic energy sources is only 24.8%, so it is dependent on oil and natural gas. By 2020, Turkey's demand for energy is expected to double (222.4 Mtep) compared to today's. (IEA 2016; Taşkın and Yılmaz 2018).

Table 3.1 provides information on public and private investment and energy use in Turkey's energy sector between 2010 and 2018. When the data related to the investments in the table are examined, it is seen that there is a fluctuating course of investment between years and it is seen that overall energy investments have decreased in total from 2010 to 2018. When the data on energy consumption in Turkey are analyzed, it is seen that energy consumption has increased over the years. Given that Turkey's growth rate has been on the rise for years, it can be said that energy use is expected to follow an increasing trend.

Information on the different types of energy consumption and the imported energy in Turkey between 2010 and 2017 is given in Table 3.2. According to this information, the rate of energy imported by Turkey has increased over the years. It is understood from the information in the table that the amount of imported energy increased with the decrease in investments in the energy sector and therefore Turkey became dependent on foreign sources in terms of energy. In terms of energy consumption, the use of nuclear and alternative energy has increased over the years, while the use of renewable energy and fossil fuels has declined.

When the data in Table 3.3 is analyzed, it is seen that Turkey's total energy production generally follows an increasing trend in some periods, although it is fluctuating. Energy production from coal and renewable energy sources increased over the years, while there has been a slight increase in the level of energy from natural gas. Furthermore, there has been a decrease in the level of energy provided by hydro and liquid fuels.

Table 3.1 Energy investment and energy use in Turley (2010–2018)

	Investment in energy with private participation (Current US $)	Public private partnerships investment in energy (Current US $)	Energy use (Kg of oil equivalent per capita)
2010	5.427.030.000	4.213.030.000	1474,67
2011	8.551.020.000	5.403.520.000	1545,48
2012	3.360.000.000	3.360.000.000	1583,63
2013	13.591.100.000	9.102.100.000	1540,10
2014	11.050.450.000	143.000.000	1573,72
2015	1.578.070.000	1.578.070.000	1651,36
2016	2.836.000.000	376.100.000	…
2017	751.500.000	727.600.000	…
2018	957.210.000	747.210.000	…

Sources: World Bank, Databank, World Development Indicators

Table 3.2 Energy depletion and energy imports of Turkey

	2010	2011	2012	2013	2014	2015	2016	2017
Adjusted savings: energy depletion (% of GNI)	0,10	0,13	0,11	0,09	0,08	0,04	0,03	0,05
Alternative and nuclear energy (% of total energy use)	5,33	5,38	5,73	6,12	4,95	6,65	...	...
Energy imports, net (% of energy use)	69,62	71,61	74,02	73,07	74,21	75,21	...	...
Fossil fuel energy consumption (% of total)	89,12	89,97	89,41	88,16	89,54	86,84	...	...
Renewable energy consumption (% of Total final energy consumption)	14,33	12,78	12,83	13,85	11,61	13,37	...	...

Sources: World Bank, Databank, World Development Indicators

Table 3.3 Electricity production sources and shares in Turkey

Year	Total (GWh)	Coal (%)	Liquid fuels (%)	Natural gas (%)	Hydro (%)	Renewable energy and wastes (%)
2000	124.922	30,6	7,5	37,0	24,7	0,3
2001	122.725	31,3	8,4	40,4	19,6	0,3
2002	129.400	24,8	8,3	40,6	26,0	0,3
2003	140.581	22,9	6,6	45,2	25,1	0,2
2004	150.698	22,8	5,0	41,3	30,6	0,3
2005	161.956	26,6	3,4	45,3	24,4	0,3
2006	176.300	26,4	2,4	45,8	25,1	0,3
2007	191.558	27,9	3,4	49,6	18,7	0,4
2008	198.418	29,1	3,8	49,7	16,8	0,6
2009	194.813	28,6	2,5	49,3	18,5	1,2
2010	211.208	26,1	1,0	46,5	24,5	1,9
2011	229.395	28,8	0,4	45,4	22,8	2,6
2012	239.497	28,4	0,7	43,6	24,2	3,1
2013	240.154	26,6	0,7	43,8	24,7	4,2
2014	251.963	30,2	0,9	47,9	16,1	4,9
2015	261.783	29,1	0,9	37,9	25,6	6,5
2016	274.408	33,7	0,7	32,5	24,5	8,6
2017	297.278	32,8	0,4	37,2	19,6	10,0

Sources: TURKSTAT, Main Statistics, Environment and Energy

3 Research Methodology

The methodology applied in this study is based on the input-output model developed by Leontief (1951). The total production of an economy is the result of the total demand and final production of intermediate consumption by different sectors operating in the economy. In this context, countries need to establish the inter-industrial relations correctly in order to determine the economic structure and to make more effective planning. In this process, the Input-Output analysis is an important tool. Input-Output analysis examines the economy as an accounting system and provides us with the basic information needed by an economy by considering both the production and consumption processes.

3.1 Literature Review

The Input-Output model can be one of the commonly accepted research methodologies that can be used to investigate both the impact of the energy industry on other sectors and its importance in the economy. Although many studies have been conducted in the literature on the identification of key sectors and multiplier analysis of the Input-Output model, the number of studies directly used to analyze the energy industry is scarce. In his study, Yildiz and Akdugan (2014) examine the economies of developed and developing countries. In the analysis, G7 countries for developed countries are included in the model. For developing countries, Poland, China, Argentina, Brazil, India, Indonesia, Greece, and Turkey are included in the analysis. As a result of the analysis, it is observed that the sector with a high employment multiplier was the service sector in all countries, while the sector with a high value added multiplier was the manufacturing industry for the developing countries.

Han et al. (2011) analyze Turkey's manufacturing industry with the input-output model. As a result of the analyzes made according to Chenery-Watanabe (1958) and Rasmussen (1957) methods, it was determined that the main metal industry, re-evaluation, food and beverage and textile products manufacturing industries are the key sectors.

Alp et al. (2017) in their study analyze the effects of inter-industry linkages and key sector for Turkey. According to the results of the analysis, the production of chemical products, metal goods industry, and re-evaluation industries are found to be key sectors in 2002 and lost their status as a key sector in 2012. Instead of these sectors, it is found that the base metals industry gained the key sector character in 2012.

Ersungur et al. (2017) compare economic terms between BRICS countries and Turkey via WIOD data for 1995, 2000, 2005, 2008 and 2011.According to the results, in the 2000s, Turkey has become dependent on foreign intermediate goods used in production. For Brazil, Russia, China and India, it has been found that this situation has been partially observed in different sectors by years. Key sectors in Turkey; domestic transport and textile products, in Brazil; mining sector in Russia; nuclear fuel and refined oil, in China; chemical products and in India; basic metals and metal production are key sectors.

Ersungur and Ekinci (2015) examine the economies of Turkey, South Korea, Taiwan, Japan and China according to the Input-Output model using the WIOD data of 1995, 2000, 2005, 2008 and 2011. As a result of the analysis, it shows that Turkey's dependence on imports in terms of imported inputs is high. According to the analysis findings; Basic Metals and their production, Quarrying and Mining,

Chemicals and Chemical products, Nuclear Fuel and Refined Petroleum, Domestic Transportation, Machinery and Other Business Activities leasing, Transportation Equipment and Machinery sectors and Electrical and Optical Equipment sectors were determined to be the key sectors for these five countries.

Botric (2013) tries to identify the key sectors in the economy by using different methods in the Input-Output analysis for the Croatian economy with the data for 2004. As a result of the analysis, while the construction sector was found to be the key sector, it is determined that the service sector was an important sector not only for export but also for the whole economy.

In their study, Jahangard and Keshtvarz (2012) identify the key sectors using the 1999 Input-Output tables for Iran, 2005 for South Korea, and 2002 for the Turkey. According to the findings of the analysis, the key sectors of the three countries were found to be similar. Key sectors; Other manufacturing sectors are Food, Beverages and Tobacco Manufacturing, Construction, Communication, Manufacture of textiles and related products.

Tounsi et al. (2013) in their studies compare 1998 and 2007 with the Input-Output analysis for the Moroccan economy, found that the food and tobacco industry sectors were the key sectors in both years.

Marconi et al. (2016) in their study examine the Brazilian economy with Input-Output analysis. It is tried to determine the levels of forward and backward linkages between the sectors and the key sectors in the economy. According to the results of the research, it is determined that manufacturing related sectors are key sectors in Brazilian economy.

Markaki et al. (2013) examine the effects of clean energy investments in Greece with the Input-Output model. For this purpose, they have included the direct and indirect macro-economic effects of the transition to new energy technologies, which affect the measurement of energy resources, within the framework of Input-Output Analysis. Thus, they obtained the necessary investment information and found the effect of this on national production.

Kucera and Milberg (2003) analyze the effects of foreign trade on employment using the Input-Output Model for ten OECD countries. For this, they took the manufacturing industry. At the end of the study, they found that changes in trade in manufacturing industry decreased employment in this sector. Ten Raa and Mohnen (2001) examine Canadian and European economies in terms of effective use of resources on the basis of input-output model in terms of foreign trade.

Duarte et al. (2017) aim to establish a social accounting matrix (SAM) for the Spanish economy for 2013 in their work on the construction and analysis of a disaggregated input-output model. As a result of the structural analysis carried out, the role of the electrical industry in Spain and its importance in the activities carried out were revealed. According to the analysis conducted according to Rasmussen model, it is found that none of the electricity producing industries are key sectors and that electricity generation and distribution activities were capital intensive activities. In addition, it is found that most of the demand is met by domestic production and the external dependence in the electricity industry is low in Spain.

In their study, Ozdemir and Yuksel (2006) calculate the forward and backward linkage effects of Turkey's energy sector by using the Input-Output tables of TURKSTAT for 1985, 1990, 1996 and 1998. As a result of the analyzes, it is found that the forward linkage effects of the sub-industries of the energy sector had higher values than the backward linkage effects in all years. Considering that the sectors with both forward and backward linkage impacts are called key sectors. It can be said that the energy sector is one of the key sectors according to the analysis results for Turkey.

In their study, Ozdemir and Mercan (2012) examine key sectors in the economy and inter-sectoral exchanges of goods by using the TURKSTAT for 2002. As a result of the analysis, it is found that the energy sector is the key sector according to Hirschman method, although Turkey is dependent on foreign sources in terms of energy. In addition, it is observed that the energy sector has a high input-output relationship with its own sub-sectors and other industries in the economy.

In their study on South Africa, Chang et al. (2014) analyze the role of the port sector in the economy by input-output analysis. In this study, forward and backward linkage effects, price change effects and employment effects on how the port sectors effect the economy from the South African case have been revealed. As a result of the analysis, although the backward linkage effect of the port sector was low, the forward linkage effect was found to be high.

Alises and Vassallo (2016) show that there is less need for transportation services in service-oriented economies with input-output analysis for Spain and UK countries between 2009–2011 to reveal the impact of the transportation sector on the economy.

3.2 General Framework of the Input-Output Model

The inter-industry flow of goods and services has a versatile and complex structure. In order to plan a country's economy, it is necessary to know the structure of the economy and the relations between the industries. In the determination of these relationships, it is necessary to make analyzes with various models. Input-Output analysis is a model that examines the relationships between industries with the help of mathematical and statistical analyzes (Todaro 1987).

The input-output analysis developed by Leontief is an important econometric tool used to determine the macroeconomic impacts of industries on an economy (Rua and Lechon 2016; Miller and Blair 2009; Leontief 1941). The input-output model describes the activities and interdependence of industries in an economy. In doing so, it uses input-output tables, where industries are both producers and consumers in the model.

The data in the columns of the input-output table represent the monetary value of the resources that a sector demands or purchases from other industries of the economy to produce goods. The data in the lines of the Input-Output table shows the distribution of the supply or goods sold by one sector among other sectors (Rua and

Lechon 2016). One of the main objectives of input-output analysis is to determine the interdependence and inter-industry relations of industries in a given economy. Input-Output analysis is a system of equations that describes the distribution of a product produced by industry among other industries in the economy and its effects on the economy. In addition, by calculating various multipliers, the impact of a sector on employment, household income and activity levels of all industries in the economy can be measured. It should be taken into consideration that this measurement is very important in terms of economic development forecasts.

The Input-Output model is a linear and inter-sectoral model that shows the relationships between sectors operating in the production process of a given economy. The basic equilibrium equation of the Input-Output model of an economy consisting of N industries can be shown as follows (Chiu and Lin 2012):

$$X_i = \sum_{N}^{j=1} X_{ij} + F_i = \sum_{N}^{j=1} a_{ij} X_j + F_i \tag{3.1}$$

$$a_{ij} = X_{ij} / X_j \tag{3.2}$$

$$r_{ij} = X_{ij} / X_i \tag{3.3}$$

X_i is the total gross output in sector i = 1, 2, …, N. a_{ij} is technical coefficients, r_{ij} is the direct output coefficients and F_i is the final demand in sector i. Eq. (3.1) explains the input-output table according to the demand-side model. The matrix consisting of input coefficients belonging to all sectors in an economy is called "input coefficient matrix and is indicated with A. The expression Xij in the equation refers to the intermediate input level that the sector j receives from the sector i during the production process.

The solution of the production equation of the input-output method can be shown mathematically as follows (Özyurt 2007):

$$X = AX + Y \tag{3.4}$$

$$X - AX = Y \tag{3.5}$$

$$(I - A)X = Y \tag{3.6}$$

$$X = (I - A)^{-1}.Y \tag{3.7}$$

In the equation, X is the production vector, Y is the final demand vector and $(I–A)^{-1}$ is Leontief inverse matrix.

In the input-output analysis, the production of one industry has two different effects on the other industries. If the "g" industry increases its production by one-unit, it is expected that other industries using "g" as input in the production process will increase the demand for the g industry. This constitutes the causality aspect of the demand-sided model of Input-Output analysis. The concept of backward linking describes such links with other sectors where one sector buys input. There will also be an increase in the input from the g industry for other industries that use the products of the g industry in the production process. This is the causality aspect of the supply side model. The forward linkage effect refers to the link between a sector and other sectors where its outputs are sold (Miller and Blair 2009). Briefly, backward linkages express the relations of a sector with input-providing sectors, while forward linkages can be defined as the relationship between activity in a sector and sales of that sector (Matallah 2007).

The Leontief inverse matrix represents the total production of each industry or sector required to meet for final demand (Rua and Lechon 2016). The Leontief inverse matrix is of great importance for the solution of the Input-Output model. The total linkage effects of sectors in an economy are calculated with the Leontief inverse matrix. The values in the row of the Leontief inverse matrix show the total forward linkage effect, and the values in the column indicate the total backward linkage effect. Formulas of total forward and backward linkage effects can be shown as follows (Aydoğuş 2010):

$$TLF_i = \sum_j r_{ij} \tag{3.8}$$

$$TLB_j = \sum_i r_{ij} \tag{3.9}$$

Sectors that have a high forward and backward linkage effect in an economy are called key sectors. Since the key sectors have high forward and backward connections with other sectors in the economy, any investment in these sectors is expected to maximize economic growth. This is important for developing countries with limited resources (Choi et al. 2014).

When the literature on input-output model is examined, it is seen that there are two different methods for calculating backward and forward connections. These methods are the method used by the input coefficient matrix developed by Chenery and Watanabe (1958) and the Leontief inverse matrix developed by Rasmussen (1957) and then developed by Hirschman (1958). The approach developed by Chenery and Watanabe is called direct linkage connections, while the methods of Rasmussen and Hirschman are expressed as total linkage connections. (Gül and Çakaloğlu 2017).

The direct linkage effect (F_i) of an industry is the share of the intermediate demand for the production of that industry in the total demand. The direct forward linkage effect of an industry can be shown as follows: (Özyurt 1982):

$$F_i = \left(\sum_j X_{ij} \right) / F_i \tag{3.10}$$

As an indicator of the magnitude of the demand of a sector for the outputs of other sectors, the ratio of the sum of the intermediate inputs directly received from this sector to other sectors and the ratio of the sector to the output gives the direct backward linkage effect (Bi) of that sector. In this context, the direct backward linkage effect for sectors i and j can be defined as follows (Aydoğuş 2010):

$$B_i = \left(\sum_i X_{ij}\right) / X_j = \sum_i a_{ij} \tag{3.11}$$

3.3 Multiplier Effects

One of the important benefits of input-output analysis is that it allows the assessment of both macro and micro impacts in the economy. Thus, it is possible to identify the total impacts that may arise in an economy (Ersungur et al. 2017). While an industry needs output as input to other industries to produce; the output of one industry provides input for other industries. This implies the interdependence between the industries in the production process. Therefore, input-output analysis is important in production planning and structure analysis of the economy (Chiang 2005). All industries in the economy are dependent on each other to produce raw materials or final products. In addition, the sectors in the production process, while doing so, have significant impacts both on the other industries and on the whole economy. Increasing the amount of production of an industry in the economy may lead to a further increase in employment, output or income. This is called the multiplier effect.

The concept of multiplier generally refers to the extent to which a model will be affected for a response to an external change (Robison 2009). In economic terms, multipliers are measures used to estimate the overall impact of an economy in response to external changes in the production of a sector (McLennan 2006). Multiplier calculations can be made for several economic factors which as income, employment, and output through the Input-Output analysis (Burrows and Botha 2013). Output, Employment and income multiplier are useful, important and guiding analytical tools for economic calculations. In addition, the calculation of these multipliers also plays an important role for public policies and decision-makers. Because with the investments to be made in sectors that have a multiplier effect, serious developments can be achieved in the important areas of the economy such as employment, income and production and also the growth rate can be increased. In this context, it can be said that it is important to calculate the multiplier effect for the countries implementing the planned development policy.

The Leontief inverse matrix allows for the calculation of multipliers for all industries operating in the economy and to see total macroeconomic effects of the potential increase or decrease in external demand (Leontief 1974; Grealis et al. 2017). Multiplier calculations for all industries can be made for any year according to the Input-Output tables. The Leontief inverse matrix is the basis of the multiplier calculations in the input-output model (Sissoko and Cruyce 2009).

The output multiplier indicates the increase in output for a response to a one-unit increase in the final demand. The total value of each column in the Leontief inverse matrix calculated using the Input-Output table represents the output multiplier value for that industry. The calculation of the output multiplier can be shown as follows (Ersungur and Kızıltan 2008):

$$Z_j = \sum_{n}^{i=1} A_{ij} \tag{3.12}$$

According to Eq. (3.12), "n" represents the number of all industries in the economy and "Z_j" represents the output multiplier of the industry in that economy.

The employment multiplier indicates the employment used in the production process and helps to define the use of labor belonging to a particular economy (Sissoko and Cruyce 2009). The employment multiplier indicates the total impact of a rise in the final demand on employment (Schaffer 1999). The simple employment multiplier of any industry can be shown as follows (Schaffer 1976; Özyurt 1982):

$$e_j = \sum_{n}^{i=1} A_{ij} \frac{L_i}{X_i} \tag{3.13}$$

In Eq. 3.13, L_i shows us the amount of labor in industry i and X_i shows the amount of production. L_i/X_i is the labor force coefficient of industry i. If we express L_i / X_i as L_e ($L_e = L_i$ / X_i) and consider L as a vector, the employment multiplier can be calculated as follows:

$$[e] = [L_e].[\mathrm{I} - \mathrm{A}]^{-1} \tag{3.14}$$

In Eq. 3.14, $(\mathrm{I}-\mathrm{A})^{-1}$ shows the Leontief inverse matrix, while the elements of the vector [e] constitute the employment multiplier. Each element of the vector [e] expresses the labor input increases necessary to increase the production of sectors against the final demand increases. Briefly, it shows how much an increase in the final demand will increase the employment of sectors.

The income multiplier represents the change in total earnings from the entire economy as a result of the change in earnings from a particular sector (Fjeldsted 1990). Income multiplier refers to the increase in household income because of the unit increase in final demand. As the value of the income multiplier increases, the degree of structural cohesion in that economy increases and the dependence of the economy on imports decreases (Özyurt 1982). The simple income multiplier can be shown as follows (Bocutoglu 1990):

$$g_j = \sum_{n}^{i=1} A_{ij} \frac{V_i}{X_i} \tag{3.15}$$

The term X_i represents the amount of production of the i industry, while the term V_i indicates the net added value of the i industry. If the value added coefficient of industry i is shown as $v_i = V_i / X_i$, the income multiplier is calculated as follows:

$$[g] = [v].[I - A]^{-1} \tag{3.16}$$

In Eq. 3.16, the vector [v] represents the value-added sequence vector, while the vector [g] represents the income multiplier. These formulas will be used in the calculation of production, employment and income multipliers for the Turkish economy.

3.4 Data and Analysis

The data used in the study are Input-Output tables prepared by TURKSTAT for 2002 and 2012. The Input-Output Table for 2002 prepared by TURKSTAT consists of 59 sectors, while the Input-Output Table for 2012 includes 64 sectors. When the similar studies in the literature is examined, it is seen that either the industry is formed by aggregating the industries or the industries are handled separately. Since the number of industries in the Input-Output tables are different, it is not possible to make one-to-one comparisons.

In this context, since the number of industries in the Input-Output tables of 2002 and 2012 are different, it is considered that the aggregation method would not yield a healthy result and it is decided to handle the data for both years separately. In this study, firstly, each year's tables are compared in the year that they belong. In the continuation of the study, a comparison is made between similar industries between the years and it is aimed to show how the development took place especially in the energy sector (Table 3.4).

Table 3.4 Energy industries in 2002 and 2012

Sector code	Sector name
Energy industries in 2002	
4	Extraction of coal, lignite and peat
5	Excluding exploration and exploration, crude oil and natural gas
17	Manufacture of coke, refined petroleum products
32	Electricity, gas, steam and hot water supply
33	Collection, treatment and distribution of water
Energy industries in 2012	
10	Coke and refined petroleum products
24	Electricity, gas, steam and air conditioning
25	Natural water; water treatment and supply services

After examining all the industries in the Input-Output tables and similar studies in the literature, the energy industries are identified in the Input-Output table for 2002 and 2012. The industries and industry codes of the energy sector are given in Table 3.4

In this study, all industries in the economy are analyzed according to the demand – side Leontief model with the help of Input – Output tables for 2002 and 2012. In this study, direct and total forward and backward linkage values of all industries are calculated. With this calculation, the structure and status of inter-industrial relations can be determined. Output, employment and income multiplier analyzes are performed. The data obtained as a result of the multiplier analysis contain the basic information necessary for a planned development model to be implemented and to help policy makers in their national planning. After these analyzes, developments in all industries, especially energy sector, are compared.

4 Empirical Results

In this part of the study, inter-industry linkage effects analyses, key sector analyses and calculations related to output, employment and income multipliers for the Turkish economy based on 2002 and 2012, and evaluations and interpretations related to these calculations are given.

4.1 Inter-Industry Linkage Effects

Since the input – output analysis has been present, it has followed a continuously developing course and has been used in many different fields. Input-Output analysis, which was only used in economic matters in its first days, is used in many fields such as employment, social criteria, international factor and commodity trade, energy consumption and environmental pollution calculations. Since input-output analysis can be used in economic planning in addition to these calculations, it has become an important analysis tool used in all countries (Corona et al. 2016).

The input-output model is a quantitative econometric model representing the exchange of goods between different sectors/industries of a national economy and the interdependencies between industries (Dai and Yang 2013). The input-output table generated from the observed data for a particular economic area (country, region, local, etc.) shows a detailed flow of goods and services between all economic units. (Feng and Hubacek 2015). The basis of the input-output analysis is the input-output coefficients that reveal the relationships between industries for the national economy (West 1995; Sun 2007).

For a sector to produce goods or services, it must be associated with other sectors and obtain the raw materials and production factors that it needs in the production process from other industries. In this context, it can be said that the industries operating in the economy are interdependent with each other in the production process. Inter-industrial forward and backward linkage effects and identification of key sectors are important indicators for a particular country's economy.

The Input-Output tables prepared by TURKSTAT in 2002 consist of 59 industries and the Input-Output Tables for 2012 cover 64 sectors. Table 3.5 shows the first 20 sectors with the highest direct forward and backward linkage effects in 2002 and 2012. The data for the direct forward and direct backward linkage effects are calculated using Eqs. 3.10 and 3.11.

In 2002, the industries with high direct forward linkage effects in Turkey were as follows; Quarrying and other mining, Reassessment, Ancillary activities for financial intermediaries, Production and distribution of electricity, gas, steam and hot water, Water collection, treatment and distribution and Manufacture of coke, refined petroleum products and nuclear fuel. The industries in which direct forward linkage effects in Turkey were of high value in 2012 are: Printing and recording services, Employment services, Advertising and market research services, Electricity, gas, steam and air conditioning, Coke and refined petroleum products. When the direct forward linkage industries of 2002 and 2012 are examined, it is seen that the electricity, gas and steam industries and coke and refined petroleum products industries belonging to the energy industry are included in the ranking. According to this

Table 3.5 Turkey direct forward and backward linkage in 2002 and 2012 (top 20 sectors)

Direct forward linkage				Direct backward linkage			
2002		2012		2002		2012	
Sector no.	Row totals	Sector no.	Row totals	Sector no.	Column totals	Sector no.	Column totals
8	1,42077	9	1,03389	31	0,78022	24	0,78197
31	1,11995	52	0,99725	9	0,71009	49	0,67633
46	1,00000	49	0,99567	12	0,70263	5	0,65826
50	0,99071	54	0,88883	14	0,68025	42	0,63431
15	0,92392	7	0,87609	32	0,67973	53	0,63199
51	0,85493	4	0,87020	50	0,66679	27	0,60491
21	0,85160	46	0,86278	11	0,65796	9	0,60241
20	0,83577	34	0,85074	10	0,64733	13	0,59021
7	0,82982	51	0,84588	30	0,63431	6	0,58945
5	0,82774	26	0,82791	13	0,63091	33	0,58462
32	0,80018	11	0,81272	41	0,61190	14	0,57526
17	0,79944	14	0,80962	19	0,60678	60	0,55771
16	0,79664	24	0,80439	20	0,59135	63	0,54141
19	0,76718	47	0,79520	16	0,57122	15	0,53028
42	0,73702	38	0,77750	17	0,57056	18	0,52409
14	0,73113	10	0,75115	46	0,56982	7	0,52116
18	0,71293	43	0,73287	22	0,56292	22	0,51900
48	0,71264	50	0,73222	15	0,55650	16	0,51120
22	0,69754	35	0,71077	34	0,54342	38	0,51005
33	0,64970	8	0,70624	38	0,53495	50	0,50451

result, both in 2002 and 2012, it can be said that the energy sector and energy industries have high forward linkage effect.

The industries with a high direct backward linkage effect in 2002 were as follows: Reassessment, Manufacture of food products and beverages, Manufacture of wearing apparel, Manufacture of wood and cork products (except furniture) And Manufacture of coke, refined petroleum products and nuclear fuel. In 2012, electricity, gas, steam and air-conditioning was the industry with the highest direct backward linkage impact, followed by Advertising and market research services, Food, beverages and tobacco products, and Insurance, reinsurance and pension funds. When the direct backward linkage data for 2012 are analyzed, it is seen that energy industry is in the first place. This is an indication of significant progress in the energy sector in Turkey.

Table 3.6 shows the top 20 industries in Turkey with the highest total forward and backward linkage effects for 2002 and 2012. The total forward and total backward linkage effects in the table were calculated using Eqs. 3.8 and 3.9.

The industries with the highest value of total forward linkage effects in 2002 are respectively; Manufacture of chemicals and chemical products, Basic metal industry, Electricity, gas, steam and hot water supply, Extraction of crude oil and natural gas and related service activities and Manufacture of coke, refined petroleum

Table 3.6 Turkey total forward and total backward linkage in 2002 and 2012 (top 20 sectors)

Total forward linkage				Total backward linkage			
2002		2012		2002		2012	
Sector no.	Row totals	Sector no.	Row totals	Sector no.	Column totals	Sector no.	Column totals
18	4,2032	24	4,36008	12	2,5106	24	2,70324
21	4,0965	4	3,92733	31	2,4710	49	2,34314
32	3,8230	11	3,40502	11	2,3171	53	2,15893
39	3,4705	31	3,33463	13	2,3058	27	2,13970
51	3,2983	15	2,97869	32	2,2974	5	2,13744
36	3,2573	29	2,81488	9	2,2417	6	2,13593
44	3,1744	10	2,77535	50	2,2367	9	2,09153
15	2,8779	44	2,60247	30	2,2124	42	2,06435
11	2,8100	54	2,21428	14	2,2026	60	2,00195
1	2,7172	8	2,21204	10	2,1430	13	1,99772
5	2,6069	41	2,19220	19	2,0834	14	1,97247
42	2,5588	1	2,17930	16	2,0745	33	1,96337
37	2,3552	34	2,14196	20	2,0727	38	1,95163
17	2,1857	38	2,09737	41	2,0684	63	1,95157
35	2,0684	6	2,06772	15	2,0351	22	1,95041
19	2,0289	49	2,02853	22	2,0349	18	1,94698
9	1,9417	27	2,02216	34	2,0182	16	1,93679
20	1,8774	5	1,95749	28	1,9973	20	1,90619
47	1,8643	46	1,90070	38	1,9910	7	1,88572
23	1,7942	26	1,83672	46	1,9666	8	1,88436

products and nuclear fuel. The industries where the total forward linkage effect has a high value are as follows in 2012: Electricity, gas, steam and air conditioning, Mining and Quarrying, Chemicals and chemical products and Coke and refined petroleum products.

The industries with the highest value of total backward linkage effects in 2002 are respectively: Manufacture of wearing apparel, Reassessment, Manufacture of textile products and Electricity, gas, steam and hot water supply. In 2012, Electricity, gas, steam and air conditioning, Advertising and market research services, Travel agency, tour operator, other reservation services and related services and Constructions and construction works industries are the industries with high total backward linkage effect. When the data in Table 3.6 is examined, it is seen that the total forward and total backward linkage effect of the energy industry has a high value.

If the backward linkage effect of industry/sector x is greater than that of sector j, it can be said that the increase in the production of sector x will have more beneficial economic effects than an equal expansion in the production of sector j. Furthermore, if the forward linkage effect of sector x is higher than the forward linkage effect of sector j, it can be said that the positive effects of the growth of sector x will be more than the positive effects of sector j, for the economy. If the total forward linking effect and the total backward linking effect are considered together for an industry operating in the economy, the industries where these two effects are high are defined as the key sector (Miller and Blair 2009; Choi et al. 2014).

In this context, it is seen that electricity, gas, steam and air conditioning services, in other words energy industry, are the most important key sectors for 2012. Other key sectors in 2012 are: Advertising and market research services, Paper and paper products, Film, video, music and television related programming, publishing and production services, and Constructions and construction works. The key sectors in 2002 are respectively: Production and distribution of electricity, gas, steam and hot water, Manufacture of paper and paper products, Manufacture of textile products, Manufacture of plastic and rubber products and Manufacture of food products and beverages. When the key industries of both years are examined, it is seen that the same industries are the key sectors in general, but the energy industry is the most important key sector especially in 2012.

Similarly, for the energy industry in 2002, electricity, gas, steam and hot water production and distribution is the key sector. In addition, in 2002, the crude oil and natural gas extraction and related service activities and Coke, refined petroleum products and nuclear fuel manufacturing industries had a high total forward linkage effect. As the total forward linkage values of these industries are high, it can be said that the positive effects of the growth of these industries on the economy will be higher than the other sectors.

4.2 *Output Multiplier*

The output multiplier refers to the change in the production of all sectors against the change in the production of a particular sector (Fjeldsted 1990). In other words, the production multiplier indicates the increase in the output level of the whole economy in response to a one-unit increase in final demand. The total value of each column in the Leontief inverse matrix calculated using the Input-Output table represents the production multiplier value for that industry (Ersungur and Kızıltan 2008). The output multiplier is a useful indicator to show the dependencies between industries in an economy (Heringa et al. 2013)

The degree of structural correlation between each industry operating in the economy and other industries indicates the value of the output multiplier. In this context, as the value of output multiplier belonging to an industry increases, so does the degree of structural correlation in that industry. Industries use both the outputs produced by themselves and the outputs produced in other sectors as intermediate inputs in the production process. This mutual trade relationship between the industries in the production process shows industrial correlation.

Output multiplier values for the years 2002 and 2012 calculated using Eq. 3.12 are given in Table 3.7. When the data in Table 3.7 are examined, the industries with high production multipliers in 2002 are respectively: Manufacture of apparel,

Table 3.7 Output multiplier values for Turkey in 2002 and 2012

2002		2012	
Sector no.	Output multiplier	Sector no.	Output multiplier
12	2,5106	24	2,70324
31	2,4710	49	2,34314
11	2,3171	53	2,15893
13	2,3058	27	2,13970
32	2,2974	5	2,13744
9	2,2417	6	2,13593
50	2,2367	9	2,09153
30	2,2124	42	2,06435
14	2,2026	60	2,00195
10	2,1430	13	1,99772
19	2,0834	14	1,97247
16	2,0745	33	1,96337
20	2,0727	38	1,95163
41	2,0684	63	1,95157
15	2,0351	22	1,95041
22	2,0349	18	1,94698
34	2,0182	16	1,93679
28	1,9973	20	1,90619
38	1,9910	7	1,88572
46	1,9666	8	1,88436

Reassessment, Manufacture of textiles, Electricity, gas, steam and hot water production and distribution. In 2012, the industries with high production multipliers were: Electricity, gas, steam and air conditioning, Advertising and market research services, Travel agency, tour operator and related services, Constructions and construction works, Food, beverages and tobacco products.

When output multiplier values are examined, it is noteworthy that the values of the energy industry are high in both years. Furthermore, the fact that the multiplier values are close to 2 in most of the industries shows that the structural correlation is high in most of the industries. When the output multiplier values of the industries in the table are examined, it is observed that the output multiplier values of the industries operating in the services and industry sectors are generally higher. In this context, it can be said that the share of services and industry-related industries will be higher in the production increase that will emerge in the economy in response to an increase in final demand due to their high structural correlation.

4.3 Employment Multiplier Effect

Assuming that there is a close relationship between employment levels in a sector/industry and the level of production, the employment/output ratio can be defined for all output levels. In this context, inputs from each industry in the input-output table can be adjusted to give employment multipliers (Bekhet 2011). The employment multiplier of an industry is the employment that will occur in all industries if the last demand of that industry increases by one unit (Tariyal 2016). The employment multiplier measures the total change in employment due to a change in the labor force in a given industry (Surugiu 2009).

The employment multiplier is an important information and indicator for policy makers as it shows information about the total employment resulting from an economic activity. (Heringa et al 2013). Furthermore, the value of the employment multiplier is of great importance for economic planning. Because, in sub-regions and provinces where population growth is high, employment increasing sectors should be identified correctly. In this context, if increasing employment is chosen as a goal of economic planning, the employment multiplier will be an important indicator (Ozyurt 1982).

The employment multiplier values in Table 3.8 are calculated via Eq. 3.14. When the data in the table is examined, the industries with high employment multiplier values in 2002 are as follows: Domestic services, Educational services, Public administration and defense, Compulsory social security, Coal mining, Lignite and peat extraction.

In 2012, the industries with high employment multiplier values are as follows: Services of households as employers, Public administration and defense services, compulsory social security services, Education services, Services provided by member organizations and Human health services. When the employment multiplier data for both years are analyzed, it is noteworthy that the multiplier value is high in similar industries and these industries are in the services sector.

Table 3.8 Employment multiplier values for Turkey in 2002 and 2012

2002		2012	
Sector no.	Employment multiplier	Sector no.	Employment multiplier
59	1,00000	64	1,00000
53	0,71352	55	0,69063
56	0,66975	56	0,67705
52	0,61765	61	0,64489
54	0,40710	57	0,55551
57	0,37311	54	0,49625
4	0,35626	43	0,44681
7	0,28154	52	0,43640
50	0,27050	60	0,39967
12	0,26796	35	0,39282
43	0,24897	48	0,36013
20	0,23937	41	0,35047
44	0,23404	42	0,34764
34	0,23172	49	0,32255
45	0,23079	38	0,31781
2	0,22657	47	0,29444
55	0,22562	29	0,29379
38	0,22472	36	0,27788
10	0,21603	58	0,27713
42	0,21423	37	0,26695

Given that the services sector has a labor-intensive structure, it can be said that this is an expected result. In this context, it can be said that the best response to the policies aimed to increase employment is the new investments to be made in these sectors. Another factor that draws attention in the table is that the employment multiplier values of public services are high. This situation makes it necessary to increase public expenditures in order to increase employment.

When the data in the table is evaluated in terms of energy industry, it is observed that the employment multiplier value is generally low. It can be said that the reason of this situation depends on the capital-intensive structure of the energy industry and the intensive technology instead of labor.

4.4 *Income Multiplier Effect*

Income multiplier refers to the increase in household income due to the unit increase in final demand. The simple income multiplier is the increase in revenue caused by a one-unit increase in final demand for any industry. As the value of the income multiplier increases, the degree of structural cohesion in that economy increases and the dependence of the economy on imports decreases (Özyurt 1982). Accordingly,

Table 3.9 Income multiplier values for Turkey in 2002 and 2012

2002		2012	
Sector no.	Income multiplier	Sector no.	Income multiplier
59	1,00000	64	1,00000
41	0,93647	40	0,95695
26	0,93166	56	0,95496
6	0,92600	48	0,94479
52	0,91729	59	0,94207
13	0,91314	52	0,93848
47	0,89877	46	0,93030
23	0,88716	43	0,92954
11	0,88550	44	0,92764
28	0,88182	41	0,90751
25	0,88125	61	0,90390
56	0,87314	30	0,90337
21	0,87213	54	0,89722
15	0,86946	58	0,89492
50	0,86848	39	0,89065
45	0,86536	2	0,89060
20	0,86164	55	0,88445
44	0,85038	51	0,88331
31	0,84979	57	0,88069
40	0,84154	29	0,87819

it can be said that there is an inverse relationship between the income multiplier and the economy's dependence on imports. The income multiplier values for the years 2002 and 2012 calculated using Eq. 3.16 are given in Table 3.9.

According to the data in Table 3.9, the sectors with the highest income multiplier in 2002 are as follows: Home works, Air transport, Radio, television, communication equipment and equipment manufacturing and Public administration and defense, compulsory social security. In 2012, the industries with a high income multiplier were: Services of households as employers, Computer programming, consultancy and related services, Educational services, Scientific research and development services. In these industries with high income multiplier value, it can be said that Turkey's low dependency rate on foreign exports.

5 Conclusions and Discussions

Input-Output analyses are used in this study, including the demand-driven model, the inter-industry linkage effect analysis, total forward and total backward linkage analysis and multiplier analysis, to investigate the role of the energy sector in Turkey's national economy in 2002 and 2012. The results of the study provide some guidelines for policymakers to make proposals for the energy industry's investment and construction.

According to the data analysis, the energy industry has high values in total forward and total backward linkage effects both in 2002 and 2012. If an evaluation is made between years, it is seen that the total forward and total backward linkage effects in 2012 are higher. Since both the total forward link and total back link values of the energy industry are high, it has been found to be a key sector in the Turkey economy in both years. Key sectors are capable of affecting the overall economy due to their high forward and backward connections. Therefore, as a result of investments in key sectors, it is possible to achieve positive progress and growth in the economy. From this point of view, it can be said that the economic growth in the Turkish economy can be realized with the investments to be made in this industry because the energy industry is a key sector.

Multiplier analysis is an important economic indicator especially for policy makers and planners due to the revival of the economy in general and the clues it gives to the future. Especially, output, employment and income multipliers are important tools in this regard. Energy industries have high values in terms of output multiplier in 2002 and 2012. This shows that the structural connection in the energy industry, in other words, inter-industrial relations is high. It has been found that energy industries have low values in terms of employment multiplier and income multiplier values. It can be said that the reason of this situation for employment multiplier, depends on the capital-intensive structure of the energy industry and the intensive technology instead of labor. In this context, it can be said that the results are as they are expected. As for the low value of the income multiplier in the energy industry shows that Turkey is dependent on foreign energy. This is also true when Turkey's data on energy imports are analyzed. For a country with a high growth rate like Turkey, it is noteworthy that the energy sector is economically dependent on the foreign trade. It is a fact that the energy sector will have an important place in the development and growth process of Turkey if it gets rid of this foreign-dependent structure. In this respect, each investment in the energy industry, which is a key sector for the country with high forward and backward links, will have a significant impact on Turkey's socio-economic development process.

References

Alises A, Vassallo JM (2016) The impact of the structure of the economy on the evolution of road freight transport: a macro analysis from an input-output approach. Transp Res Proc 14:2870–2879

Alp E, Kök R, Ozan BM (2017) Türkiye ekonomisinde sürükleyici endüstri analizi: 2002–2012 karşılaştırması. Kafkas Üniversitesi İktisadi ve İdari Bilimler Fakültesi Dergisi 8(16):211–241

Aydoğuş O (2010) Girdi-çıktı modellerine giriş. Efil Yayınevi, Ankara

Baumol WJ (2000) Leontief's great leap forward: beyond quesnay, marx and von bortkiewicz. Econ Syst Res 12(2):141–152

Bekhet HA (2011) Output, income and employment multipliers in Malaysian economy: input-output approach. Int Bus Res 4(1):208–223

Bocutoğlu E (1990) Endüstrilerarası iktisat, teori ve türkiye uygulamaları. Trabzon, Karadeniz Teknik Üniversitesi Basımevi

Botric V (2013) Identifying key sectors in Croatian economy based on input-output tables. EIZ working papers, EIZ-WP-1302, Zagreb

Burrrows LR, Botha AP (2013) Explaining the changing input-output multipliers in South African: 1980–2010. Paper presented at the biennial conference of the Economic Society of South Africa, 25–27 September 2013

Chang Y, Shin S, Lee P (2014) Economic impact of port sectors on South African economy: an input–output analysis. Transp Policy 35:333–340

Chenery HB, Watanabe T (1958) International comparisons of the structure of production. Econometrica 26(4):487–521

Chiang AC, Wainwright, K (2005) Fundamental methods of mathematical economics. Mcgraw Hill: New York

Chiu RH, Lin YC (2012) Applyıng input-output model to investıgate the inter-industrıal linkage of transportatıon industry in Taıwan. J Mar Sci Technol 20(2):173–186

Choi S, Ji H and Zhao X (2014) Identifying key sectors using regional input-output model at sub-national level. Paper presented at the 54th Congress of the European Regional Science Association: regional development and globalisation: best practices. 26–29 August 2014, St. Petersburg, Russia, European Regional Science Association (ERSA), Louvain-la-Neuve

Corona B, De La Rua C, Guillermo SM (2016) Socio-economic and environmental effects of concentrated solar power in Spain: a multiregional input-output analysis. Sol Energy Mater Sol Cells 156:112–121

Dai Q, Yang J (2013) Input-output analysis on the contribution of logistics park construction to regional economic development. Procedia Soc Behav Sci 96:599–608

Ersungur ŞM, Ekinci ED (2015) Türkiye ve Doğu Asya ülkeleri arasındaki dış ticaret ilişkileri: girdi – çıktı yöntemi ile bir analiz. Atatürk Üniversitesi İktisadi ve İdari Bilimler Fakültesi 29(4):721–748

Ersungur ŞM, Kızıltan A (2008) Türkiye ekonomisinde sektörlerarası yapısal bağınlaşma- girdi-çıktı yöntemiyle bir uygulama. Atatürk Üniversitesi İktisadi ve İdari Bilimler Fakültesi Dergisi 22(2):17–31

Ersungur ŞM, Bayramoğlu T, Pabuçcu H (2017) Türkiye'nin Almanya, Rusya ve Çin ile ticari ilişkileri: girdi-çıktı analizi ile bir uygulama. Gaziantep Univ J Soc Sci 16(1):252–267

Feng K, Hubacek K (2015) A multi-region input-output analysis of global virtual water flows. In: Ruth M (ed) Handbook of research methods and applications in environmental studies. Edward Elgar Publishing, Northampton, pp 225–247

Fjeldsted BL (1990) Regional input-output multipliers: calculation, meaning, use and misuse. Utah Econ Bus Rev 50(10):1–20

Grealis E, Hynes S, O'Donoghue C, Vega A, Van Osch S, Twomey C (2017) The economic impact of aquaculture expansion: an input-output approach. Mar Policy 81:29–36

Gül ZB, Çakaloğlu M (2017) Inşaat sektörünün dinamikleri: Türkiye için 2000-2014 girdi-çıktı analizi. Akdeniz Üniversitesi İ.İ.B.F. Dergisi 36:130–155

Han E, Tosunoğlu T, Özsoy C (2011) Türk imalat sanayinde geri ve ileri bağlantılar: g-ç tablosuna dayalı yapısal bir çözümleme. TİSK Akademi 6(11):104–129

Heringa PW, Heide CMV, Heijman WJM (2013) The economic impact of multifunctional agriculture in Dutch regions: an input-output model. NJAS – Wageningen J Life Sci 64–65:59–66

Hirschman AO (1958) The strategy of economic development. Yale University Press, New Haven

IEA (2016) Energy policies of IEA countries: Turkey review. www.iea.org/publications/freepublications/publication/EnergyPoliciesofIEACountriesTurkey.pdf. Accessed 20 July 2019

Jahangard E, Keshtvarz V (2012) Identification of key sectors for Iran, South Korea and Turkey economies: a network theory approach. Iran Econ Rev 16(32):41–63

Karadaş A, Koşaroğlu M, Salihoğlu E (2017) Enerji tüketimi ve ekonomik büyüme. Cumhuriyet Üniversitesi İktisadi ve İdari Bilimler Dergisi 18(1):129–141
Kucera D, Milberg W (2003) Deindustrialization and changes in manufacturing trade: factor content calculations for 1978–1995. Rev World Econ (Weltwirtschaftliches Archiv), Springer 139(4):601–624
Leontief W (1941) The structure of American economy: 1919–1939. Oxford University Press, New York
Leontief W (1951) The structure of the American economy, 2nd edn. Oxford University Press, New York
Leontief W (1974) Structure of the world economy. Am Econ Rev 64:823–834
Marconi N, Rocha IL, Magacho GR (2016) Sectoral capabilities and productive structure: an input-output analysis of the key sectors of Brazilian economy. Braz J Polit Econ 36(3):470–492
Markaki M, Belegri-Roboli A, Michaelides P, Mirasgedis S, Lalas DP (2013) The impact of clean energy investments on the Greek economy: an input-output analysis (2010-2020). Energy Policy, Elsevier 57(C):263–275
Matallah KE (2007) A multiplier and linkage analysis: case of Algeria. J North Afr Econ 1:287–300
McLennan W (2006) Australian national accounts: introduction to input-output multipliers, Information paper. no.5246.0. Australian Bureau of Statistics, Canberra
Miller RE, Blair PD (2009) Input-output analysis: foundations and extensions, 2nd edn. Cambridge University Press, Cambridge
Ozdemir A, Yuksel F (2006) Türkiye'de enerji sektörünün ileri ve geri bağlantı etkileri. Yönetim ve Ekonomik Dergisi 13(2):1–18
Ozdemir A, Mercan M (2012) Türkiye'de enerji sektöründe yapısal bağınlaşma: girdi-çıktı analizi. Bus Econ Res J 3(2):111–133
Ozyurt H (1982) Trabzon alt bölgesi girdi-çıktı modeli ve alt bölge planlamasında kullanılması. Trabzon, KTÜ Temel Bilimler Fakültesi
Ozyurt H (2007) İktisadi planlama. Derya Kitabevi, Trabzon
Rasmussen PN (1957) Studies in inter-sectoral relations. North-Holland, Amsterdam
Robison RS (2009) Input-Output guidebook: a practical guide for regional economic impact analysis. Moscow, EMSI
Duarte R, Langarita R, Choliz JS (2017) The electricity industry in Spain: a structural analysis using a disaggregated Input-Output model. Energy 141:2640–2651
Rua C, Lechon Y (2016) An integrated multi-regional input-output (MRIO) analysis of miscanthus biomass production in France: socio-economic and climate change consequences. Biomass Bioenergy 94:21–30
Schaffer WA (1976) On the use of input-output models for regional planning. Springer, Boston
Schaffer WA (1999) Regional impact models. Regional Research Institute, West Virginia University. Http://Www.Rri.Wvu.Edu/Webbook/Schaffer/Regionalgt.Pdf. Accessed: 12 July 2019
Sissoko AA, Cruyce BVD (2009) Qualitative employment multipliers for the Belgian environmental industry. Working Paper 13-9. Federal Planning Bureau, Brussels
Sun Y (2007) Adjusting input-output models for capacity utilization in service industries. Tour Manag 28:1507–1517
Surugiu C (2009) The economic impact of tourism: an input-output analysis. Institute of National Economy 29(2(38)):142–161
Tariyal P (2016) Industry specific multipliers to identify key industries of Indian economy: an application of input output analysis. Glob J Hum Soc Sci E 16(4)
Taşkın E, Yılmaz M (2018) Türkiye'de elektrik enerjisi üretiminde doğal gazın yeri ve enerji ekonomisi açısından önemi. Paper presented at the international geography symposium on the 30th anniversary of TUCAUM, 3–6 October 2018, Ankara
Ten Raa T, Mohnen P (2001) The location of comparative advantages on the basis of fundamentals only. Econ Syst Res 13(1):93–108
Todaro MP (1987) Kalkınma planlaması (modeller ve yöntemler), (Turkish Edition Translator: Orhan Sezgin), İstanbul

Tounsi S, Ezzahid E H, El Alaoui A, Nihou A (2013) Key sectors in the Moroccan economy: an application of input-output analysis. Economics 7(18). https://doi.org/10.5018/economics-ejournal.ja.2013-18. Accessed 7 June 2019

TÜİK (2019) Girdi-Çıktı Tabloları. http://tuik.gov.tr/PreTablo.do?alt_id=1021. Accessed 27 June 2019

West G (1995) Comparison of input–output, input–output econometric and computable general equilibrium impact models at the regional level. Econ Syst Res 7(2):209–227

Yildiz N, Akdugan U (2014) Girdi-çıktı tabloları ile sektörler arası ilişkilerin analizi: seçilmiş ülke örnekleri. Akademik Bakış Dergisi 44

Yılmaz A, Kelleci SÜ, Bostan A (2016) Türkiye ekonomisinde sektörel enerji tüketiminin ayrıştırma yöntemiyle analizi. Dokuz Eylül Üniversitesi İktisadi ve İdari Bilimler Fakültesi Dergisi 31(2):1–27

World Bank (2019) Databank, World Development Indicators. https://databankworldbankorg/source/world-development-indicators. Accessed 8 June 2019

Chapter 4
Impacts of Electricity Subsidies Policy on Energy Transition

Asma Dhakouani, Essia Znouda, and Chiheb Bouden

Abstract Energy subsidy programmes are socio-economically designed to offer a modern and affordable energy, accessible for specific social groups, protect domestic industry, simulate economic development and protect the environment. However, in several countries, energy subsidies have deviated from their objectives and become an energy budget burden and a sustainable development barrier. Many questions arise: what are exactly energy subsidies? How are they implemented in a country mechanism? And what are their real effects?

This Chapter presents a review of energy subsidies: definitions, typologies, measurement approaches and effects. Facing harmful energy subsidies, reforms are also examined to help decision-makers phase out energy transition barriers related to subsidies. The case of the Tunisian power system is displayed in deeper details, characterised with a heavy burden of end-users electricity subsidies and an energy transition aiming 30% of renewable energies by 2030, against 3% in 2019. Using a holistic approach, based on hybrid energy systems modelling, has allowed presenting insights on reforming electricity subsidies and achieving sustainable development. This approach links subsidies, pricing, emissions, demand and supply of the power system through the advanced version of OSeMOSYS. Dynamics between energy, economics and environment are appealed within an integrated analysis of electricity subsidies policy.

Keywords Energy subsidies · Electricity pricing · Hybrid modelling · OSeMOSYS · Tunisian system · Econometric approach · Optimisation modelling · Reform policy

A. Dhakouani (✉) · E. Znouda · C. Bouden
Université de Tunis El Manar, Ecole Nationale d'Ingénieurs de Tunis, LR-11-ES16, Laboratoire de Matériaux, Optimisation et Énergie pour la Durabilité, Tunis, Tunisia
e-mail: asma.dhakouani@enit.utm.tn; essia.znouda@enit.utm.tn; chiheb.bouden@enit.utm.tn

H. Qudrat-Ullah, M. Asif (eds.), *Dynamics of Energy, Environment and Economy*, Lecture Notes in Energy 77, https://doi.org/10.1007/978-3-030-43578-3_4

1 Introduction

Local energy pricing policies is a governmental intervention in energy markets targeting social, economic and environmental objectives. Commonly known as energy subsidies (ES), this policy facilitates access to basic fuels, e.g. diesel, electricity and liquefied petroleum gas, etc. Subsidy programmes are originally designed to make modern energy affordable and accessible for specific social groups or in rural areas, protect domestic industry from competition, simulate economic and regional development and protect the environment (Fattouh and El-Katiri 2012). Taking into consideration these numerous benefits that Governments try to reach through energy subsidies, what are exactly energy subsidies? How are they implemented in a country mechanism? And what are their real effects? This chapter pleads energy subsidies by bringing answers to these questions. Section 2 reviews energy subsidies, followed by Sect. 3 analysing renewable energy (RE) subsidies. Then, Sect. 4 will overview the benefits, challenges and costs of ES reform, and present methodologies for the identification of *harmful energy subsidies* and designing their reform. Ultimately, in Sect. 5 the case study about the power system of Tunisia is analysed. Applicable electricity subsidies definition in Tunisia are identified, existing types and their effects are reviewed, and different Government interventions for electricity, electrification and RE subsidies are cited. In order to present holistic insights on the power system, this chapter relies on an advanced version of Open Source energy MOdelling SYStem "OSeMOSYS", i.e. a hybrid cost-based model relating electricity demand and supply through long-term optimisation.

Based on simulation results, recommendations to policy makers will be presented on how to reform electricity subsidies in Tunisia to achieve energy transition.

2 Review of Energy Subsidies

2.1 Definitions

Given the areas that subsidies are covering, there is no fit-all definition; and even within the energy sector the same problematic is faced. Many works have defined ES, where they agree about common points and differ about others. Given the developed literature of international organisations in terms of energy economics, their definitions of ES are taken into consideration in this work, listed in the Box 4.1.

The common point in these definitions is built around the principle that ES is a Government intervention. However, definitions differ about the target (consumer and/or producer), type of Government intervention (financial or in-kind) and effects on market prices.

Box 4.1: Energy Subsidies Definitions

United Nations (UN) Statistical Office definition based on the System of National Accounts (SNA)

Subsidies are current unrequited payments that Government units, including non-resident Government units, make to enterprises on the basis of the levels of their production activities or the quantities or values of the goods or services which they produce, sell or import (Von Moltke et al. 2004).

World Trade Organisation (WTO) Agreement on Subsidies and Countervailing Measures (ACMS)

A "subsidy" exists when there is a "financial contribution" by a Government or public body that confers a "benefit". A "financial contribution" arises where: (i) a Government practice involves a direct transfer of funds (e.g. grants, loans, and equity infusion), potential direct transfers of funds or liabilities (e.g. loan guarantees); (ii) a Government revenue due which is foregone or not collected (e.g. fiscal incentives such as tax credits); (iii) goods or services other than general infrastructure or purchased goods provided by the Government; or (iv) an entrusted or directed private body by the Government to carry out one or more of the above functions.

All WTO members accept this subsidy definition (Sovacool 2017).

Organisation for Economic Co-operation and Development (OECD) definition

ES are measures that keep prices for consumers below market levels or for producers above market levels, or reduces costs for consumers and producers (Kleinbard 2010). Within the European Union, there isn't any common designation of ES.

United Nations Environment Programme (UNEP) definition

During the twenty-first century, UNEP considers ES any Government action that influences energy market outcomes by lowering the cost of energy production, raising the price received by energy producers or lowering the price paid by energy consumers. Governments like to keep subsidies "off-budget" for political reasons, since "on-budget" subsidies are an easy target for pressure groups interested in reducing the overall tax burden (Steenblik 2003).

US Energy Information Administration (EIA) and International Energy Agency (IEA) definition

Both institutions have defined an energy subsidy as any Government action designed to influence energy market outcomes, whether through financial incentives, regulation, research and development or public enterprises, which lowers the cost of energy production, raises the price received by energy producers, or decreases the price paid by energy consumers (Brown 2001).

International Monetary Fund (IMF) definition

Energy subsidies comprise both consumer and producer subsidies. Consumer subsidies arise when the prices paid by consumers, including both firms (intermediate consumption) and households (final consumption), are below a benchmark price, while producer subsidies arise when prices received by suppliers are above this benchmark price (Fisher and Rothkopf 1989).

2.2 *Typology*

The classification of energy subsidies is highly related to their definitions. Some definitions include parameters and ignore others. These parameters are generally:

- Pathway benefits (direct or indirect, explicit or implicit, on-budget or off-budget)
- Target (consumer or producer, input or output)
- Instrument (budget, tax, transfer, assets)
- Purpose (regional development or energy conservation).

Previous research works where based on one of these parameters to classify subsidies mechanisms. However, the same mechanism could be classified under different categories. For example, a loan could be a financial transfer, direct and for a consumer. Table 4.1 summarises the different mechanisms.

2.3 *Measurement*

The task of measuring and comparing ES across countries or on a national basis might be highly complex and complicate their reform. This is due to the differences in definitions, classifications, the degree of transparency of fiscal systems and the complexity of energy systems. The attempt studies about quantifying energy subsidies are very limited given the lack of data and the complexity of the tasks (Von Moltke et al. 2004).

Sections 2.3.1, 2.3.2, 2.3.3 and 2.3.4 are the most common approaches for measuring ES which have been mentioned and applied by international organisations.

2.3.1 The Price Gap Approach

It is the most commonly applied method to calculate subsidies and is based on a comparison of the average end-use prices paid by consumers with reference prices presenting the full cost of the supply. In a given economy, the basic calculation of subsidies for a product is expressed in Eq. 4.1:

$$Subsidy = \left(Reference\ price - End\_user\ Price\right) \times Units\ consumed \tag{4.1}$$

Where:

- Reference price represents the supply cost — in electricity field: generation, transmission and distribution costs— plus the value added taxes (VAT);
- End_user price represents the price paid by the final consumer;
- Units consumed are the amount of energy consumed per unit; for e.g. quantity of kWh consumed.

Table 4.1 Classification of energy subsidies The Price Gap Approach

Government interventions	Definitions/schemes	Sources	Effects on prices	Classification parameters
Direct financial transfers	Grants to producers, low-interest or preferential loans to producers	Sovacool (2017)	Lower production's cost	Transfer, direct, consumers and producers
	Grants to consumers		Lower consumer's price	
Preferential tax treatments	Rebates or exemption on royalties, duties, producer levies and tariffs	Sovacool (2017)	Lower production's cost	Tax, consumers and producers
	Tax credit		Lower production's cost and consumer's price	
	Accelerated depreciation allowances on energy supply equipment		Lower production's cost	
Tax expenditures	Related to final consumption, typically targeted at households	Kleinbard (2010)	Lower consumer's price	Tax, consumers and producers
	Related to energy as inputs to production: Targeted at fuels or electricity, used as input to the final production of another good or service.		Lower production's cost	
	Related to energy production: Targeted at the actual extraction and production of energy, including refining and transport.		Lower production's cost and consumer's price	
Government revenue foregone	Governments also forego revenue by offering the use of non-depletable (e.g. land) or depletable assets (e.g. fossil-fuel resources) under their control to a private company (or individuals) to exploit them for their own use or for sale	Steenblik (2003)	Lower production's cost or consumer's price	Asset, consumers and producers, regional development

(continued)

Table 4.1 (continued)

Government interventions	Definitions/schemes	Sources	Effects on prices	Classification parameters
Regulation of the energy sector	Demand guarantees and mandated deployment rates	Sovacool (2017)	Lower production's cost and raise price to producer	Energy conservation
	Price controls		Raise the price to producers and lower it to consumers	
	Market-access restrictions and preferential planning consent and controls over access to resources		Raise price to producers	
Energy-related services provided by Government at less than full cost	Direct investment in energy infrastructure	Sovacool (2017)	Lower production's cost	Development, direct, explicit, budget
	Public research and development		Lower production's cost	
Trade restrictions	Quota, technical restrictions and trade embargoes	Sovacool (2017)	Raise price to producer	Producers, indirect
Failure to impose external costs	Environmental externality costs	Brown (2001)	Lower consumer's price	Off-budget
	Energy security risks and price volatility costs		Lower consumer's price	
Depletion allowance	Allow gross income deduction up to ~27% for depletion of exhaustible resources (oil, gas, minerals).	Fisher and Rothkopf (1989 and Mead (1979)	Raise price to producers	Direct, assets
Transfer of risk to Government	Governments assume some of the risk taken by energy producers through all kinds of measures.	Mills (2003)	Lower production's cost	Off-budget and indirect

In general to avoid overestimation, electricity reference prices rely on the LCOE from a combined-cycle gas turbine plant as a ceiling (Olejarnik 2013; Lin and Jiang 2011).

This methodology captures the support for end-users, which is the major part for fossil fuel energies, and doesn't take into consideration other types of subsidies such as industry development. This entails an underestimation of ES and their economic impact on the Government budget (Koplow 2009).

2.3.2 The Effective Rate of Assistance

The Effective Rate of Assistance (ERA) calculation considers local price direct and indirect effects. This consideration makes it practically difficult to apply because of the intensive required data such as the upstream industry of the estimated product (Plunkett et al. 1992; Anderson 1980). Originally, it derives from the effective rate of protection developed in 1960s. It measures net incentives for activities producing tradable goods.

ERA's formula is in Eqs. 4.2 and 4.3:

$$Net\,Subsidy\,Equivalent = Assisted\,Value\,Added - Unassisted\,Valued\,Added \tag{4.2}$$

$$ERA = \frac{Net\,Subsidy\,Equivalent}{Unassisted\;\,Value\;\,Added} \times 100 \tag{4.3}$$

Where:

- Assisted value added is the fully subsidised added value of an activity. It considers domestic pricing arrangements, export subsidies or taxes, local content schemes, production subsidies, quantitative import restrictions, tariffs, variable levies and voluntary export restraints. Moreover, it includes land, labour or capital returns.
- Unassisted value added representing real value of an output without any intervention.

2.3.3 Producer Subsidy Equivalent and Consumer Subsidy Equivalent

The Producer Subsidy Equivalent is the nominal cash allocated to local producers corresponding to the total support for a certain level of outputs, consumption and trade. Therefore, it focuses on the supply side and does not capture all subsidies. On the other hand, the Consumer Subsidy Equivalent (CSE) rather focuses on consumption and end-uses and is presented in Eq. 4.4 (Cahill and Legg 1990):

$$\begin{aligned} CSE = \Sigma(&\text{domestic prices} - \text{international prices}) \times consumed\,quantity \\ &+ \text{direct financial payment to consumers} \end{aligned} \tag{4.4}$$

Where:

- Domestic prices are prices paid by consumers;
- International prices reflect the cost of energy including transportation and distribution costs;
- Direct financial payment to consumers are direct subsidies paid by the Government to consumers to afford access to energy.

2.3.4 The Pass-Through Approach

Equation 4.5 is another method for measuring the magnitude of energy subsidies developed by the International Monetary Fund (IMF), named the pass-through approach (IMF 2013).

$$P.T = \frac{\Pr(t) - \Pr(t-1)}{\mathrm{Pw}(t) - \mathrm{Pw}(t-1)} \quad (4.5)$$

Where:

- *Pw(t)* is the international price or the supply cost;
- *Pr(t)* is the actual retail price paid by consumers, including taxes.

Thus, PT factor reflects the changes in international prices for energy goods on retail prices in the local market. There are three possible cases:

- **PT** = 0, then prices are fixed or virtually fixed.
- **PT** < 1, then there is exemptions of subsidies or tax.
- **PT** ≥ 1, then there is full pass-through or no subsidies, meaning that all the fluctuations in the international prices of energy goods are reflected in the domestic market.

However, this methodology has some limitations since the indicator can be negative around an inversion of the trend in international prices; it is as well sensitive to the choice of start and end-dates; and it does not account for initial price levels.

2.4 *Effects*

ES effects have economic, social and environmental dimensions, connected to each other. Indeed, ES accompanied with other policies conditioning the supply and demand and protecting the environment affect primary marginal revenues, production and incomes. Subsequently, they affect the composition of production and the outputs related to each sector. Further, ES have direct and indirect effects on emissions and resource use causing other consequences on environmental damage. Hence, ES effects should be interpreted based on a holistic approach (Ogarenko and Hubacek 2013; Sdralevich et al. 2014).

2.4.1 Economic Effects

On the one hand, several studies have identified the high costs associated to ES and the appearance of economic efficiency losses.

Firstly, by lowering blindly end-use prices there is an augmentation in energy's utilisation. This causes a reduction in incentives designated for more efficient energy and technologies, which was the case of the former Soviet Union. Those technologies have a high investment cost comparing to conventional technologies. The subsidies targeting consumption may decrease energy suppliers' profits and their ability to invest in new infrastructure, causing severe energy shortages.

Moreover, direct subsidies in the form of grants and tax exemptions present a heavy burden on Government's budget especially when international prices are high. Price caps, a subsidy form, have led to physical shortages and have required costly administratively arrangements.

Additionally, the increase in energy use due to subsidised prices entails an augmentation of imports or a decrease in exports. This engenders a high dependency on other nations or a spread of smuggling fuels, the case in many African countries. This illegal trade causes budgetary costs for originating countries, and ability limitation to tax domestic consumption for receptor countries.

Specifically, electricity subsidies could present a heavy burden on the state budget and weaken public funds. Therefore, they limit the development on other sectors. Other problems might occur (Narula et al. 2012; USAID SARI 2004):

- Overconsumption and waste of electricity due to distorted prices;
- Equipment negligence, subverting production and limited innovation;
- High interest and focus on subsidising one kind of sources creating barriers to diversification and to private sectors to flourish;
- Lifeline rates poorly designed and can be serving political considerations leading to disastrous financial and environmental damages.

On the other hand, there are several economic advantages from energy subsidies. For instance, subsidies targeting producers will reduce charges on investors and participate in industrial development. The energy related services provided by the Government such as investment in infrastructure or public R&D represent a tool of improving the value chain of the energy sector and increasing related added-value. Moreover, the transfer of risks to the Government allows providing a safer and a more favourable environment for investors.

2.4.2 Social Effects

Social effects could be deduced based on the type of subsidy. For this matter, let's take into consideration the division consumer/producer subsidies.

Consumption subsidies allow:

- Providing an affordable and accessible energy;
- Promoting a switch from traditional, i.e. wood energy, to modern forms of energy, i.e. conventional or renewable.

With subsidies' policies, Governments aim also to improve the living conditions of people. An affordable energy allows indeed a creation of more productive activities either in rural or urban areas affecting thus the phenomenon of exodus. Unfortunately, in many developing economies the social benefits don't reach rural areas. It is the rich and high consumer who are most likely to benefit, leaving the poor in a worse situation since the costs of subsidies are shared by the whole population (United Nations Environment Programme and International Energy Agency 2002). As a matter of fact, the situation results on:

- The poor is unable to afford subsidised energy;
- The financial value to the poor is very low due to their modest consumption and price caps;
- The middle and high income classes get more advantages from ES.

In many cases, ES ended up in capital intensive projects causing, on the one hand, a displacement of communities or affecting the health of poor neighbours unable to move away. On the other hand, ES boosted improving conditions of power and infrastructure in neighbouring areas of projects investments. This reflects hence the contradictory effects of E.S.

2.4.3 Environmental Effects

The environmental effects are dependent on the type of subsidy and energy sources. ES to end-users lower the prices, increase the consumption, and cause therefore an augmentation of emissions and pollution (especially with the utilisation of conventional energy like oil). ES to producers, increase production and therefore raise pollution. However in some developing countries, ES for fuel lower deforestation phenomenon and carbon emissions, making the switch to modern forms (fuel and electricity) highly favourable. Moreover, subsidising renewable energies will avoid emissions and decrease deforestation.

With the increase of concerns lately about climate change, many research works have tried to solve the dilemma (Dipa et al. 2015): *"Is subsidy worthy or harmful for mitigating climate change?"* Many have concluded that fossil fuels contribute to higher GHG emissions. This was based on the observations about the removal of coal subsidies in OECD countries, which have generated important environmental benefits. Cert, subsidising renewable energies and energy efficiency initiatives will further decrease emissions. However, it requires high supports to make it competitive and therefore it negatively affects the Government's budget. Some types, such as biofuels or non-consideration PV panels recycling, might cause also severe damages on the natural resources. In conclusion, emissions have to be assessed during the whole life-cycle from production, to conversion, transportation and end-use (UNEP-DTIE 2008).

3 Renewable Energy Subsidies

3.1 Framework

Among the challenges encountered during the integration of RE, Governments need to (Been 2013):

- Adapt the energy balance;
- Implement policies which promote cost competitiveness and avoid windfall profits;
- Identify appropriate policy funding mechanisms;
- Ensure that decisions are made transparently and are accountable;
- Achieve wide-scale political willing and social acceptance;
- Map institutional discrepancies, policy implementation and uncertainty of decision making roles;
- Overcome bottleneck situations related to the infrastructure of conventional energy sources, long-term planning and funding and qualified human capacity.

Governments have then adapted different policies to promote and implement renewable energies.

RE policies go along legal, institutional and regulatory frameworks, are country-specific and inspired from international experiences.

3.2 Typology

Table 4.2 summarises Government interventions for RE development, where each type is divided into different forms alongside the energy supply chain.

4 Energy Subsidies Reforms

Abolishing ES might be complex and difficult because of their entrenchment in institutional barriers and lock-in mechanisms. Indeed, they generally entail economic rents for industrials and certain consumer classes who are the prime beneficiaries, offer incentives and make political actions.

Table 4.2 Government interventions for renewable energies

Government interventions	Type	Definitions/ Schemes	Sources
Direct financial transfers	Consumer subsidies	Given directly to consumers to encourage them switching from fossil fuel to RE, for e.g. through the acquisition of an equipment part of the installation to decrease investment cost.	Saidur et al. (2010)
	Feed-in tariffs (FiT) and power purchase agreements (PPAs)	FiTs offer investors in RE a preferential price, often guaranteed by long-term PPAs and grid access. FiT allows investors calculating the period to recoup their investments. Prices include profits as well.	
	Preferential credit	For many commercial or development banks, investments in RE projects entail technology and policy risks. Such institutions could offer low-cost credit lines, i.e. loans with lower interest rates, and partial risk guarantees handled by the Government. The Government encourages these institutions to offer preferential lending to RE project developers, for e.g. over 3 years grace period.	
	Production subsidies for equipment manufacturing	Capital grants or low-interest loans to RE equipment manufacturers decrease the cost of equipment production or help to expand manufacturing capacity.	
	Export subsidies	Could be direct grants or concessional loans, to encourage local manufacturers and other firms offering clean energy products and services.	
Preferential tax credits	Tax credits for consumption	Tax exemptions, could be a reduction from personal taxes, used to encourage consumers to adopt RE. This could be applied through electricity bill in case consumer has a RE installation.	Haas et al. (2011)
	Accelerated depreciation	Project developers could use higher depreciation rates on their RE assets and thus receive related tax breaks. In fact, it will lead to high deductions in the earlier years of the lifecycle of the RE power plant and later on investors could have higher profits.	
	Investment tax credits	Attract more foreign capital in RE. Income tax breaks offer investors the attraction of higher profits. Another form is perceived when the Government decrease taxes for an investor in order to invest further in RE projects.	
	Production tax credits	The credits, or generation-based incentives, are paid per kWh of electricity produced over the guaranteed power tariff. The amount paid is dependent on the actual amount of power produced. Then, there is an incentive for producers to generate more.	
	Excise duty rebates	Rebates on sales, royalties and other levies are targeted at increasing RE production or manufacturing RE capacity.	
	Export tax rebates	Like export subsidies, tax concessions could be used to encourage exports of RE products and services.	

Regulation	Grid connections	Access to the grid: One of the most difficult challenges for scaling-up RE infrastructure. Government could assure the availability of the grid and related reinforcement.	Toke (2007)
	Renewable Purchase Obligations (RPOs)	It represents a demand guarantee: a guarantee of ever-increasing demand for RE helps bringing prices down. In fact large consumers are obliged while purchasing electricity to assure that it is originated from RES.	
	Trading via RE certificates	Instruments promoting shares of RE used by a utility based on an integrated and efficient national or even cross border electricity grid. It consists on receiving a certificate when the producer has achieved RES objective; once electricity is sold, the producer could sell the certificate as a commodity.	
	Government procurement	Regulatory policies obligating the Government to procure more energy-efficient products or have more shares of RE. This is achieved through national strategies and engagement of the Government in international agreement, for e.g. Protocol of Kyoto.	
	Compulsory licensing of intellectual property	It ensures that companies have access to the best technologies. This will help promote and access to innovative RE technologies.	
Infrastructure support	Grid access	Extension of grid access could enable people living in rural areas to benefit from larger RE projects. The Government should then invest in electrifying the whole territory to assure investments in rural areas.	Sovacool (2009)
	Land acquisition and access to other natural resources	Infrastructure support offered to build RE generation capacity and local manufacturing bases or to promote exports.	
Investment and trade restrictions	Market access restrictions and investment measures	Regulations could prohibit foreign firms from participating as project developers in setting up RE generation capacity. Tariff barriers could also favour domestic equipment suppliers. This will help local small investors to survive and promote RE activities.	Menanteau et al. (2003)
	Quotas	A form of trade restriction promoting domestic production that has an economic impact on lightening the firm's budget to tax payers. Therefore, part of produced goods, i.e. RE technologies, should be injected in the local market to assure their utilisation and promote indirectly RE investments.	
	Technical standards	Restriction of imports by promoting domestic manufacturing technical standards	
	Local content requirements	The purpose is to progressively stimulate local industries to familiarise and create jobs	
Potential risks with subsidy schemes	Governments and stakeholders have to be aware of the risk that would be generated from undesirable effects of subsidies and programmes, and in case of non-transparency and ignorance of the participatory and consultancy approach.		Sovacool (2009)

4.1 Benefits and Challenges of Reforms

During the last years, ES reform received much attention internationally especially with the financial and economic crises. Many Governments believe that phasing out ES would be highly beneficial for their budgets. This occurs exclusively when subsidies are not really serving the objectives that they have been implemented for and causing economic, financial and environmental damages (Sovacool 2017; Sdralevich et al. 2014; UNEP-DTIE 2008).

With the continuous interest in green growth strategy, both developed and developing economies subsidising energy goods have been concerned with the reform of their energy subsidies. In the view of reducing GHG emissions and initiating green development, Governments have to examine environmentally harmful subsidies (EHS). The success of a reform is dependent on the implication of the Government, stakeholders and society. Analysts have identified the benefits and challenges of ES reform through research studies in (Von Moltke et al. 2004; Lin and Jiang 2011; Sdralevich et al. 2014).

On the one hand, the transformation of subsidies could result in:

- Reducing the intensive use of resources and thus polluting less;
- Increasing competitiveness by supporting other sectors;
- Making polluters pay for their pollution;
- Overcoming technological lock-in;
- Improving the cost-effectiveness of financial mechanisms by re-allocating budget.

On the other hand, people behaviours have become dependent on subsidies mechanisms[1] since goods and services are at low prices. Reforming ES would have to deal with the income distribution and social acceptance.

ES reform faces numerous challenges whether economic, institutional or political; known that in most cases the main barrier is not economic but rather political.

The lack of political will to reform subsidies could occur. This is due to the strong special interests and rent seeking behaviours. And then, it makes ES benefits concentrated in a small group not presenting the whole population, while the costs are spread within the population. Politicians fear the change and social disruption, which was the case in India, Iran, Malaysia and Nigeria. The increase in energy prices engendered an increase in related goods and even people losing their jobs.

Institutional constraints are due to either the institution benefits from ES or the lack of vision. ES mechanisms are highly connected with other financial policies entailing holistic transformation of the economy and finance of a country. Worthwhile, the lack of transparency, information and data represents a major barrier (Sdralevich et al. 2014).

[1] For e.g. low subsidised fuel prices, in Middle Eastern countries, have resulted in the increase of car purchases and utilisation.

Eliminating ES without a clear planning will lead to a decline in households' welfare due to the increase of energy, related goods and services prices. Furthermore, eliminating ES has different generated effects in the short and long runs, as follow:

- ES reform considerations in the short-run: In the case of an augmentation of prices, the poor classes won't be able to pay for basic needs such as food. This situation was not captured during the modelling made by OECD in (IEA, World Bank, OPEC, and OECD 2010) using a Computable General Equilibrium (CGE) model. Contrary, OECD found income gains. Moreover, the removal of ES can also affect industrial competitiveness because the price of energy as input will rise. Furthermore, effects on profit margins engender an inability to secure finding and to plan for long-term, especially the case for industries with high energy intensity and competition. In addition, the removal of ES decreases demand, which affects as well employment. However, impacts on Government budget are seen annually impacting positively budget deficits.
- ES reform considerations in the long-run: The reform might lead to an improvement of the economy's growth, deflation of energy prices. This situation could be reached if during the medium term there isn't any increase in workers' wages and the increase in prices is rather seen as an augmentation of taxes indirectly (Sovacool 2017; IEA, World Bank, OPEC, and OECD 2010). The removal impacts as well the environment, since energy will be used more efficiently generating a decrease in GHG emissions and pollution in general. However, there is a high risk of failure of the reform on the long-run because of poor management of saved Government interventions.

4.2 Benchmarking of Reform Methodologies

4.2.1 International Monetary Fund Method

Based on the IMF research and case studies (Von Moltke et al. 2004; IMF 2013; Been 2013) about reforming ES in numerous developed and developing countries, a successful reform could be divided into two main parts:

- Identification of barriers to reform;
- Design of a reform strategy.

Identification of Barriers to Reform The barriers encountered to reform ES is country-specific. However, there are some common obstacles in most of the cases that decision makers should take into consideration. These obstacles are primarily related to social opposition (Fattouh and El-Katiri 2013), due to:

- The lack of information to the public renders the population unaware of the size of fiscal costs that the Government is covering and counter-effects of subsidies on poverty;

- The lack of Government credibility and transparency because of corruption make people opposed to the reform since they ignore and are suspicious on how the savings from ES reform will be spent;
- The short-run social impact on low income classes will increase their poverty due to the increase of fuel prices;
- The periods of low economic growth and unstable prices favour oppositions. Therefore, decision makers should consider the timing of reform to be able to recompense the increase in prices, e.g. reforming ES when rising household incomes.

In an economic perspective, the Government should be aware while reforming ES of fuel inflation effects in the short-run, international volatile oil prices and energy-intensive sectors suffering from a loss of competitiveness.

Finally, except population's resistance to the increase in energy prices, there are oppositions from interest groups such as politically vocal groups, labour unions, middle/upper class, losers from the reforms such as State-Owned Enterprises in the energy sector or exporters (Painuly 2001).

Design of a Reform Strategy In order to overcome the barriers mentioned above and to successfully implement a reform of ES, IMF has developed a list of key ingredients that decision makers should follow to trace a clear strategy for the short and long-runs described in (IMF 2013). To summarise, setting up a price mechanism requires four main steps (De Broeck and Kpodar 2014):

(i) Relating the retail price to import costs, distribution costs and taxes;
(ii) Specifying the time step to update prices;
(iii) Establishing a price smoothing mechanism as a periodic lagged rate of price increase;
(iv) Specifying the smoothing mechanism and creating an authority in charge.

4.2.2 Organisation for Economic Cooperation and Development and World Bank Methods

Both the Organisation for Economic Cooperation and Development (OECD) and the World Bank (WB) have developed various tools, for instance quick-scan and adapted models, to estimate the impact of subsidies removal on the economic and environmental levels. Commonly, they relied on decision trees (OECD 2013). Decision trees are support tools that use a tree-like graph for decisions. They identify the possible consequences of decisions through providing analytical frameworks to understand the effectiveness of a subsidy scheme. They do not measure the subsidy's scale or impacts of removal on political and economic spheres. We could find OECD checklist, OECD integrated assessment framework and World Bank checklist, as quick-scan methods. Within checklists, each answer launches a chain of continuous consequences until a final decision is reached.

Conversely, models, such as OECD ENV-linkages, are used to assess and quantify the effects of ES on environmental, economic and social spheres. They follow either an economic or econometric vision and have been developed on a local, regional or global scale. They point out the combination of demand and supply elasticities and energy prices (Château et al. 2014).

For this matter while analysing electricity subsidies, Sect. 5 touched upon Energy Systems Models (ESM) as a tool of analysing and forecasting energy mechanisms based on a holistic approach considering economic, econometric and technological perspectives.

5 Case Study: Electricity Subsidies in Tunisia

The Government has been since the 1960s supporting the industrial sector and assuring an affordable access to the energy for the Nation. The Government intervention is practiced through the fixed pricing policy assuring acceptable economic and social standards. However, in the twenty-first century the country has known an energy deficit due to the decline of national fossil resources and the incremental increase of demand. This deficit has affected the financial situation and caused a budget deficit (Dhakouani et al. 2017).

Precisely, electricity in Tunisia is produced at 97% by natural gas (NG) in 2017; and most of NG demand is reflected through electricity consumption. Roughly speaking, over 50% of electricity costs is subsidised for different end-users. Following the engagement of Tunisia to the COP21 in 2015 for climate change reduction, the Government launched a new strategy to assure a sustainable energy transition. Amongst major axes, the integration of renewable energies for the production of electricity is reflected through the national plan intending to reach 30% of Renewable Energy Sources (RES) in 2030 in the electricity mix. Other climate change engagements consist on decreasing carbon intensity by 41% comparing with the rates in 2010 and decreasing primary energy demand by 30% in comparison with the actual trend scenario. However, the financial situation plays a major role in postponing renewable projects implementation due to their high costs and the lack of financial resources (Dhakouani 2018).

A reform of the pricing mechanism of energy and specifically electricity could be of practical value in light of minimising the energy budget deficit. This reform should also target the maximisation of renewable energies integration. Nevertheless, the investment cost of renewable technologies is not competitive comparing to conventional technologies. On the one hand, costs could be treated with a long-term cost-based optimisation of the supply, which appeals technologies competitiveness. On the other hand, electricity pricing—*subsidies*—could be treated following one of the reform methodologies introduced in Sect. 4.

Analysing the case study of the Tunisian power system entails the following steps:

(i) Identification of electricity system subsidies in Tunisia;
(ii) Description of undesirable subsidies effects;
(iii) Measurement of subsidies;
(iv) Identification of the reform methodology;
(v) Optimisation of electricity costs produced;
(vi) Integration of the smoothing mechanism and reshaping electricity demand;
(vii) Reconfiguration of technologies competitiveness.

5.1 Applicable Definition

In the field of electricity, subsidies target principally the consumer since the market follows a vertical integrated monopoly. However, the objectives of subsidising power system components differ:

- Electricity subsidies have been introduced to offer to end-users lower prices than supply costs. These subsidies are the most important, proportional to the consumption of electricity and are variable.
- Electrification subsidies is introduced to all end-users. These subsidies are at fixed prices.
- Electricity produced by RES is subsidised since 2010, through a national programme targeting self-production regime using solar roofs, on investment.

Referring to definitions in Box 4.1, IMF definition is the most applicable to the case of Tunisia. It allows capturing consumer-based subsidies in function of electricity consumption, which is the predominant part in electricity subsidies.

5.2 Types of Electricity Subsidies in Tunisia

Taking into consideration the typology introduced in Sect. 1, Table 4.3 exhibits power system subsidies in Tunisia.

In regard to RE subsidies, Fig. 4.1 shows the different Government interventions (Tunisian Parliament 2015, 2016, 2017).

5.3 Measurement: Relating Costs to Prices

Measurement of electricity subsidies in Tunisia is assessed referring to Sect. 2.3. Firstly, ERA cannot be applied since it considers local price direct and indirect effects. This involves other related sectors than electricity which are out of the scope of this work. A part, the intensiveness of data could increase error rates and impact

Table 4.3 Government interventions in the Tunisian power system

Government interventions	Case schemes	Effects on prices
Tax expenditures	Electricity produced by the state-owner enterprise Société Tunisienne de l'Electricité et du Gaz (STEG) either from conventional or renewable technologies, and distributed for final consumption.	Lower consumer's price
Government revenue foregone	Government allows using land by STEG or independent power producer (IPP) for the implementation of power plants.	Lower production's cost
	The contract between STEG and IPP is regulated considering the notion of depletable assets	
Regulation of the energy sector	Price control of electricity	Lower price to consumers
	Market-access restrictions and preferential planning consent	–
Energy-related services provided by Government at less than full cost	Direct investment energy infrastructure: Transmission and distribution of the grid	Lower production's cost
	Collaboration with internal resources in public institutions instead of a full dependency on private consulting firms	Lower production's cost
Failure to impose external costs	Even though emissions are regulated and the excess is penalised, the Government is not applying environmental externality costs.	Lower consumer's price
	The Government by almost fixing electricity prices is handling price volatility costs.	Lower consumer's price
Transfer of risk to Government	The Government assumes a part of risks instead of IPPs, for e.g. exchange rate risks	Lower production's cost

insights. Secondly, even though CSE focuses on consumption and end-uses subsidies which are the predominant in the case of Tunisia, the direct financial payments to consumers are not applicable. Thirdly, P-T approach does not quantify subsidies but rather measures their magnitude. In regard to the *price gap approach*, this method captures principally the support for end-users based on the LCOE as a reference price. In addition, the availability of data and its extensiveness represent a major feature for its application in Tunisia.

Considering an integration of 30% RES by 2030 in the electricity production would be processed through a cost-based modelling— LCOE. The *price-gap approach* allows then relating electricity subsidies to supply costs through supply shares, demand and prices. Moreover, using the *price-gap approach* will allow quantifying subsidies related to consumption.

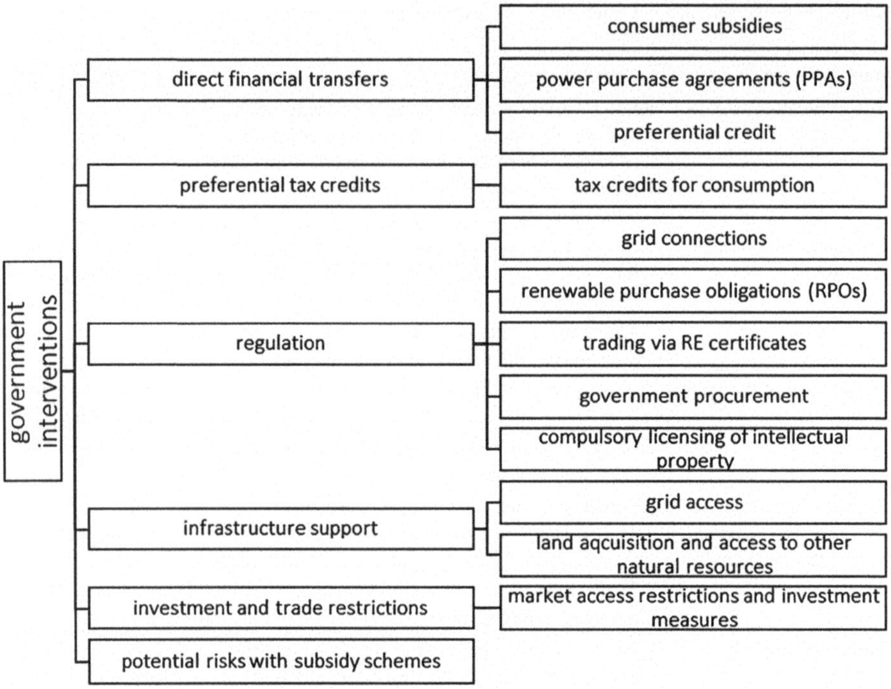

Fig. 4.1 RE subsidies in Tunisia

5.4 Reforming Electricity System Subsidies

In order to identify subsidies requiring a reform, it is essential to study the different effects engendered by subsidies. Following, Table 4.4 depicts the negative effects of electricity subsidies. RE subsidies are very limited compared to conventional electricity subsidies; then they are not included. Moreover, electrification subsidies have allowed reaching high rates of access to electricity above 99% in rural and urban areas by 2017, and have reached the objective that have been implemented for: *access to electricity* (Société tunisienne de l'électricité et du gaz*)*.

In regard to the identification of the most suitable methodology for the reform of electricity subsidies, this Chapter considers the IMF methodology as the most appropriate since it is based on the *price-gap approach*. Integrated with other methodologies, it allows treating different issues related to renewable energies integration in Tunisia, such as linking supply cost with energy subsidies. Indeed, the *price gap-approach* and the related method of reform could be modelled using an econometric model or integrated in a cost-based model to offering partial equilibrium features.

Per contra, OECD checklist targets only environmentally harmful energy subsidies. It is noticeable in Table 4.4 that undesirable economic effects are the main ones. In addition, OECD integrated assessment framework is a qualitative method

Table 4.4 Undesirable effects of electricity subsidies in Tunisia

Types	Effects
Economic	Augmentation in electricity and NG utilisation;
	Energy balance deficit;
	Deficit in energy Government budget;
	Decrease in tax revenues due to international prices fluctuations;
	Augmentation of NG imports and decrease of energy security;
	Reduction in incentives designated to more efficient energy;
	Decrease of investments in economically more attractive technologies;
	Decrease in STEG profits and debt ratio;
	Decrease of public infrastructure investment.
Social	High consumers benefitting most;
	Economic activities benefitting;
	Low income classes not affording subsidised prices.
Environmental	Augmentation of emissions and pollution;
	Decrease of opportunity to access to green funds.

and focuses on the linkages of energy and other sectors. Considering that the scope of this work is the electricity system, this method wont' satisfy the objectives to reach. In regard to the World Bank checklist, this method is based on a cost-effectiveness analysis and comparison of alternatives which is out of the scope of this work. Finally, with respect to developed models treating energy subsidies, the choice of the model depends on other technical, economic and regulatory criteria to assure a high integration of renewable energies too.

5.5 Hybrid Modelling of the Power System

The developed countries have been using Energy Systems Modeling (ESM) (Kern and Smith 2008) since the 1950s for energy transitions. This has led to a high number of approaches and types. The different types of analyses which were primarily developed target the supply side of an energy system, where the objective was to meet a given exogenous energy demand (Bhattacharyya 2011). Due to the oil crisis in the 1970s, demand side energy systems analyses have appeared and have kept evolving through years, leading to forecasting approaches, and later on translated to top-down models. Meanwhile, supply-oriented focus has kept evolving and has led to integrated bottom-up models which are technologically oriented to identify the needed investments or to operate short-term solutions (Hoffman and Jorgenson 1977; Hoffman and Wood 1976). Lastly, hybrid models, i.e. a combination of bottom-up and top-down models, have resulted in better insights for decision makers (Gargiulo and Gallachóir 2013; Bhattacharyya and Timilsina 2009). Considering the review articles on ESM such as (Urban et al. 2007), the choice of the appropriate modelling framework is identified from the context and the challenges depicted in major developing countries. In fact, it has been proven that the problem of optimally designing the electricity system in such a way as to match the demand with available resources and supply technologies can be solved by a bottom-up optimisation modelling framework. Given the easy access to the source code, the different developed versions, and the accessibility to manuals, the advanced version of *Open Source energy MOdelling SYStem* (OSeMOSYS) has been chosen to model the power system (Dhakouani et al. 2017; Howells et al. 2011). OSeMOSYS is originally a bottom-up, dynamic and linear optimisation model applied to the integrated assessment and energy planning.

5.5.1 Advanced Version of "OSeMOSYS"

Long-Term Optimisation of Electricity Supply OSeMOSYS aims to determine the lowest net present cost of an energy system to meet given demands and constraints. The original code of OSeMOSYS consists of several blocks of functionality, computing balances for costs, storage, capacity adequacy, energy balances and emissions. Recent works added blocks of functionality for reserve capacity dispatch and costs of flexible operation, in order to refine the analysis of short-term implications due to the penetration of intermittent RES (Welsch et al. 2015; Gardumi 2016).

Long-Term Simulation of Electricity Demand In order to present insights and to analyse the impacts of electricity end-users subsidies on the integration of renewable energies in the electricity mix, the integrated functionality block in the core code of OSeMOSYS related to electricity demand is appealed. The variation of prices using an automated smoothing mechanism impacts the evolution of demand. Thus, this version allows analysing the impact of subsidies on the demand and on the supply sides.

In regard to electricity demand modelling, the advanced version of OSeMOSYS relied on two approaches, i.e. decomposition and econometric.

Econometric Approach

The econometric approach allows connecting the demand to prices. It is a standard quantitative approach for economic analysis that establishes a relationship between the dependent variables (prices) and independent variables (income or economic growth rates) by static or dynamic analysis of historical data (Ryan and Plourde 2009; Bhattacharyya and Timilsina 2010; Momani 2013). The log-linear form of econometric specifications provides imbedded estimation of price and income elasticities and is better suited to energy demand than a simple linear specification (Bhattacharyya 2011). The advanced version of OSeMOSYS relies on the dynamic log-linear equation of energy demand in function of prices and income/ economic growth rates variables, developed by *Cobb Douglas* function in Eq. 4.6. The double-log regression model is a standard approach in the energy consumption studies (Madlener et al. 2011; Zaman et al. 2015).

Considering a time period of 1 year, demand block of OSeMOSYS stands on Eq. 4.6:

$$lnDy = \ln a + b * \ln GDPy + c * \ln Py + d * lnDy - 1 \tag{4.6}$$

Where:

- *y:* year
- GDP_y/GNI_y*:* gross domestic production per capita or gross national income per capita during the year y
- D_y*:* energy demand during the year y; D_{y-1}: energy demand during the previous year
- P_y*:* price of the energy during the year y
- *a:* scaling factor
- *b:* short-term GDP or GNI elasticity
- *c:* short-term price elasticity
- *d:* lagged-demand coefficient

The embedded prediction of prices during a year is processed using IMF method (Kaltenbacher et al. 2008) and presented in Eq. 4.7:

$$\ln P_y = (y - yo) * \ln(1 + s) + \ln P_{y0} \tag{4.7}$$

Where:

- y_0: the start-up year of modelling;
- P_{y0}: the price in the start-up year;
- P_y: the price in the year y; and P_{y-1}: the price in the previous y;
- *s:* the smoothing band. It is the percentage of subsidies decrease.

Decomposition Approach

The repartition of electricity demand by sector is appealed since subsidies differ by end-users. Moreover, demand types are differently sensitive to the smoothing band; depending on their elasticities to prices. The disaggregation follows Eq. 4.8:

$$\begin{aligned} Total\,demand &= low\,voltage\,residential\,demand \\ &+ (High + medium + low\,voltages)\,industry\,demand \\ &+ (High + medium + low\,voltages)\,services\,demand \\ &+ (High + medium + low\,voltages)\,agriculture\,demand \end{aligned} \tag{4.8}$$

Where:

- *Low voltage residential demand:* residential electricity demand. Generally residential demand is associated to low voltage power.
- *High, medium and low voltages industry demand:* electricity demand by voltage type in the industry sector
- *High, medium and low voltages services demand:* electricity demand by voltage type in the service sector
- *High, medium and low voltages agriculture demand:* electricity demand by voltage type in the agriculture sector

The short-run price and income elasticities, the coefficients of the regression, the scaling factor (intercept), logarithmic GDP or GNI per capita factor, logarithmic price factor, and logarithmic lagged demand factor are obtained from (Dhakouani 2018).

5.5.2 Main Input Data

Main input data of the model, not scenario dependent, are from (Dhakouani et al. 2017; Dhakouani 2018) and are summarised in Table 4.5. These parameters cover technical, economic and environmental characteristics of modelled technologies. The principal characteristics of considered scenarios are taken from (Dhakouani et al. 2017) and are related to the energy transition objectives, where the Business As Usual (BAU) scenario and RE scenario target respectively 5% and 30% of RES penetration by 2030 in the electricity mix. The rate of 5% of RES represents the natural technological integration. The 30% of RES represents the national objective of energy transition. The time horizon of modelling is from 2010 to 2030.

5.6 Electricity Subsidies vs. RES Penetration

In order to promote sustainable development and RE penetration in the power system, this modelling proposes to replace eliminated variable consumer (end-users) subsidies by production subsidies. As demonstrated in Sect. 4, production subsidies are more beneficial than consumer subsidies since they are targeted, on-budget and explicit. Moreover, these subsidies aid the industry development. To align with the energy transition, production subsidies will be targeted to promote sustainable and renewable technologies investments.

The *injection* of eliminated electricity subsidies on renewable technologies investments will lower investment costs of renewable technologies. Since LCOE from RES is primarily impacted by investment costs, subsidising renewable technologies will render these technologies more competitive. The *injection* reflects an incentive mechanism to promote RES penetration.

Table 4.5 Main input data

Parameters[a]	Units	Conventional technology[b]		Renewable technology							
		fuel fired Steam PP	CCGT	OCGT Large	OCGT Small	Wind	Cen PV	Dec PV	CSP	Hydro	PSH
Capital cost	USD/kW	220.5	200.9	705.0	428.0	1523.8	3312.8	5114.5	9800	1628.0	2618
Variable costs	M-USD/PJ	0.0	0.4	0.7	0.0	0.1	0.1	0.0	1.24	0.0	3.85
Fixed costs	USD/kW	21.0	12.1	20.0	15.0	12.5	29.0	29.0	67.26	14.4	18.27
Availability factor	%	91.8	94.9	97.0	94.3	97.5	99.0	99.0	96.0	91.0	92.0
Life cycle	Years	37	28	30	45	20	25	25	30	40	40
Efficiency	%	36.36	44.84	31.85	23.87	100.00	100.00	100.00	100.00	100.00	90.00
Unitary cost of the starts	USD/MW	58.16	53.32	47.86	29.08	–	–	–	90	–	
Ramping rate	MW/min	8	11	15	10	–	–	–	17	–	15
Minimum technical load	%	37.3	41.2	6.6	27.9	0.0	0.0	0.0	1	0.0	6.6
Max prim. Upward reserve	%	20.0	14.0	13.0	12.0	0.0	0.0	0.0	20.0	0.0	13.0
Max prim. Downward reserve	%	20.0	14.0	13.0	12.0	5.0	5.0	5.0	20.0	5.0	13.0
Max sec. upward reserve	%	12.0	6.0	70.0	75.0	0.0	0.0	0.0	60.0	0.0	70.0
Max sec. downward reserve	%	12.0	6.0	70.0	75.0	100.0	100.0	100.0	40.0	100.0	70.0
CO_2 generation	ton/PJ	1.93	1.46	2.38	2.38	0.00	0.00	0.00	0.00	0.00	2.38
NOx generation	10^{-3} ton/PJ	9.36	6.84	10.44	10.44	0.00	0.00	0.00	0.00	0.00	10.44

Reported from Dhakouani et al. (2017)

[a]Dashed cells: not applicable, no data available, out of research scope / All data is relative to 2010 – some parameters are evolving during the time horizon

[b]Conventional technologies characteristics where primarily extracted from the existing generating capacities. This data might be different from data sheets presented by manufacturers

Based on literature, the applicable smoothing mechanism bands in the energy sector and specifically electricity are 3%, 5% or 7% (IMF 2013; Sdralevich et al. 2014; Hanieh 2015). It has been demonstrated that greater smoothing mechanism values could generate a deterioration of the demand.

Due to the sensitivity of residential electricity demand to prices in Tunisia, the increase in prices through the application of above mentioned rates could lead to a deterioration of the demand — applied without any direct incentives to consumers. Based on research works elaborated in (Dhakouani 2018), a smoothing mechanism band for the residential sector of 0.24% annually will be applied. It has been demonstrated that this band allows a healthy growth of residential electricity demand. A smoothing mechanism band of 3% annually will be applied to the economic sectors, i.e. services, agriculture and industry, since they are less elastic to prices. Actually, economic sectors are more sensitive to GDP. Applying a 7% smoothing band to economic activities sectors has already resulted to an ill-conditioning and instability of the model. Moreover, eliminated energy subsidies in case of 5% smoothing band has much increased renewable technologies investments above 70% which is illogic for the considered time horizon.

Then, the *injection* of eliminated subsidies followed these hypotheses:

- A smoothing mechanism of 0.24% for residential sector;
- A smoothing mechanism of 3% for economic activities sectors.

Simulations included the application of hypotheses to:

- BAU scenario;
- RE scenario;
- No scenario: The power system without any scenario targeting a RES penetration rate. In other words, the constraint of achieving a certain rate of RES penetration was released and technologies were *freely* competing.

The results of simulations of the Tunisian power system, using the advanced version of OSeMOSYS, point out the impact of *injection* and are summarised in Figs. 4.2 and 4.3. Presented results are supply oriented showing the capacity shares evolution from 2010 to 2030 of the power system and the electricity mix in 2030.

As displayed in Fig. 4.2, the evolution of demand installed capacities is varying in the BAU, RE and No scenarios, achieving respectively by 2030, 8 GW, 11 GW and 12 GW.

The BAU scenario is characterised with high rates of conventional technologies because of the constraint limiting RES penetration to 5% even though renewable technologies are subsidised. Referring to Table 4.6, the BAU scenario achieved a total discount cost for the whole time horizon of 12.58 billion USD. In fact with subsidised investments of RE technologies, the limit of 5% of RES penetration represents a constraint that higher costs.

The low rate of total installed capacities is due to the elevated capacities of invested conventional technologies, their high availability factors and efficiency rates, referring to Table 4.5. By 2030, electricity demand would be satisfied at 70% from combined cycle (CCGT) because of its high efficiency as shown in Fig. 4.3a.

Gaz Turbines (OCGT) are primarily used to respond to the demand variation. Pumping Storage Hydro (PSH) represents 11% of electricity generation in 2030 since it allows responding to load changes by smoothing the load curve. PSH has originally a high investment cost, subsidising such investment would higher its chances to be more competitive independently from RES penetration.

In Fig. 4.2b, OCGT are more installed than CCGT even though their efficiency is lower. This is explained by their short-term characteristics — ramping characteristics as displayed in Table 4.5. OCGT are able to respond to quick variation either caused by the RES intermittency, especially wind participating at 18% in 2030, or the load curve variation. The highest share amongst renewable technologies is Centralised Photovoltaic because of the matching between generation production and the daily demand profile, where the summer peak is the most important in Tunisia. The integration of RES has led to a higher utilisation of PSH installed capacities than in the BAU scenario, as exposed in Fig. 4.3b. Where PSH is appealed to respond to RES intermittency and load curve changes. The RE scenario even though had 3 GW higher installed capacity, the subsidised capital costs of renewable technologies allowed decreasing the total discounted costs, reaching thus 12.53 billion USD. By 2030, primarily *renewable electricity* will be composed of 18% wind energy and 11% from Centralised PV.

The released scenario has led to high rates of renewable installed capacities by 2030 as shown in Fig. 4.2c. This is due to not constraining RES penetration and subsidising related investments. In fact, the model resulted in high investments in RE technologies, around 6 GW since they are becoming more competitive. However, due to their intermittency, the model should assure back-up conventional technologies for power system flexibility. Therefore, rates of 25% and 19% of the electricity mix were satisfied respectively by CCGTs and OCGTs in 2030 as displayed in Fig. 4.3c. By 2030, 50% of the electricity demand is satisfied by renewable technologies. It is noticeable that CSP by the end of the modelled period is invested, differently from the BAU and RE scenarios. Due to its flexibility in comparison with other renewable technologies, it will participate with 3% to electricity demand satisfaction. The integration of CSP has impacted PSH investment and utilisation, falling down to 350 MW in 2030. Subsidising RE investments without imposing a long-term objective to achieve allowed decreasing costs to 12.42 billion USD from 2010 till 2030.

In regard to the BAU and RE scenarios crossed with smoothing bands hypotheses, from 2010 to 2030 simulations have resulted respectively in 2.13 kton CO_2 and 1.92 kton CO_2. Accumulated NO_x passed respectively from 9.45 ton to 8.38 ton. It is remarkable that the RES penetration has impacted the decrease of NO_x emissions more than CO_2 emissions. This is due to the types of conventional technologies invested to respond to RES intermittency, i.e. OCGTs, and PSH— see Table 4.5. The lowest levels of emissions registered correspond to 50% of RES penetration, 1.77 kton CO_2 and 7.68 ton NO_x. Independently from the shares of conventional technologies types, CO_2 and NO_x emissions are proportional to the RES penetration rates. Referring to the agreements "Intended Nationally Determined Contributions", that the Tunisian Government has signed during the COP21, the RE scenario will

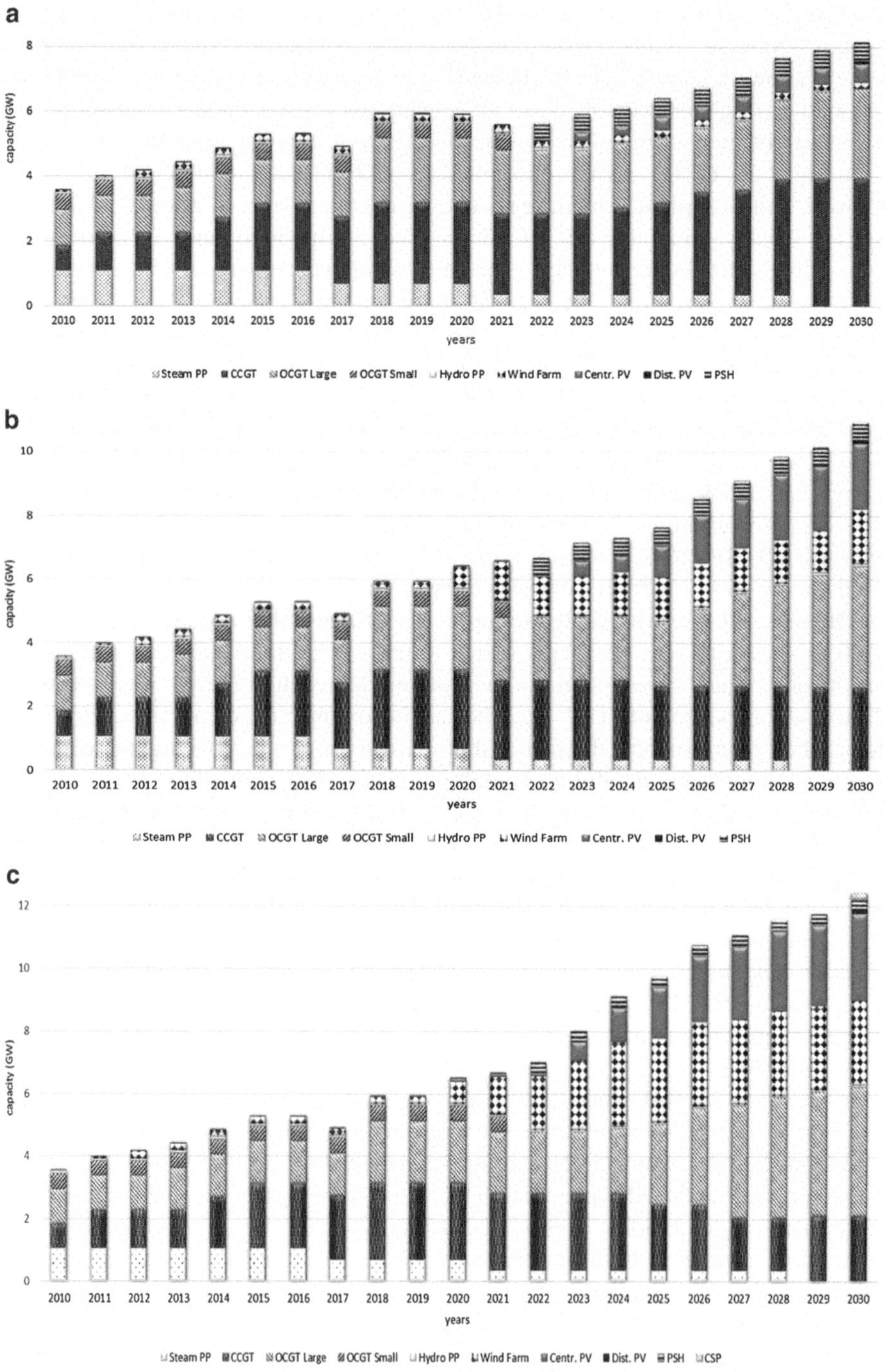

Fig. 4.2 Capacity shares evolution (**a**) BAU scenario; (**b**) RE scenario; (**c**) No scenario

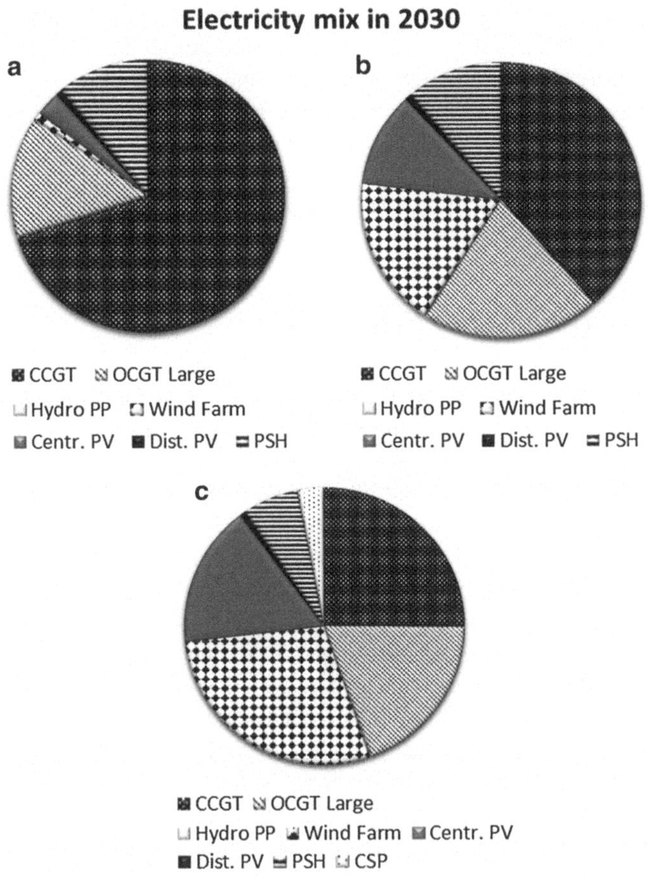

Fig. 4.3 Electricity mix by 2030 (**a**) BAU scenario (**b**) RE scenario (**c**) No scenario

Table 4.6 Total discounted costs – Subsidised RE technologies

	Cost (billion USD 2010)
BAU scenario	12.58
RE scenario	12.53
No scenario	12.42

highly contribute to achieving the target of reduction of carbon intensity to 41% compared to 2010, alongside changes within industry and transport sectors. It's important to mention that in 2010, Tunisia has reached 0.541 ton equivalent CO_2 per 1000TND expenditure. National efforts and INDC target to achieve 0.320 ton equivalent CO_2 per 1000TND expenditure by 2030 (Boisgibault 2015). Based on the hypothesis of emission penalty (3.3USD/tCO2), the impact on total emissions, total discounted costs and supply shares would have changed in function of emission penalties. Higher penalties of emissions would definitely increase the chances of having a supply share rich in terms of renewable technologies.

Finally, it is important to mention that the number of accumulated direct jobs created has reached 9153 if 50% of RES penetration was achieved by 2030. The identification of jobs could be simulated using the developed add-on of job creation introduced in (Dhakouani et al. 2017).

5.7 Electricity Subsidies Reform Policy

After all, the compound smoothing mechanism, i.e. 0.24% for residential sector and 3% for economic sectors activities applied to the advanced version of OSeMOSYS without imposing a RE objective, has lowered costs by 0.16 billion USD, in comparison with the BAU scenario, and allowed an integration of 50% of RES in the electricity mix by 2030. Injecting eliminated end-users electricity subsidies for subsidising renewable technologies investments has highly improved the competitiveness of renewable technologies. It is important to mention that contrary to conventional technologies where the LCOE is primarily impacted by fuel costs, generally renewable technologies has a high LCOE due to investment costs. In addition, subsidising RE technologies has generated a decrease in emissions. The effect of applying high penalties on emissions is reciprocal to subsidising RE investments.

The advanced version of OSEMOSYS has allowed presenting hybrid insights on the evolution of the power system in Tunisia. The balance of the supply and demand is mainly based on the connection between costs and pricing.

In fact, the Government is just targeting and orienting subsidies while i) decreasing end-users electricity subsidies at a rate that won't impact electricity demand evolution and ii) injecting eliminated subsidies into renewable technologies investments. Subsidies will be direct, for e.g. under the form of a financial contribution for producers. Yet, producers are not only limited to the field of renewable energies. Producers receiving renewable energies subsidies might be local investors in renewable energies projects, manufacturers of renewable technologies components, any other industrial willing to integrate renewable technologies to satisfy its needs in electricity consumption, or even within households for self-consumption (e.g. solar roofs). Through this mechanism, decision makers will avoid non-targeted and indirect electricity subsidies and will assure targeted and direct energy subsidies; switching then from harmful to beneficial incentive mechanisms. Moreover, the development of this field would eventually have social positive impacts such the creation of employment. Additionally, subsidising renewable technologies investments will higher the supply of these technologies because of an increase in renewable technologies demand. Referring to the market law, this will cause naturally a further decrease in the market price. When the market achieve a certain maturity, the Government could gradually eliminate even beneficial subsidies oriented to renewable technologies. Other long-term benefits that could be observed from this reform are the improvement of economic growth and energy security.

Finally, in order to avoid the aftermath of political instability from the switch of electricity subsidies, it should be appropriate to (i) identify and settle clear detailed objectives of an energy strategy, (ii) clarify public private partnerships to share risks and, (iii) assure social acceptance.

6 Conclusions

This chapter focuses on one of the major barriers of the integration of renewable energies in numerous developing countries: *energy subsidies*. Firstly, the literature review carried about energy subsidies, i.e. definitions, typology, measurement methods and effects, has shown that international organisations that have mostly brought this topic on the table. However, the notion of energy subsidies has differed. Therefore, different definitions were found which resulted in different typologies and methods of measurement. It is noteworthy that the price-gap approach is the mostly applied approach to quantify energy subsidies due to its simplicity, data extensiveness and relating supply cost to market price and retail prices. In addition, a special focus has been given to renewable energies subsidies. Based on the observation about the deviation of energy subsidies from the objectives that they have been settled for, the benefits, challenges and methodologies for reforming harmful energy subsidies have been reviewed. Significantly, a holistic approach for modelling an energy system taking into consideration the supply and demand would present useful insights to decision makers for energy transition. Energy subsidies and renewable energies are then introduced in a macro level based on the supply/demand equilibria of the power system.

Based on the context characterised by an energy budget deficit due to end-users electricity subsidies and an energy strategy targeting 30% of renewable energies by 2030, while it is currently at 3% (2019), the power system of Tunisia has been considered as a case study. The overview displays the applicable energy subsidies definition, the different Government interventions in terms of electricity, electrification and RE subsidies. To present insights on the power system reform, the advanced version of OSeMOSYS has been used as a tool of an integrated modelling. This version is a hybrid model, i.e. bottom-up with ends-users partial equilibrium based on decomposition and econometric approaches. The pricing system follows the smoothing mechanism approach developed by IMF to eliminate gradually subsidies. The decomposition approach pointed out separately economic sectors and residential sector, and voltage types.

A consideration of a smoothing mechanism of 0.24% for residential sector and 3% smoothing mechanism annually for economic activities sectors injected in renewable technologies investments as direct subsidies allowed achieving naturally 50% of renewable energies integration by 2030 and decreasing emissions.

Switching from end-users subsidies to production subsidies with a clear planning on a long-term perspective allows decisions makers to achieve high RES penetration, decrease emissions, diminish energy budget deficit and assure a sustainable energy transition.

References

Anderson K (1980) The political market for government assistance to Australian manufacturing industries. Econ Rec 56(153):132–144

Been H (2013) Meeting renewable energy targets: global lessons from the road to implementation. World Wide Fund Nat, p. 82

Benjamin K. Sovacool, (2009) The importance of comprehensiveness in renewable electricity and energy-efficiency policy. Energy Policy 37 (4):1529–1541

Bhattacharyya SC (2011) Understanding and analysing energy demand. In: Energy economics. Springer, London, pp 41–76

Bhattacharyya SC, Timilsina GR (2009) Energy demand models for policy formulation: a comparative study of energy demand models. Social Science Research Network, Rochester, NY, SSRN Scholarly Paper ID 1368072

Bhattacharyya SC, Timilsina GR (2010) Modelling energy demand of developing countries: are the specific features adequately captured? Energy Policy 38(4):1979–1990

Boisgibault L (2015) COP21 objectives: towards a joint energy transition in the Mediterranean? In COP21. Paris, pp 8

Brown MA (2001) Market failures and barriers as a basis for clean energy policies. Energy Policy 29(14):1197–1207

Cahill C, Legg W (1990) Estimation of agricultural assistance using producer and consumer subsidy equivalents: theory and practice. OECD Econ Stud 13:13–42

Château J, Dellink R, Lanzi E (2014) An overview of the OECD ENV-Linkages model. Organisation for Economic Co-operation and Development, Paris, OECD Environment Working Papers

De Broeck M, Kpodar R (2014) Mali: Automatic fuel pricing mechanism. International Monetary Fund, Washington, DC, Technical Assistance Report 14/31

Dhakouani A (2018) Multi-dimensional integrated modelling of the power system: towards renewable energy integration. UNIVERSITE DE TUNIS EL MANAR, Tunis

Dhakouani A, Gardumi F, Znouda E, Bouden C, Howells M (2017) Long-term optimisation model of the Tunisian power system. Energy 141:550–562

Dipa DD, Abedin K, Khan MM, Hasan MM (2015) Impacts of energy subsidy in Bangladesh: an analysis. ABC J Adv Res 4(1):39–56

Fattouh B, El-Katiri L (2012) Energy subsidies in the Arab world

Fattouh B, El-Katiri L (2013) Energy subsidies in the Middle East and North Africa. Energy Strateg Rev 2(1):108–115

Fisher AC, Rothkopf MH (1989) Market failure and energy policy A rationale for selective conservation. Energy Policy 17(4):397–406

Gardumi F (2016) A multi-dimensional approach to the modelling of power plant flexibility. Italy
Gargiulo M, Gallachóir BÓ (2013) Long-term energy models: principles, characteristics, focus, and limitations. Wiley Interdiscip Rev Energy Environ 2(2):158–177
Haas R, Panzer C, Resch G, Ragwitz M, Reece G, Held A (2011) A historical review of promotion strategies for electricity from renewable energy sources in EU countries. Renew Sust Energ Rev 15(2):1003–1034
Hanieh A (2015) Shifting priorities or business as usual? Continuity and change in the post-2011 IMF and World Bank engagement with Tunisia, Morocco and Egypt. Br J Middle East Stud 42(1):119–134
Hoffman KC, Jorgenson DW (1977) Economic and technological models for evaluation of energy policy. Bell J Econ 8(2):444–466
Hoffman KC, Wood DO (1976) Energy system modeling and forecasting. Annu Rev Energy 1(1):423–453
Howells M et al (2011) OSeMOSYS: the open source energy modeling system: an introduction to its ethos, structure and development. Energy Policy 39(10):5850–5870
IEA, World Bank, OPEC, and OECD (2010) Analysis of the scope of energy subsidies and suggestions for the G-20 initiative. G-20 Summit Meet Tor Can 5:81
IMF (2013) Energy subsidy reform: lessons and implications. *IMF eLIBRARY*. [Online]. Available: http://www.elibrary.imf.org/page/120. Accessed: 01 July 2017
Kaltenbacher B, Neubauer A, Scherzer O (2008) Iterative regularization methods for nonlinear ill-posed problems, vol 6. Walter de Gruyter, Berlin
Kern F, Smith A (2008) Restructuring energy systems for sustainability? Energy transition policy in the Netherlands. Energy Policy 36(11):4093–4103
Kleinbard ED (2010) The congress within the congress: how tax expenditures distort our budget and our political processes. Ohio North Univ Law Rev 36:1
Koplow D (2009) Measuring energy subsidies using the price-gap approach: what does it leave out? Social Science Research Network, Rochester, NY, SSRN Scholarly Paper ID 1631668
Lin B, Jiang Z (2011) Estimates of energy subsidies in China and impact of energy subsidy reform. Energy Econ 33(2):273–283
Madlener R, Bernstein R, González MÁA (2011) Econometric estimation of energy demand elasticities, vol 3. E. ON ERC, Aachen
Mead WJ (1979) The performance of government in energy regulations. Am Econ Rev 69(2):352–356
Menanteau P, Finon D, Lamy M-L (2003) Prices versus quantities: choosing policies for promoting the development of renewable energy. Energy Policy 31(8):799–812
Mills E (2003) Risk transfer via energy-savings insurance. Energy Policy 31(3):273–281
Momani MA (2013) Factors affecting electricity demand in Jordan. EPE 5(1):1–9
Narula K, Nagai Y, Pachauri S (2012) The role of decentralized distributed generation in achieving universal rural electrification in South Asia by 2030. Energy Policy 47:345–357
OECD (2013) Analysing energy subsidies in the countries of Eastern Europe, Caucasus and Central Asia. OECD
Ogarenko I, Hubacek K (2013) Eliminating indirect energy subsidies in Ukraine: estimation of environmental and socioeconomic effects using input–output modeling. J Econ Struct 2(1):7
Olejarnik P (2013) World energy outlook 2013. International Energy Agency, Paris
Painuly JP (2001) Barriers to renewable energy penetration; a framework for analysis. Renew Energy 24(1):73–89
Plunkett H, Wilson R, Argy S (1992) The measurement of effective rates of assistance in Australia-Industry
Ryan D, Plourde A (2009) Empirical modelling of energy demand. In: International handbook on the economics of energy. Edward Elgar Publishing, Cheltenham
Saidur R, Islam MR, Rahim NA, Solangi KH (2010) A review on global wind energy policy. Renew Sust Energ Rev 14(7):1744–1762
Sdralevich MCA, Sab MR, Zouhar MY, Albertin G (2014) Subsidy reform in the Middle East and North Africa: recent progress and challenges ahead. International Monetary Fund, Washington, DC

Sovacool BK (2017) Reviewing, reforming, and rethinking global energy subsidies: towards a political economy research agenda. Ecol Econ 135:150–163

Société Tunisienne de l'Electricité et du Gaz (2017) Rapport d'Activité. STEG, Tunis, https://www.steg.com.tn/fr/institutionnel/publication/rapport_act2017/Rapport_Activite_steg_2017_fr.pdf.

Steenblik R (2003) Subsidy measurement and classification: developing a common framework. Environmental Harmful Subsid Policy Issues Chall OECD Paris. pp. 101–142

Toke D (2007) Renewable financial support systems and cost-effectiveness. J Clean Prod 15(3):280–287

Tunisian Parliament (2015) Production d'électricité à partir des énergies renouvelables, vol 38, pp 926–932

Tunisian Parliament (2016) les conditions et les modalités de réalisation des projets de production et de vente d'électricité à partir des énergies renouvelables, vol 71, pp 2764–2774

Tunisian Parliament (2017) Model validation for the transfer of electricity produced from renewable energy contract, vol 13, p 546

UNEP-DTIE (2008) Reforming energy subsidies: opportunities to contribute to the climate change agenda. UNEP

United Nations Environment Programme and International Energy Agency (2002) Reforming energy subsidy. Oxford

Urban F, Benders RMJ, Moll HC (2007) Modelling energy systems for developing countries. Energy Policy 35(6):3473–3482

USAID SARI (2004) Subsidizing rural electrification in South Asia: an introductory guide. Energy Program

Von Moltke A, McKee C, Morgan T (2004) Energy subsidies: Lessons learned in assessing their impact and designing policy reforms, vol. 15. Emerald Group Publishing Limited

Welsch M et al (2015) Supporting security and adequacy in future energy systems: the need to enhance long-term energy system models to better treat issues related to variability. Int J Energy Res 39(3):377–396

Zaman M, Shaheen F, Haider A, Qamar S (2015) Examining relationship between electricity consumption and its major determinants in Pakistan. Int J Energy Econ Policy 5(4):998–1009

Chapter 5
ZEMCH Strategic Framework for Low Carbon Solutions in Sustainable Housing Delivery

Masa Noguchi

Abstract In the light of Sustainable Development Goals acknowledged by the United Nations, homes today need to be socially, economically, environmentally and humanly sustainable. This diverse sustainable housing production challenge necessitates multidisciplinary stakeholders' effective collaboration on the R&D actions. To form the global collaboration platform, the *ZEMCH Network* was established in 2010. ZEMCH is an acronym of *Zero Energy Mass Custom Home* which was conceptualised with the aim to function as a new domain for sustainable housing development in global contexts. For the purpose of ZEMCH design, production and marketing knowledge generation, collection and dissemination, the global network initiated and operated the academic conferences, industry-focused technical study tours and the design training workshop. In consequence, several housebuilders and housing manufacturers, who participated in the knowledge transfer activities, embarked on ZEMCH delivery in their local contexts. This chapter crystallised ZEMCH strategic framework for low carbon solutions in sustainable housing delivery through reviewing the design, production and marketing innovations applied to ZEMCH practices selected in Japan, Canada and Scotland.

Keywords Zero energy homes · Sustainability · Environment · Energy · Buildings

---

M. Noguchi (✉)
ZEMCH EXD Lab, Faculty of Architecture, Building and Planning,
The University of Melbourne, Melbourne, Australia
e-mail: masa.noguchi@unimelb.edu.au

H. Qudrat-Ullah, M. Asif (eds.), *Dynamics of Energy, Environment and Economy*,
Lecture Notes in Energy 77, https://doi.org/10.1007/978-3-030-43578-3_5

1 Introduction: ZEMCH Concept and Research Advancement

Dated back to 1987, "Sustainable Development" concept was first advocated by the World Commission on Environment and Development and it described the challenge as "a process of change in which the exploitation of resources, the direction of investments, the orientation of technological development, and institutional change are all in harmony and enhance both current and future potential to meet needs and aspirations" (WCED 1987). In 1992, this view was given additional impetus at the United Nations Conference on Environment and Development (or the Earth Summit) held in Rio de Janeiro where an initial international treaty on environment was produced. Nonetheless, this had neither a limit on greenhouse gas emissions nor legal enforcement provisions for individual nations. In 1997, the context of the Kyoto Protocol to the United Nations Framework Convention on Climate Change was adopted eventually at the 3rd Conference of the Parties held in Kyoto, Japan. As of April 2008, 178 states signed and ratified the Protocol. In consequence, most industrialised nations and some central European countries agreed to legally binding the reductions of greenhouse gas emissions of an average of 6–8% below 1990 levels between the years 2008 and 2012. In response to growing global warming issues and the constant increase of energy prices, housebuilders and housing manufacturers today are becoming more aware of the necessity for the delivery of zero energy and carbon dioxide (CO_2) emission sustainable homes than ever. Within this context, the sustainability may also embrace housing economy and adequacy beyond the legitimacy in which the quality barely coincides with individuals' dynamic various needs, desires and expectations. With due consideration of the United Nations' acknowledged 17 "Sustainable Development Goals", homes today need to be socially, economically, environmentally and humanly sustainable (United Nations 2019).

Housing needs and demands are dynamic having been changed in the course of time. The shifts in socio-demographics may be escalated by the emergence of non-traditional households and influence housing configurations and performances. Thus, homes today need to be *customisable* in a way to embrace the human and social sustainability in the built environment where users' individual requirements, desires and expectations are well accommodated. In the light of "fuel poverty" issues arising in various countries in which the drastic hike of energy cost is a serious concern, the notion of housing affordability may be reviewed to encompass both initial and operating costs over the lifespan (Boardman 1991). To secure the economic sustainability that impacts on individuals' financial capacity for daily activities, homes need to be *affordable* essentially. Affected by local climatic and living conditions, the amount of energy consumption in housing differs from one house to another, yet normally, it is transformed into CO_2 emission that contributes to global warming before and after the occupancy. Housing is indeed a system of energy and environment and in consideration of environmental sustainability, it needs to be designed in a way to reduce or eliminate CO_2 emission over the lifespan. The advancement of carbon neutral (or *zero energy*) homes is in need. Nevertheless, the construction industry's business operation tends to follow routines and the close system mode often hinders enterprises from adopting unfamiliar building innovations. In theory, homebuilders and housing manufactures are sensitive to societal needs and demands, yet in practice, traditional builders generally tend to follow routines in their way of doing business and cut down information search for determining whether to adapt unfamiliar design challenges, and innovative building materials and systems (Roberts 1970; Noguchi 2008a). Then, a question arises: how can the conventional practices of housebuilders and housing manufacturers be shifted in a way to adopt a new business operation required for the delivery of socially, economically, environmentally and humaly sustainable homes, whose design, production and marketing approaches may not be akin to those to which they are accustomed?

With due consideration of housing sustainability needs and demands applicable to both developed and developing countries, a "Zero Energy Mass Custom Home" or "ZEMCH" concept emerged, and the advanced sustainable design and lean construction approaches have been studied and practiced globally – the distinctive features and strategies are articulated in a book entitled "ZEMCH: Toward the Delivery of Zero Energy Mass Custom Homes" and edited by the author of this chapter (Noguchi 2016). In order to crystallise a wide spectrum of hopes and fears around the design, production and marketing approaches to the ZEMCH delivery in global contexts, an international multidisciplinary collaboration volunteer group called "ZEMCH Network" was formed in 2010. Today, the ZEMCH Network consists of over 750 partners from academia, industry and government based in nearly 45 nations and the number of participants in the collaboration activities is constantly on the rise. Originally, the ZEMCH Network was established by a group of academics, who participated in the 2010 industry-academia knowledge transfer technical visits to the production and sales facilities of low/zero energy mass customised housing manufacturers in Japan. The technical tour, later called "Zero-energy Mass Custom Home Mission to Japan" (or simply "ZEMCH Mission to Japan"), dates to the 2006 operation and it celebrates the 10th operational anniversary in 2018 (Noguchi 2011a). Moreover, in 2012, for the purpose of technical knowledge assortment, the ZEMCH International Conference was also launched in Glasgow, Scotland. Subsequently, it was held in 2013 (Miami, US), 2014 (Londrina, Brazil), 2015 (Lecce, Italy), 2016 (Kuala Lumpur, Malaysia) and 2018 (Melbourne, Australia). The ZEMCH 2019 International Conference is currently organised in Seoul, South Korea. In practice, ZEMCH production may hardly be realised without the complex interdisciplinary design engineering process. In order to simulate the sustainable design thinking experience and training, the "ZEMCH Workshop" was formed in 2014 and experimented at the State University of Londrina in Brazil (ZEMCH Network 2019). To date, the ZEMCH Workshop was run at the University of Sao Paulo (Brazil) in 2015, University of Melbourne (Australia) from 2016 to 2019 and the University of Bio Bio (Chile) in 2019. Linking it to the 2019 conference operation, South Korea is currently organising the ZEMCH Workshop to be held in Seoul.

In view of ZEMCH Network's pedagogical attempts and research advancement in global contexts, the following sections will recapitulate some design, production and marketing innovations studied and/or experimented through the mission, conference and workshop operations carried out to date.

2 ZEMCH Design, Production and Marketing Innovations

In Japan, a total of 942,370 new houses were built in 2018 and amongst them, 131,496 homes, or 14%, were estimated to be industrialised prefabricated homes (JPA 2019). Japanese housing manufacturers are advancing the massive supply of zero energy mass custom homes (ZEMCHs) today. To provide international housing research academics, government authorities, housebuilders and housing manufacturers in global contexts with opportunities to scrutinise Japan's ZEMCH design, production and marketing innovations, a technical study visit to the prefabricators' production and sales facilities (i.e. ZEMCH Mission to Japan) was initiated in 2006. In 2005, CANMET Energy Technology Centre (called "CanmetENERGY" today)—Varennes, Natural Resources Canada, appointed the author of this chapter and Professor Roger-Bruno Richard at the University of Montreal for the organisation of this technical study visit to Japan, which was then called "Japan Solar Photovoltaic Manufactured Housing Technical Mission 2006" (Noguchi 2011a). The knowledge transfer educational event was held from 20th to 23rd February, 2006, with the aim to provide Canadian homebuilders, housing manufacturers, building component suppliers, architects, academics and government officers with opportunities to explore not only the state-of-the-art production facilities of the then four leading net zero energy cost housing manufacturers in Japan, i.e., Sekisui Chemical Co. Ltd., Misawa Homes Co. Ltd., PanaHome Corp., and SANYO Homes Corp., but also a sales centre that displays a variety of Japanese manufacturers' housing prototypes (Richard and Noguchi 2006). Based on this experience, from 3rd to 5th September 2007, the educational event was reorganised by MEARU, Mackintosh School of Architecture, The Glasgow School of Art, in collaboration with the Centre for the Built Environment in the UK. Then it was renamed to "Zero-carbon Mass Custom Home Mission to Japan" and added a visit to Sekisui House Ltd. which had been committed to the development of a zero carbon emission house for the purpose of the G8 Hokkaido Toyako Summit exhibition that was expected to be held in the following year. In 2008, the title was again changed to "Zero-energy Mass Custom Home Mission to Japan" and was held from 10th to 12th September (Noguchi 2011a). Subsequently, to date, the ZEMCH Mission to Japan was organised almost annually.

All Japanese housing prefabricators visited and studied through the ZEMCH Mission to Japan operated to date have been practising a systematic way of *mass-customising* low/zero energy homes through their value-centric production and marketing approaches (or strategies). The following sections firstly recap the principles of a mass custom design approach and secondly, some zero energy home design features will be explored to shape fundamental techniques applicable to the ZEMCH delivery in developed and developing countries.

2.1 Mass Custom Home Design

Housing is composed of several parts and components that can be standardised. The choice of the housing design elements needs to be made carefully with due consideration of the project's initial and operational cost, quality, and time and the decision in turn affects the amount of future CO_2 emission that contributes to global warming. "Mass Customization" is an oxymoron since the term is composed of two opposite notions: mass production and customisation (Pine II 1992). This notion was anticipated by Alvin Toffler (1970) in his book entitled "Future Shock" and the term was eventually coined by Stanley M. Davis (1987) in his book entitled "Future Perfect." This paradoxical concept has been recognised as a means that helps to enhance efficiency in production and communication and to secure design customisability through economies of scope rather than economies of scale. Profoundly, Joseph B. Pine II (1992) systematised the general methods of mass-customising products and services in his book entitled "Mass Customization: The New Frontier in Business Competition." Frank T. Piller and Mitchell M. Tseng (2009) edited a "Handbook of Research in Mass Customization and Personalization" and compiled the research and development (R&D) activities and outputs delivered by a variety of industries across the globe. The handbook also includes a comprehensive universal design system approach to delivering "mass custom homes" (Noguchi 2001) and it stresses the significance of modularising house-building components. The total number of possible ordered pairs (or combination) of given standard housing components can be quantified (Noguchi and Friedman 2002). In the approach, the mass customisation (MC) has been systematised and visualised simply by making use of a conceptual analogue model as follows: MC $= f$ (PS) (Noguchi and Hernández 2005). In this model, the service sub-system (S) concerns communication platforms that leads home users to participate in customising their design output while the product sub-system (P) covers production techniques that aim to help standardise housing design components for establishment of the mass production capacity. Standardisation of building components seems to be a limited hindrance to design customisation if communication platforms are well developed. Design-consulting staff and appropriate communication interface are required to facilitate user choice of standard design components. These fundamental design service factors can also be integrated into a comprehensive model: S $= f$ (l, p, t). In this model, the service sub-system (S) admits the necessity of location (l), personnel (p) and tool (t) factors that affect design communication and they are interrelated. Basically, building components can be divided essentially into three categories: volume, exterior and interior. These can be considered main elements of the product sub-system (P) which can be described using the following conceptual model: P $= f$ (v, e, i, o). The volume (v) components are used to configure the building's usable space that determines the size and location of each room while the interior (i) and exterior (e) components serve as decorative and functional elements that customise a house. In addition, 'o' denotes other optional features such as building amenity and security systems, inclusive design components and renewable energy technologies. Generally, fabric and ventilation heat loss are associated with building volume and configuration, as well as the envelope quality whilst thermal transmittance links up with materials used in the skin.

The ZEMCH Mission to Japan generates industrial impacts and actions. In fact, some homebuilders and housing manufacturers took initiatives for the actual development of their own low/zero energy housing prototypes after the study visits. Reviewing the mission outcomes, the zero energy home design features and/or approaches will be examined below.

2.2 *Zero Energy Home Design*

"Alouette Homes" was a housing manufacturer, who attended the first ZEMCH Mission to Japan held in 2006. Before merged with the other company, the prefabricator was delivering panelised and modular homes mainly in Quebec, Canada. Immediately after returning from the first mission held in Japan, the company formed an R&D team that included the author of this chapter as the housing design lead and shaped the first architectural appearance. Later, a design charrette to embrace multidisciplinary challenges was arranged and Professor *Andreas Athienitis at Concordia University joined the team to lead* the further engineering upgrade. Eventually in 2007, the manufacturer built its own (near) net zero energy solar house prototype, later called "ÉcoTerra house" (Noguchi 2008b). In parallel, Scottish homebuilder "ROBERTRYAN Homes" attended the ZEMCH Mission to Japan 2007 and became an ardent industry advocate of zero carbon modular homes through Scottish Building Federation activities. Immediately after the mission experience, the company also established a zero energy/emission housing R&D team including the author of this chapter, when he served as a Reader at the Mackintosh School of Architecture, The Glasgow School of Art, then. In 2010, the housebuilder made their attempt to conceptualise the design of its own net zero energy housing prototype, called "Z-en house" (Noguchi 2011a). Both post-mission projects can be characterised by the application of advanced passive and active systems for energy efficiency and power generation towards zero energy housing operation. The low/zero energy housing techniques or systems applied to these projects, where the author of this chapter was involved in the detailed design, may be worthy of further review.

2.2.1 ÉcoTerra House

The ÉcoTerra house is a single detached home built on a 1.1-hectare rural lot of a new mass housing subdivision development in Eastman, Quebec (Fig. 5.1). The total heated floor area of the house was estimated at 234m^2 and the heated volume at 671m^2 including the basement. When the basement is excluded, the utilised floor area of this house is 141m^2 and 2 bedrooms are located on the first floor and semi-private spaces, such as a kitchen, dining room, and sunspace family room/lounge, are on the ground floor. The soil surrounding the house covers almost half of the basement wall. Accordingly, the space before the ground floor was called "semi-basement" that benefits the natural earth coverage which helps reduce the wall exposure to the outside climate (Noguchi 2011a). The semi-basement serves as a multifunctional space and machine room. The house was constructed by making use of the housing manufacturer's pre-engineered modular housing system that helped eliminate or reduce on-site construction nuisances, such as bad weather, theft and vandalism. A closed garage is located on the north side of the house being attached to the façade. Thus, the garage somewhat functions as an air buffer space that helps

Fig. 5.1 ÉcoTerra house, Alouette Homes

reduce the fabric heat loss from the north façade. The south wall is fitted with large triple-glazing to optimise passive solar gain for space heating. The south-glazing to floor ratio is 9.1%. The window area of south, west, east and north walls was estimated at 20.90m^2, 5.20m^2, 6.67 m^2 and 0.65 m^2, respectively. The glazing areas allocated to the exterior walls of this house were designed to ensure the comfortable level of natural daylighting and reduce the total amount of electricity required for artificial lighting.

In the ÉcoTerra house, a 3kWp building integrated photovoltaic thermal (BIPV/T) system was installed (Fig. 5.2). The PV array of the BIPV/T system is comprised of 22 amorphous silicon 136 W laminates placed on the 55m^2 (5.8 m × 9.5 m) south-facing metal rooftop. The potential annual electricity production was estimated initially at approximately 3420 kWh when the roof is sloped at 30.3° (Noguchi 2008b). The air under the PV panels is heated by solar radiation, drawn into a lower portion of the house from the roof ridge ducted. The BIPV/T system introduced to the ÉcoTerra house contributes to the drastic reduction of operational energy consumption for heating the space and water in addition to the alleviation or elimination of an electric clothes tumble dryer use. With due consideration of the utilisation of the advanced renewable energy technologies combined with passive solar techniques applied to the ÉcoTerra house, the annual space energy consumption was simulated to be as low as 1130 kWh. The annual electricity used for domestic hot water reduced from 3353 to 553 kWh due to the contributions of the heat pump desuperheater, BIPV/Thermal system and drain-water heat recovery unit. As for the yearly energy consumption of indoor and exterior lighting and other appliances, it was decreased from 3975 to 3275 kWh and this was affected partially by the reduced

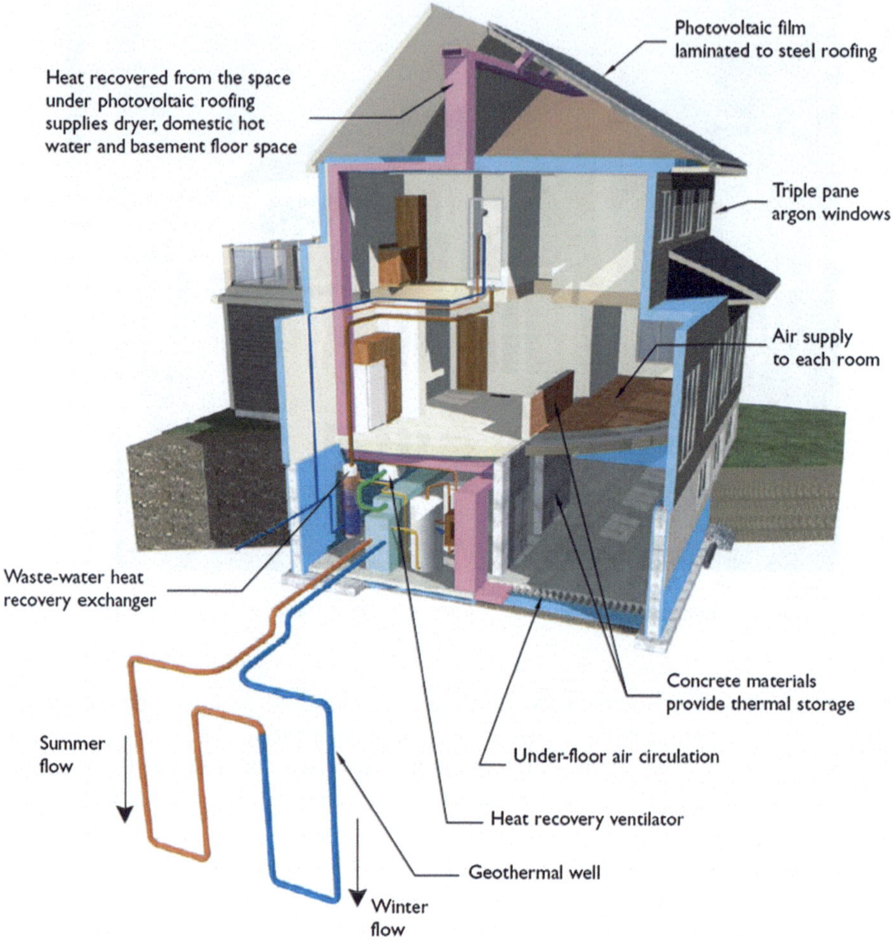

Fig. 5.2 ÉcoTerra house's renewable technology scheme

energy demand of an electric dryer. Subtracting the expected PV electricity generation of 3420 kWh from the sum of these figures, the annual energy consumption of the ÉcoTerra house can be re-estimated at 2155 kWh. However, this excludes the magnitude of user behaviour, which may further contribute to reducing or eliminating the total operational energy use.

In principle, the ÉcoTerra house was designed to optimise energy gains while minimising the total heat loss (Noguchi 2011b). Therefore, passive solar design techniques were well implemented in harmony with advanced active renewable energy technologies in order to achieve the net zero energy cost when the home is used properly. The key innovative aspects of the house can be summarised as follows:

- Effective building orientation with large lights that optimise the sun exposure for solar heat gains and daylighting;
- Triple glazing windows filled with argon gas and low emissivity coating for reduction of fabric heat loss through windows;
- Minimised openings on the north, east and west facades for reduction of fabric heat loss through the openings;
- Multifunctional outdoor garage that is equipped with a green roof terrace connected to the 1st floor main bedroom and serves as an air buffer space for reduction of fabric heat loss from the critical heat-losing wall;
- High thermal performance foam insulation that achieves U-values of 0.15 W/m^2K in walls and 0.1 W/m^2K in ceilings and ensures airtightness at approx. 0.8 air-changes per hour that reduces uncontrolled ventilation heat loss;
- Waist-high thermal mass wall that improves the house's capacity for heat-storing, daylighting and natural ventilation and distinguishes the project from others with this new environmental architectonic feature;
- Unpartitioned open staircase that allows for not only human movement and heat and daylight distribution as well as gives a sense of visual extension to the moderately sized interior aimed at reducing energy demand for space heating;
- Unfinished multipurpose semi-basement for spatial flexibility and affordability;
- Resource-efficient modern method of construction that helps ensure the high levels of product quality and waste management;
- Green landscaping (including the garage's roof terrace) that is aimed at reducing the amount of ambient CO_2 presence and rainwater runoff, as well as conserving natural habitat for sustainable community development;
- Building integrated photovoltaic thermal (BIPV/Thermal) system that generates not only electricity using free clean sunlight, but also solar-heated air applied for heating space through solar thermal mass, pre-heating domestic hot water through an air-to-water heat exchanger, and supplying warm air to the clothes dryer;
- PV and/or solar thermal capacity expansion space that allows for the growth of rooftop solar panels after occupancy;
- Integral indoor sunspace on the semi-basement and ground floors that are composed of generous solar thermal mass connected to the BIPV/Thermal system;
- Balanced heat recovery system that recycles preheated indoor air up to 76%;
- Ground-source heat pump that is used as the primary system for heating and cooling the house;
- Heat exchanger designed to apply the heat from warm drain (waste) water for heating incoming cool water from the mains;
- Energy efficient appliances and lighting; and
- Rainwater butt that harvests the natural free water for gardening and car-washing.

The ÉcoTerra sustainable house was designed to provide its occupants with comfortable and healthy indoor living environment and produces as much energy as it consumes on an annual basis. The simulation of the energy use indicates that the house experiences nearly net zero energy consumption when it comes into operation. However, the domestic energy use still depends on occupants' energy saving behaviour. Accordingly, a user manual that helps educate occupants about behaviour-related energy saving techniques may need to be developed and the effect can be calibrated through energy meters that can display both energy use and cost.

2.2.2 Z-en House

The Z-en house is a single detached home conceptualised for the company's new rural residential development in West Kilbride, Scotland (Fig.5.3). The floor area of this house is approx. 346m^2 excluding the basement floor area and the exposed wall area was estimated at 279m^2. The house contains 4 bedrooms and a study are located on the first floor and semi-private spaces, such as a kitchen, dining room, lounge, and sunspace family room, are on the ground floor. A basement is also introduced to this project, designed to serve as a multifunctional space, in which thermal mass components are installed heavily so as to capture heat from the sun and active hybrid renewable energy technologies.

Like the ÉcoTerra house in Canada, the 'Z-en house' in Scotland also attempted to install a photovoltaic thermal (PV/T) system in addition to various passive solar housing design techniques for energy efficiency. Nonetheless, this Scottish house was designed to integrate the PV/T with a Mechanical Ventilation Heat Recovery (MVHR) system (Fig. 5.3). In order words, it was aimed at examining the performance of a newly conceptualised PV/T MVHR system, instead. Thus, 20–30% of fresh air extracted from the PV/T roof will be mixed with the recycled indoor air circulated via the MVHR system. Accordingly, the fresh air supplied through the PV/T MVHR system may somewhat contribute to the space heating when desirable.

Fig. 5.3 Z-en House South-west Façade

The configuration of a BIPV/T roof (i.e. PV roof sizes, slope angles and ventilation rates) and the cell types may affect the heat and power generation performance. Thus, in order to investigate the PV/T heat and power generation potentials, a feasibility study was conducted (Table 5.1). As an initial attempt to develop a guideline for PV/T performance in Scotland, the roof slope angle was initially set to be 30°, 40°and 50°. The study confirmed that PV generates heat which makes the air running under the PV panels 10–15 °C warmer than the outside temperature even during the Scottish winter with the ventilated air velocity at 1 m/s (Fig. 5.4). More heat can be generated in the summer. Low efficient amorphous silicon PV generates more heat than high efficient PV of the same nominal power output due to the necessarily larger area of amorphous PV roof coverage as well as the less sensitivity to temperature rise as opposed to the mono/polycrystalline counterparts (Noguchi 2013).

Amongst these roof design options, the slope angle of 40° provides the best performance in terms of both heat and power generation. Due to the lowest height amongst the options given, the 30° roof pitch can be considered as most efficient in terms of the building material consumption and the associated initial cost. Nonetheless, it also contributes to lessening the amount of PV heat and power generation, but the expected outcomes will be better than the PV/T roof with an angle of 50°. Thus, amongst the given alterative arrangements, the 50° angled PV/T roof is the worst option in the heat and power generation performance and the most expensive approach to the construction. When the area of the roof coverage becomes double, both low and high efficient PV panels tend to serve nearly twice as much to generate electricity. On the other hand, albeit the vertical extension of the PV roof from 7.14 to 14.28 m (thus, the increase of the roof area from 57.14 to 114.28m^2) studied, the heat production of the amorphous PV roof with an angle of 30° can increase by 6% only when the velocity of ventilation air is limited to 0.5 m/s and about 17% when 1.0 m/s. In the case of the polycrystalline PV under the same condition, the heat production can increase by 25% when the air velocity is set to be 0.5 m/s and 43% increase with the air flow of 1.0 m/s. The ventilation rate of the PV/T roof can be considered as one of the most cost-effective influential factors that

Table 5.1 PV/T system variations simulated using EESLISM

PV Type	Conversion Efficiency	nominal power output	PV/T collector roof area
Amorphous silicone	7%	4kWp	57.14m^2 (8 m in width & 7.14 m in length)
		8kWp	114.28m^2 (8 m in width & 14.28 m in length)
Polycrystalline silicone	14%	4kWp	28.57m^2 (8 m in width:& 3.58 m in length)
		8kWp	57.14m^2 (8 m in width & 7.14 m in length)

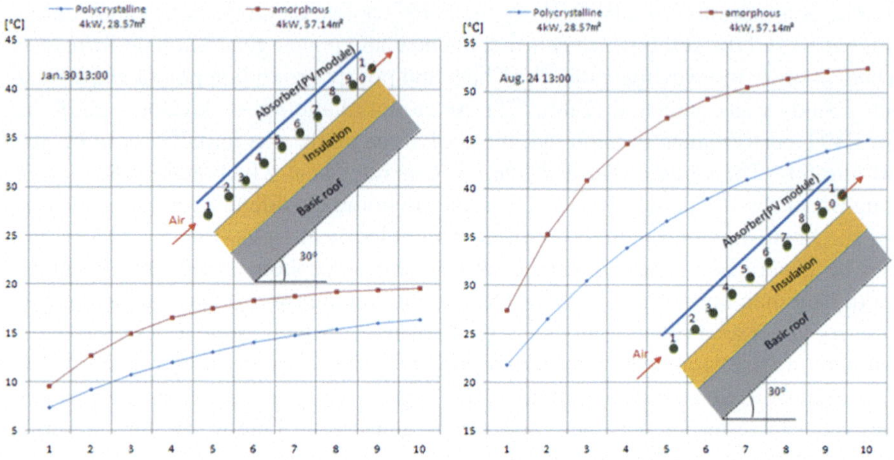

Fig. 5.4 4kWp ventilated PV/T Air temperature profile at the air flow of 300m^3/h, 30th January (left) and of 300m^3/h, 24th August (right)

help improve the heat collecting performance while contributing to cooling the temperature of PV cells. For instance, in the low efficient 4kWp PV roof with an angle of 30°, the annual rate of heat collection can be increased by 77% when the velocity is changed from 0.5 to 1.0 m/s and this approach is about 13 times more efficient than the mere increase of the PV size from 4 to 8kWp. In the case of high efficient 4kWp PV roof, 55% performance improvement can be expected and it is about twice higher than the increase achieved by enlarging the PV size itself.

2.3 *Value-Focused Production and Marketing*

In Japan, housing manufacturers bring mass custom design and production systems into play and reduce the production cost. Contrarily, they tend not to lower their selling price, yet they install advanced energy efficiency and healthy housing measures, as well as renewable energy technologies in their homes as standard features rather than options (Noguchi and Collins 2008). Between 1994 and 2003, the number of domestic PV installations in Japan drastically increased from 539 to 52,863 houses. To alleviate the eccentric appearance of their contemporary solar PV mass custom homes, Japanese manufacturers tend to well integrate the cells into the roofing and façade materials. This quality-centric production approach rather than the simple selling price reduction reflects their high "cost-performance" marketing strategy which is also applied to both the reduction of housing energy consumption for affordability and the improvement of living environments for health and wellbeing (Noguchi 2003). The cost-performance marketing strategy itself is far from new, having been applied to a variety of end user products internationally. For instance, although today's automobiles can be produced with lower production costs than those in the past, their selling price does not seem to be affected dramatically by higher productivity. New cars are still generally regarded as expensive. Nevertheless, the list of items now offered as standard features in new cars, such as air conditioning, a stereo set, airbags, remote-control keys, power steering, power windows, adjustable mirrors and a GPS, were offered only as expensive options in older models. Clearly, the quality of newer models is much higher than that of older models. The same is true for the housing industry. Quality-oriented production contributes towards the delivery of high cost-performance housing in which high-tech modern conveniences that are installed as options in conventional homes are available as standard equipment (Se). In view of this cost-performance marketing strategy, the product subsystem (P) of the mass custom design system model, as described above, can further be modified into the following conceptual model: $P = f(v, e, i, o) + Se$. In fact, Japanese housing manufacturers zero energy mass custom homes in which a variety of housing amenities and renewable energy and environmental technologies (e.g. PV, air source heat pump, home energy management system, and lithium ion battery) tend to be installed as standard features rather than options. Despite the reduction of equipment choices, volumetric, exterior and interior design components remain options from which the users can choose to customise their home. This value-focused production and cost-performance marketing strategy may be considered as effective means to introduce relatively new housing innovations (e.g. renewable energy technologies) to the market where customers may not take expensive design options easily when the buying decisions drastically increase the selling price.

3 Conclusions

This chapter was a first attempt to crystallise the ZEMCH strategic framework for low carbon solutions in sustainable housing delivery and recapitulated the design, production and marketing innovations in view of ZEMCH practices observed in Japan, Canada and Scotland. The mass custom design approach being implemented in Japan was considered as a practical approach to enhancing social, economic and human sustainability in housing delivery, while passive and active solar systems introduced to Canadian and Scottish zero energy homes were regarded as effective means to secure environmental sustainability through the drastic reduction or elimination of operational CO_2 emission that contributes to global warming issues arising today. Nonetheless, passive solar housing features scrutinised in this chapter were based solely on this chapter author's zero energy home design experiences in cool or cold climate alone, where space heating dominates the domestic energy use. Sustainable housing development is a global concern. Accordingly, passive and active cooling design strategies for reduction or elimination of domestic energy use in warm or hot regions should be studied further and integrated into ZEMCH design framework for the applicability to both developed and developing countries.

References

Boardman B (1991) Fuel poverty: from cold homes to affordable warmth. Belhaven Press, New York

Davis SM (1987) Future perfect. Addison-Wesley, New York

Japan Prefabricated Construction Suppliers and Manufacturers Association (JPA) (2019) Purehabu kenchiku no genjyo. http://www.purekyo.or.jp/3-1.html. Accessed 6 Sept 2019

Noguchi M (2001) The mass custom home: vivienda personalizada masiva: optimización de la tecnología de la vivienda canadiense y japanesa en latinoamérica. Simposium internacional: prospectiva urbana en latinoamerica. Secretaría de Desarrollo Social & la Universidad Autónoma de Aguascalientes. Aguascalientes. Mexico. 31 August 2001

Noguchi M (2003) The effect of the quality-oriented production approach on the delivery of prefabricated homes in Japan. J Housing Built Environ 18:353–364

Noguchi M (2008a) Structuring a choice model for mass customisation. Int J Mass Custom 2(3/4):264–281

Noguchi M (2008b) Net zero-energy home design strategies learned from Canadian experience. Open House Int 33(3):88–95

Noguchi M (2011a) Enhancement of industry initiative through the zero-energy mass custom home mission to Japan experience towards commercialisation. Int J Mass Custom 4(1/2):106–121

Noguchi M (2011b) The essence of low-energy affordable housing design strategies learnt from Scottish and Canadian homebuilders' attempt and experience. J Green Build 6(3):59–75

Noguchi M (2013) Calibrating air heating and power generation capacities of low and high efficient BIPV roofs for hosing application in Scotland. Int J Environ Stud 70(6):987–998

Noguchi M (ed) (2016) ZEMCH: toward the delivery of zero energy mass custom homes. Springer, Basel

Noguchi M, Collins D (2008) Commercialisation strategies for net zero-energy-cost housing in Japan. Open House Int 33(3):96–104

Noguchi M, Friedman A (2002) A mass custom design system model for the delivery of quality homes-learning from Japan's prefabricated housing industry. In: Proceedings of International council for research and innovation in building and construction. CIB W060-096 Syllabus joint conference, Measurement and management of architectural value in performance-based building. Hong Kong: CIB, 6–10 May 2002, pp 229–243

Noguchi M, Hernández C (2005) A 'mass custom design' approach to upgrading traditional housing development in Mexico. J Habitat Int 29(2):325–336

Piller F, Tseng MM (eds) (2009) Handbook of research in mass customization and personalization. World Scientific, Hackensack

Pine BJ II (1992) Mass customization: the new frontier in business competition. Harvard Business School Press, Boston

Richard RB, Noguchi M (2006) The Japan solar photovoltaic manufactured housing technical mission 2006. CanmetENERGY, Natural Resources Canada, Varennes

Roberts J (1970) Home-building U.S.A.: a system analysis. Ind Forum 1(3):35–40

Toffler A (1970) Future shock. Bantam, New York

United Nations (2019) Sustainable development goals. https://sustainabledevelopment.un.org/?menu=1300. Accessed 6 Sept 2019

World Commission on Environment and Development (WCED) (1987) Our common future. Oxford University Press, Oxford

ZEMCH Network (2019) ZEMCH workshop Londrina 2014. http://www.zemch.org/zemch_workshop_uel_2014. Accessed 6 Sept 2019

Part III
Dynamics of Energy and Environment

Chapter 6
Role of Hybrid Energy System in Reducing Effects of Climate Change

Somudeep Bhattacharjee, Uttara Das, Moumita Chowdhury, and Champa Nandi

Abstract Climate change is a very rising topic nowadays since the climate of this world is changing rapidly day by day. In the technical field, it is seen that so many things or techniques used here, which have a very bad impact on our environment like use of non-renewable energy source, emission of greenhouse gases and so on. At present electric power generation is mainly dependent upon non-renewable sources. Due to rapid uses of non-renewable energy sources, its storage reserves are decreasing rapidly. So an alternate source is required and that is the renewable energy source, nowadays renewable sources are utilized but in small amount. Renewable sources are environmentally friendly, so using of renewable energy sources are more preferable than non-renewable sources for the betterment of our environment. Due to rising environmental concerns day by day, the utilization of renewable energy need to be increased as much as possible. There are so many remote or island places in this world where huge numbers of renewable sources are available which can be used for power generation. And the most important thing is that they have no effect (or very less effect) in this environment. So our goal is to model and simulate a grid connected solar-wind hybrid energy system which is used to solve the problems regarding the power generation. In this chapter a 24 h case study analysis is done by taking the real time data of solar radiation and wind speed of a selected location. The results of this analysis indicate that the hybrid system is profitable and environmentally friendly. This analysis simply gives an idea about to what extent there will be the generation of power and how much it will be helpful to this environment. In addition, it includes detailed discussion on climate change, harmful effects of non-renewable energy sources on the environment and the need of renewable energy based hybrid energy system to combat climate change. By this explanation we will get to know more about how renewable energy sources mitigate two problems – climate change & power demand.

S. Bhattacharjee · U. Das · M. Chowdhury · C. Nandi (✉)
Department of Electrical Engineering, Tripura University, Agartala, India
e-mail: somudeep.ee@tripurauniv.in; cnandi@tripurauniv.in

H. Qudrat-Ullah, M. Asif (eds.), *Dynamics of Energy, Environment and Economy*, Lecture Notes in Energy 77, https://doi.org/10.1007/978-3-030-43578-3_6

Keywords Climate change · Renewable energy sources · Solar power · Wind power · Grid · Hybrid energy system

1 Introduction

Climate change is a very raising topic nowadays. Climate change takes place when changes happen in the climate of the earth and result in new weather patterns which last for two to three decades or maybe for millions of years (Climate Change of Earth 2019). The climate can change by some nature activities like cyclical ocean patterns, Atlantic multi-decadal oscillation, volcanism etc. Climate change can be happened by human activities too. Burning of fossil fuels and cutting trees are mainly responsible for changing the climate of this world drastically and by doing such activities; humans are impacting the climate and temperature very badly. The huge amount of greenhouse gasses is coming out of the burning of fossil fuels to the atmosphere which has a big role in global warming. These green house gasses trap the heat of the sun and don't allow them to go back to the space. Some of them are produced naturally but most of them are produced by human activities which are increasing so fast, like CO_2, CH_4, N_2O & fluorinated gasses. Some of the major impacts of climate change are extreme heat, changing patterns in rainfall, drought, decreasing ground water, glacier melting and so on (Cause of Climate Change 2019). The impacts of non-renewable energy sources not only in the form of greenhouse gas emissions, but it also affects the environment in various forms. Non-renewable energy resources emit lots of pollutants which affect human health and the environment. For example, coal fired power plants are the largest source of mercury in many countries. If mercury is emitted into the air, then it stores on the ground water which can affect our health through the food chain. In addition, the effects of non-renewable energy not only present in atmospheric temperature and air, its impact also falls on water cycle. When sulfur and other chemicals are produced from the industry and it mixed in the atmosphere which turns the rain mildly acidic. Acid rain also damages the trees and weak the forest ecosystems, it created the corrosion on the machinery equipment. In addition, the environmental impact also comes from the use of non-renewable resources and disposal of their waste. The effect of surface mining in short term, as well as long term, a clear proof of land pollution. Huge volume of excess rock and soil are discarded in the location nearby valleys also affect that ecosystem. At the end, there are other hazards are present in the environment except air, water, land pollution. An oil spill is very dangerous and destroying matter near the shores and its ecosystem (Effect of Non-Renewable Resources on the Environment 2019). Therefore, for protecting the environment from the dangerous impacts of climate change and non-renewable energy sources, the utilization of renewable energy sources needs to be increased at a very large scale. At present electric power generation is mainly dependent upon non-renewable sources. Due to rapid uses of non-renewable energy sources, its reserves are decreasing at a faster rate. So an alternate source is required and that is the renewable energy source, nowadays renewable energy sources are utilized but in smaller amounts. Since the environmental concerns are increasing day by day so the utilization of renewable energy is required as much as possible. Renewable resources (like solar, wind, biomass and hydro power) are available in many places that can easily used to generate power. But most of the renewable energy sources are fluctuating in nature. Solar power generation depends on sunlight which is not present at night. Wind power generation depends on wind speed which is very fluctuating all over the day. Thus, it is not possible to maintain continuity of power supply from individual renewable energy sources. But it is possible to maintain the continuity of power supply by utilizing the hybrid energy system concept. A hybrid energy system is an integrated form of two or more energy sources in such a way that the continuity of power supply is maintained (Bhattacharjee et al. 2018). So our goal is to model and simulate a grid connected solar-wind hybrid energy system which is used to solve the problems regarding the power generation. In this chapter a 24 h case study analysis is done by taking the real time data of solar radiation and wind speed of a selected location. It also discusses about the topic climate change in detail. In addition, it includes discussion on the harmful effects of non-renewable energy sources on the environment and the need of renewable energy to combat climate change. The results of this analysis indicate that the hybrid system is profitable and environmentally friendly. This analysis simply gives an idea about to what extent there will be the generation of power and how much it will be helpful to this environment. By this explanation we will get to know more about how renewable energy sources mitigate two problems – climate change & power demand. Increasing dangerous impacts of climate change and rising power demand are the two major issues which are necessary to be controlled for protecting environment. Therefore, this study provides a solution in terms of hybrid energy system by utilizing renewable energy sources.

This chapter consists of following sections apart from the introduction. Section 2 describes about the climate change with its causes and impacts. Section 3 describes about the harmful effects of non-renewable energy sources in the environment. Section 4 describes about the need of renewable energy based hybrid energy system to battle against climate change. Section 5 describes about the system description of grid connected solar-wind hybrid energy system. Section 6 describes about the simulation model of the hybrid power plant. Section 7 describes about the case study on Kavaratti, Lakshadweep, India. Section 8 describes about the conclusion.

2 Climate Change with Its Causes and Impacts

2.1 Introduction

Climate change is a very raising topic nowadays. Climate change takes place when changes happen in the climate of the earth and result in new weather patterns which last for two to three decades or maybe for millions of years. Earth's climate system mainly takes its all energy from the sun and a little amount of the earth's interior. Earth's climate also gives energy to the outer space. The budget of earth's energy mainly depends on the stability of incoming & outgoing energy, and the route of the energy through the climate system. If the outgoing energy is more than incoming energy then the climate system of the earth is cooling and if the incoming energy is more than outgoing energy then the climate system of the earth is warming. As the climate of the earth helps to move the energy, earth's weather is created by it. In short, the phenomenon of changes in the climate is known as climate change. The climate can change by some nature activities like cyclical ocean patterns, Atlantic multi-decadal oscillation, volcanism etc. Climate change can be happened by human activities too. Climate change is a big cause of global warming. That's why its need to be reduced as soon as possible (Climate Change of Earth 2019).

2.2 *Causes of Climate Change*

The climate can change by some nature activities like cyclical ocean patterns, Atlantic multi-decadal oscillation, volcanism etc. Climate change can be happened by human activities too. Burning of fossil fuels and cutting trees are mainly responsible for changing the climate of this world drastically and by doing such activities; humans are impacting the climate and temperature very badly. The huge amount of greenhouse gasses is coming out of the burning of fossil fuels to the atmosphere which has a big role in global warming. These green house gasses trap the heat of the sun and don't allow them to go back to the space. Some of them are produced naturally but most of them are produced by human activities which are increasing so fast, like CO_2, CH_4, N_2O & fluorinated gasses. CO_2 is a greenhouse gas, which is mostly responsible for global warming. Other gasses also discharged in fewer amounts by various processes, but catch heat in more amount than CO_2. Burning fossil fuels helps to increase the level of CO_2 & N_2O the atmosphere. Normally trees help to absorb excess CO_2, but due to huge deforestation, the level of CO_2 is rising day by day. Fertilizers which contain nitrogen are responsible for producing N_2O emission. Fluorinated gases are responsible for global warming up to twenty three thousand times more than CO_2. In short, human activities are mainly responsible for global warming. The average temperature of industrial area is 2 °C more than non-industrial area. Because of this global warming world's climates are changing drastically. Present global warming temperature is greater than earlier by 0.85 °C which has a major impact on climate (Cause of Climate Change 2019).

2.3 *Impacts of Climate Change in India*

Extreme heat is the first major impact of climate change. Climate of India is already becoming warmer than earlier, if it goes on then within a few years a big part of India will experience an unexpected high temperature. Changing patterns of rainfall are the second major impact of climate change. Since 1950, it has been noticed that a decreasing level in monsoon rainfall. But the regularity of intense rainfall is increased, because of these changes climate of such areas has also changed. A drought is the third major impact of climate change. Various reports prove that various parts of South Asia have become drier than earlier. If it goes on then the days are not so far when the drought will become a common matter in all over the world. Decreasing ground water is the fourth major impact of climate change. Indian agriculture, mostly (more than 60%) depended on ground water. India highly depended on ground water for fulfilling its needs. If ground water decreases in this way, then the day is very near when ground water level will reach to zero or near zero. Glacier melt is the fifth major impact of climate change. At 2.5 °C warming, dissolving of glaciers of different parts of the Himalayas may threaten the stability and reliability of the water level in the glacier-fed rivers (Cause of Climate Change 2019). These may cause flood like situation in nearby areas of rivers. Sea level rise is the fourth major impact of climate change. This rise of sea level causes flood like situation in areas like Mumbai, Kolkata etc., which are near to sea. It may contaminate drinking water and raise many diseases like diarrhea and cholera. Climate change may also disturb the food productivity and hydro power generation (Cause of Climate Change 2019). These are some of the major impacts of climate change in India.

3 Harmful Effects of Non-renewable Energy Sources on Environment

The usage of non-renewable energy resources has so many bad impacts on our health & environment. Human health and the wellbeing of this environment, these two are connected with the use of non-renewable energy sources so intimately. There are different kinds of harmful effects on this environment either it is in the way they are processed and extracted or in terms of how they are used. The harmful effects are variable based on the types of non-renewable resource used. Mainly there are five such ways by which non-renewable energy sources affect our health and environment (Effect of Non-Renewable Resources on the Environment 2019). These five ways are:

3.1 Greenhouse Gas Emissions

The main reason behind climate change is mainly due to increase of greenhouse gas emissions in the environment. Main greenhouse gases are carbon dioxide and methane which is mainly obtained from the use of non-renewable resources. Various types of non-renewable resources, discharge various levels of greenhouse gases like coal, which produce high amounts of carbon dioxide when it burn. It has been found from US history, 71% of carbon dioxide released from the electric power sector, which comes from the burning of coal, in 2015, where natural gas discharges 28%. On the other hand, natural gas discharges 50–60% less carbon dioxide compared to coal, also discharges less heat-trapping gases, which trapped heat when it is used to power a vehicle, which is 15–20% compared to gasoline. Methane, which is 34 times stronger than CO_2 in terms of ability to trap the heat, is found as leakage during drilling and extraction of natural gas from wells. So, natural gas is also dangerous as increasing greenhouse gas. Greenhouse gas and climate change are coordinated to each other, climate change is the reason of the change in weather patterns and rising temperatures as a result of flood and drought which further changes the living in humans. Although, climate change is directly effective for the ecosystem, change the biodiversity of the world, overall, the whole change in normal livelihood (Effect of Non-Renewable Resources on the Environment 2019).

3.2 Air Pollution

Non-renewable energy resources do not only emit greenhouse gases; they also emit lots of pollutants which affect human health and the environment. For example, coal fired power plants are the largest source of mercury in many countries. If mercury is emitted into the air, then it stores on the ground water which can affect our health through the food chain. This mercury has not only effects on the biodiversity, but also affect human health so badly as the study says that mercury can lead neurological problem in embryos and young children. Another important source of air pollutant is fossil fuel combustion, which includes sulfur dioxide, nitrogen oxides and particulate matter (Effect of Non-Renewable Resources on the Environment 2019).

3.3 Acid Rain and Water Pollution

The effects of non-renewable energy not only present in atmospheric temperature and air, its impact also falls on water cycle. When sulfur and other chemicals are produced from the industry and it mixed in the atmosphere which turns the rain mildly acidic. Acid rain also damages the trees and weak the forest ecosystems, it created the corrosion on the machinery equipment. In the change of acidic level of lakes, streams and different water resources, acid rain play a very threatening role which may be effective in the life of fish and other aquatic organisms. Another way of water pollution is thermal pollution, which means use of fossil fuel like coal, nuclear energy. Every fossil fuel, nuclear power plant uses water to running and cooling purposes of power plant. After using when plants release the water, its temperature has been changed and decayed quality. This heated water reintroduced and sometimes used in the locality for the ecosystem or agriculture, which contains lower dissolved oxygen level. This water makes stress in wildlife like increase the heart rate of fish, decreasing fertility (Effect of Non-Renewable Resources on the Environment 2019).

3.4 Land Pollution and Waste Generation

Environmental impact comes from the use of non-renewable resources and disposal of their waste. The effect of surface mining in short term, as well as long term, a clear proof of pollution. Huge volume of excess rock and soil are discarded in the location nearby valleys which affect that ecosystem. Due to the mining in terms of a long process, the soil quality becomes poor and sometimes chemical used in land near water resources make the land as well as water pollution. In extraction of nuclear energy like uranium, and disposal of its waste create some very deprecatory issues which have no long-term solution due to the radioactive nature element being mined (Effect of Non-Renewable Resources on the Environment 2019).

3.5 Oil Spills and Other Accidents

At the end, there are other hazards are present in the environment except air, water, land pollution. An oil spill is very dangerous and destroying matter near the shores and its ecosystem. In referring to this, British Petroleum's losses $2.5 Billion in oil spilling in 2010 in the Louisiana fishing market alone, where Florida losses $3 Billion from the income from tourism. Due to the rub out of oil from the Gulf of Mexico, biologists are suspicious about the species of algae, hundreds of species of animals. In Chernobyl, the devastating effects are seen as a result of the nuclear disaster in nature and humans both. Due to negligence, technology failure, lack of preparedness or a combination, these accidents are occurring (Effect of Non-Renewable Resources on the Environment 2019).

4 Need of Renewable Energy Based Hybrid Energy System to Battle Against Climate Change

At the current time, the role of electrical energy is very crucial. Without it, human beings nowadays couldn't imagine his life as it controls some of the crucial needs of human beings directly or indirectly. Electrical energy plays a crucial role in society as its accessibility lead to the progress of electrical appliances in the workplace, which include light bulbs, water pump, television, computer, mobile charging and so on. At home, these electrical appliances establish a special place as they are time-savers, reliable, safer and easy to use. So human beings couldn't even think of surviving without electrical energy. In the absence of electrical energy, the life of human beings becomes difficult and boring. The societal and financial progress in the history of human beings is decided by the energy availability and accessibility. Due to this reason, the answers were investigated for delivering the energy demand for the requirement of the societal and financial progress. In the current time, a large of countries, including India generates a large amount of power using thermal power plant which is a dangerous issue as it increases the dangerous impacts of climate change. Due to this, the necessity of utilization of renewable energy sources for supplying the electrical energy is highly vital for protecting the environment from the utilization of non-renewable energy sources. Due to increasing global warming and rising energy demand, the interest of people now moves towards those energy sources which are environmentally friendly for saving the earth for the future and currently good thing is that there are never-ending sources of energy which are being developed and modernize. For instance, solar energy which is generated from the sunbeams, wind energy which is generated from the high speed moving winds, hydro energy which is generated from the intensity of running water and so on. The utilization of renewable energy sources in large scale can amplify the energy security for both the current and future generation. Renewable energy utilization is rising every year with the focus on various types of renewable energy sources, but the speed of progress is very low. It is a fact that it was not possible to maintain continuity of power supply by utilizing individual renewable energy sources. But it is possible to maintain the continuity of power supply by utilizing the hybrid energy system concept. A hybrid energy system is an integrated form of two or more energy sources in such a way that the continuity of power supply is maintained. One of the most popular hybrid energy systems is photovoltaic (PV) -wind turbine-diesel generator based hybrid energy system. The utilization of renewable energy technologies diminishes the use of non-renewable energy sources. Some of the major advantages of hybrid energy system are:

- It saves fuel upto 50%.
- It is a silent system and its maintenance cost is low
- It is connected to other energy supplies (solar panels, wind turbines and so on)
- It has low unpredictability of system efficiency and its results are more consistent.
- Its life is long as it provides electrical energy for more than 20–25 years.
- It keeps the environment clean by producing very less or no emission of greenhouse gases in the atmosphere.
- It is a cost-effective system as it provides high power quality and reliability.

It is a true fact that the hybrid energy system also possesses some drawbacks, but still a hybrid energy system is better as its usage reduces the dangerous impacts of climate change. Some of these drawbacks are: It requires a large initial cost to setup the project, it is a complex system as the integration of two or more energy sources is difficult and so on. The drawback of the initial cost is easy to manage since the lifetime earnings of hybrid system easily cover the initial investment and provide an extra large profit. In addition, the technologies of renewable energy and the technologies of renewable energy integration are improving day by day, which can easily solve the problems of complexity in the integration of two or more renewable energy sources (Importance of Electricity – How It Changed People's Lives 2012) (Marisarla and Kumar 2013) (The Impact of Electricity on Society 2019) (Vuc et al. 2013) (Paska et al. 2009) (Soon 2015) (Bhikabhai 2005) (Advantages of the Hybrid System 2019) (Nandi et al. 2019). Therefore, hybrid energy system is fully capable of solving the problem of rising dangerous impacts of climate change and the excessive utilization of non-renewable energy sources by integrating two or more renewable energy sources in such a way that the continuity of power supply is maintained.

5 System Description of Grid Connected Solar-Wind Hybrid Energy System

5.1 Block Diagram

The block diagram of grid connected solar wind hybrid energy system is shown in Fig. 6.1. The block diagram of hybrid system consists of five different parts: PV array, wind farm, converter, electric load and grid. In this figure, the inverter is represented by converter as shown in Fig. 6.1. In this hybrid system, a protection system is utilized which is used to break the circuit in case of low voltage. The grid is connected to the hybrid energy system for consuming excess energy generated from the hybrid energy system or to supply the required energy for fulfilling the load demand during requirement. In this hybrid system, PV array and wind farm are the two energy sources which are used to fulfill the load demand.

5.2 Mathematical Modelling of PV Array System

The circuit diagram of solar cell is shown in Fig. 6.2. In designing the PV array system, a fixed number of solar cells are connected in series to build one solar module. Then a fixed number of series connected solar modules are connected in a string and then a fixed number of modules connected strings are connected in parallel to

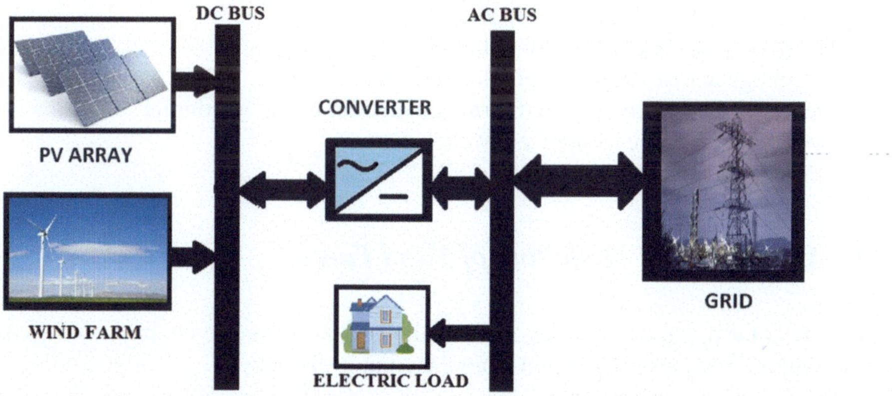

Fig. 6.1 The block diagram of grid connected solar wind hybrid energy system

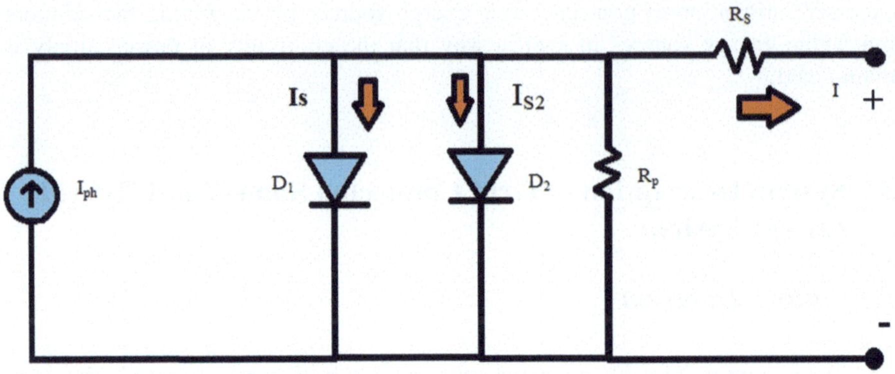

Fig. 6.2 The circuit diagram of solar cell

obtain the desired power output. This PV array system transforms the sunbeams into electrical energy which is increased using MPPT (Maximum Power Point Tracker) with a boost converter.

The output current of the solar cell is given by.

$$I = I_{PH} - I_S * \left(e^{[(V+I*Rs)/(N*Vt)]} - 1\right) - I_{S2} * \left(e^{[(V+I*Rs)/(N2*Vt)]} - 1\right) - \left[\left(V + I * R_S\right) / R_P\right]$$

Where, I is the output current (ampere),

I_{PH} is the photo generated current (ampere),
I_s is the saturation current of the first diode,
I_{S2} is the saturation current of the second diode,
V_t is the thermal voltage, $V_t = (kT/q)$,
k is the Boltzmann constant,
T is the device simulation temperature,
q is the elementary charge on an electron,
N is the quality factor of the first diode,
N_2 is the quality factor of the second diode,
V is the voltage across the solar cell electrical ports,
R_S is the series resistance and wind farm, a fixed number of windmills
R_P is the shunt resistance (Solar Cell Model 2019).

5.3 Mathematical Modelling of Wind Farm

In this wind farm, a fixed number of windmills are connected to obtain the desired power output. The power extracted from the wind is given by.

$$P_{WIND} = C_P(\lambda, \beta) .\bullet \rho \bullet A \bullet v^3$$

Where, ρ is the air density in atmosphere (kg/m^3),

A is cross-sectional area of a turbine blade (m^2),
V is wind velocity (m/s) and.
C_p is the wind turbine energy conversion coefficient. The rotor power coefficient is usually given as a function of two parameters: the tip speed ratio λ and the blade pitch angle β (in degrees). The blade pitch angle is defined as the angle between the plane of rotation and the blade cross-section chord. And the tip speed ratio is defined as the ratio between the speed of the tips of the blades of a wind turbine and the speed of the wind (Dave and Sinha 2017).

In the wind farm, a fixed number of windmills are connected and in one windmill, a wind turbine is present, which is connected to a permanent magnet synchronous machine to obtain the desired AC power output. The wind turbine transforms the kinetic energy of wind into mechanical power which is later converted into AC electrical power using a permanent magnet synchronous machine. This AC electrical power is converted into DC power using rectifier which is increased using MPPT (Maximum Power Point Tracker) with a boost converter.

5.4 Literature Review

A literature review of several research works related to the integration of different renewable energy sources and its protection is carried out to understand the progress. In (Comodi et al. 2015), the authors proposed an on-grid PV and microgas turbine hybrid system with management policy in the year 2015. In this paper, the energy management system has been designed by the linear programming method. The load demand has been considered to be fixed. The proposed algorithm mainly solves the problem which immerged because of uncertainty of solar irradiance in PV panels. The paper also aims at reducing the emission of toxic pollutants which emerged from a traditional power generating system. The strategy helps in injecting more power to the grid (Comodi et al. 2015). In (Abushnaf and Rassau 2018), the authors developed a novel energy management system for grid supported PV/Battery system in the year 2018. In this paper, genetic algorithm is applied to energy management strategy. Here, the system is applied only for fixed types of load. The scheme reduces the cost of energy production and moreover the optimal use of the components likes: solar panel, battery, DC-AC converter is also computed. The scheme easily fulfills the load demand (Abushnaf and Rassau 2018). In (Aktas et al. 2018), the authors proposed an energy management for grid connected PV system with hybrid storage facility comprising of battery and ultra-capacitor in the year 2018. The energy management is designed by linear programming in MATLAB environment. The proposed process can satisfy the load and can sell excess power to the grid. The paper introduces the concept of Phase Locked Loop in renewable energy management Scheme (Aktas et al. 2018). In (Nge et al. 2019), the authors proposed a dynamic management system of energy for grid supported PV and battery storage during the year 2019. The energy management system maximizes the energy production from renewable and injects excess to the grid while keeping the battery fully charged. In this paper solution to the optimization predicament is computed by Lagrangian technique. In this paper the load demand is taken as fixed. The whole method of EMS is verified with dynamic programming of brute-force. The algorithm easily solves the problems occurred in energy management due uncertainty in PV generations (Nge et al. 2019). Microgrid is also a type of hybrid energy system where two or more local energy sources are integrated with the energy storage system for fulfilling the electric load demand of a small area. Therefore, some research works related to microgrid are also discussed here. In (Mirsaeidi et al. 2017a), the author proposed the concept of hybrid AC/DC micro-grid with the advantage of AC and DC architectures in the year 2017. These AC and DC sub-grids of micro-grid are connected in the same distributed grid, which gives the facility to direct integration of both AC- and DC-based DG sources, energy storage systems (ESSs) and loads. To minimize the modifications of the current distributed grid and reducing the total cost of upcoming renewable energy sources (RES) or electric vehicles (EV), this feature of AC-DC micro-grid provides an efficient way (Mirsaeidi et al. 2017a). In (Liu et al. 2011), for reducing the processes of multiple DC–AC–DC or AC–DC–AC conversions in an individual AC or DC grid, the author proposed a hybrid AC/DC micro-grid in the year 2011. Both AC and DC network are connected together with multi-bi-directional converters are making the hybrid grid. Although, DC sources and loads are connected in the DC network, whereas AC sources and loads are ties in the AC network. The suggested hybrid grid can operate in a grid-tied or self-governing mode. To stable the system operation under various generations and load conditions for smoothing the power transfer between AC and DC links, the coordinate control algorithm has been proposed (Liu et al. 2011). In (Gaur and Singh 2017), the author represents shortcoming confronted and different technical issue in installing of micro-grid, especially in rural areas in the year 2017. It highlights some aspects of stability in micro-grids and discusses the technical challenges involved in grid integration of micro-grid. This paper reviewed the different issues in implementations of micro-grid technology. It focuses on active and current progress in micro-grid research, especially in Europe and America. Voltage-frequency control, islanding and protection are the three key issues of technical challenges that must be overcome for effective implementation of micro-grid. Improvement of power quality as seen by the end user is contributed by the integration of small scale production in the form of micro-grid that supports the inclusion of power electronics devices (Gaur and Singh 2017). In (Lai et al. 2015), the author suggests a secure operation and comprehensive protection strategy for insuring dependable of an island micro-grid system in the year 2015. To prevent unnecessary loss of critical load and distributed generators (DGs), the paper implemented the use of microprocessor based relay. To clear the way of plug and play of distributed generators (DGs), different improvements are proposed. To tackle the exclusive high impedance fault in distributed system, different recommendations are present. Moreover, different analyses are carried out to demonstrate the protection strategy (Lai et al. 2015). In (Mirsaeidi et al. 2017b), the author propounds a review paper on different comprehensive strategies for protection of hybrid AC/DC micro-grids in the year 2017. Apart from this, it classifies them into specific groups and critically analyzes the main challenges of each approach. Finally, some recommendations are found out to solve the protection problems (Mirsaeidi et al. 2017b). In (Hussain and Kim 2016), the author represents a hybrid adaptive protection scheme for the protection of micro-grids in the year 2016. Although, a gateway is proposed to maintain the communication between the serial interface device and the IEC 61850 process bus. Finally, a framework is introduced by using IEC 61850-based intelligent electronic devices (IEDs) for the protection of micro-grid (Hussain and Kim 2016). In (There et al. 2011), for delivering a micro-grid frequency response-as an ultra-capacitor and inverter for delivering short-term energy response, the author proposed an energy storage system in the year 2011. Depending on measured condition, a control system is created to monitor the grid frequency and active an inverter to either charge or discharge an ultra-capacitor. Finally, a simulation result is described with their major components (There et al. 2011). In (Patil et al. 2015), the author propounds the application of super capacitor as an energy storage system connected to micro-grid in the year 2015. This proposal explored here to address carbon and energy emission concerned into global issues. It discusses the design and the process of development of a dynamic support system for specific micro-grid applications. It also represents the hardware and software design and development of 7.8 kW single-phase converter connected to the 125 V, 100 Wh super capacitor bank (Patil et al. 2015). In (Khorasani et al. 2017), for improving the power quality and to compensate the reactive power of an AC micro-grid using DC bus capacity, the Author suggests a new global solution with new design of hybrid AC/DC micro-grid in the year 2017. This paper also proposed new design of back to back connection of two series and parallel converters, as well as new controllers and simultaneous uses of earthing switches. Under different power quality problem (e.g., interruption, sag, harmonics, and any variation of voltage/current signal from pure sinusoidal), their proposed method guarantees to supply quality voltage and drawn current to consumers from the network. Furthermore, even in the worst load quality condition, the proposed design of new AC/DC micro-grid will help to achieve the operation of network under islanded mode in accordance with power quality standard (Khorasani et al. 2017). In (Hosseini et al. 2016), the author reviewed the structure and topology of micro-grid and proposed a summary of studies in the field of protection in the year 2016. Afterward, the challenges and solving methods of protection of micro-grid through DG-equipped network are discussed. In this paper, six main categories of solving method of protection problem are described. Finally, it also identifies the factor of micro-grid type and topology, DG type, communication type and delay type, method of fault detection and analysis, relay type, fault type, method of grounding and use of smart transformer in micro-grid (Hosseini et al. 2016).

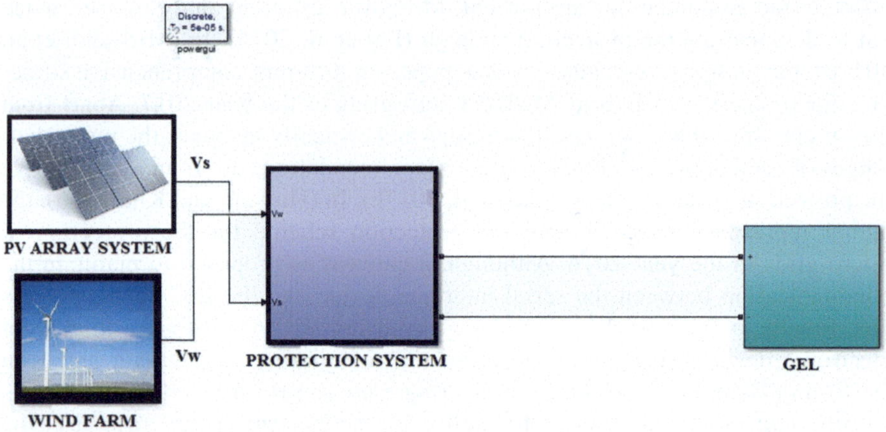

Fig. 6.3 The simulation model of hybrid energy system

6 Simulation Model of the Hybrid Power Plant

A simulation model of grid connected solar–wind hybrid energy system is designed, which is shown in Fig. 6.3. The simulation model of hybrid energy system consists of four main parts: PV array, wind farm, protection system and GEL. The full form of GEL is Grid and Electric load. GEL represents a DC grid and it consists of an inverter, electric load and main AC grid connection.

6.1 PV Array System

In the simulation model of the PV array system, 120 solar cells are connected in series to build one solar module. Then 600 series connected solar modules are connected in a string and then 400 modules connected strings are connected in parallel to obtain the desired power output. This PV array system transforms the sunbeams into electrical energy which is increased using MPPT (Maximum Power Point Tracker) with a boost converter. The total rated capacity of the PV array system is 13 MW and the final output voltage Vs of PV array system is sent to the protection system.

6.2 Wind Farm

In the simulation model of wind farm, 30 windmills are connected. In one windmill, a wind turbine is present, which is connected to a permanent magnet synchronous machine to obtain the desired AC power output. The wind turbine transforms the kinetic energy of wind into mechanical power which is later converted into AC electrical power using a permanent magnet synchronous machine. This AC electrical power is converted into DC power using rectifier which is increased using MPPT (Maximum Power Point Tracker) with a boost converter. The total rated capacity of wind farm is 600 kW (with MPPT) and the final output voltage Vw of wind farm is send to the protection system.

6.3 GEL (Grid and Electric Load)

GEL represents a DC grid and it consists of an inverter, electric load and main AC grid connection. In the GEL simulation model, the DC input voltage is converted into AC voltage using IGBT two-level inverter. Then this AC voltage is used to fulfill 50 Hz electric load of 50 kW. For maintaining grid stability, a voltage source (indicating grid) of 50 Hz is connected near to electric load, which is used to feed the electric load whenever required. The grid is connected to the hybrid energy system for consuming excess energy generated from hybrid energy system or to supply the required energy for fulfilling the load demand during requirement.

6.4 Protection System

The protection system is based on a protection algorithm. Based upon the protection algorithm, the protection system decides its control actions. The protection system takes input of solar DC voltage Vs and wind DC voltage Vw which is used to calculate total voltage V (Vs + Vw = V). The flowchart of protection algorithm is shown in Fig. 6.4.

The working of protection system is based on protection algorithm where the total voltage V is checked that it is more than or equal to Vreq or not. Vreq (440 V) is the minimum DC grid voltage below which, if the voltage enters into GEL then grid stability problem arise. If the total voltage V is more than or equal to Vreq, then at that time, total DC voltage is sent to GEL where it is converted into AC voltage using an inverter and then send to fulfilled the load demand of 50 kW and excess energy send to AC grid. If the total voltage V is less than Vreq then at that time, total DC voltage is disconnected from GEL (i.e., disconnect wind farm and PV array system from GEL). In this situation, if the AC grid is available, then it is used to fulfill the load demand of 50 kW, otherwise load shedding occurs. Since this protection algorithm managing the flow of energy so it can be considered as an energy management algorithm.

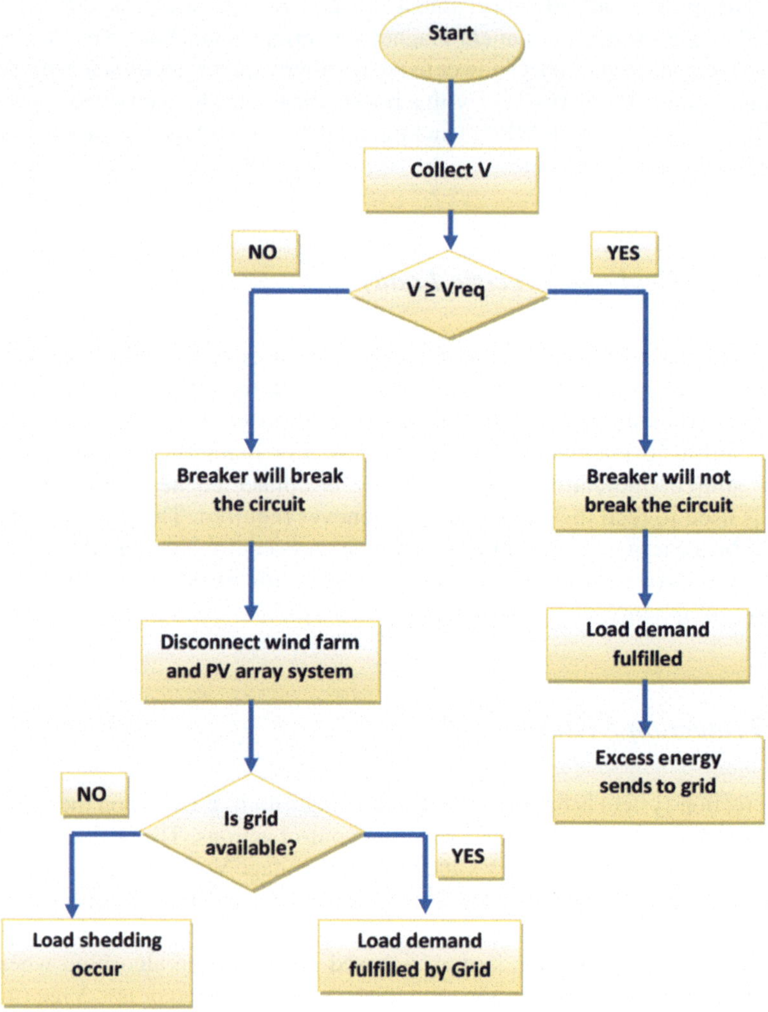

Fig. 6.4 The flowchart of protection algorithm

7 Case Study on Kavaratti, Lakshadweep, India

The feasibility of the proposed hybrid energy system plays a crucial role and so for performing feasibility analyzing, a 24 h case study analysis is done on Kavaratti, Lakshadweep, India by utilizing the real time data of solar irradiance and wind speed. The latitude and longitude of the chosen location of Kavaratti, Lakshadweep, India, are 10 degrees 33.6 minutes north, 72 degrees 38.6 minutes east.

7.1 Solar Radiation, Clearness Index, Wind Speed and Ambient Temperature

This section describes the climatic conditions of the selected location which is necessary to be consider for proper analyzing of the feasibility of the hybrid energy system. These climatic conditions are solar radiation (kWh/m^2/d), clearness index, wind speed (m/s) and ambient temperature (°C) of the chosen location of Kavaratti, Lakshadweep, India. Figure 6.5 indicates the average monthly solar radiation (kWh/m^2/d) and average monthly clearness index of the chosen location. Figure 6.6 indicates the average monthly wind speed (m/s) of the chosen location. Table 6.1 indicates the monthly ambient temperature in terms of minimum, daily low, average, daily high and maximum temperature in °C. All these real time data on the chosen location are obtained from the database of NASA Prediction of Worldwide Energy Resources (NASA Prediction of Worldwide Energy Resources 2019).

The average solar radiation in Kavaratti area is 5.76kWh/ m^2/day for an annum as shown in Fig. 6.5. The speed of wind is taken 50 m above the ground level of Kavaratti area. The range of the wind speed in Kavaratti area lies between 3.49 and 8.99 m/s which shown in Fig. 6.6. The maximum speed of the wind is found in the month of June at a speed of 5.21 m/s. The average annual temperature in Kavaratti area is 27.3319 °C.

All these real-time data (Figs. 6.5 and 6.6 and Table 6.1) simply indicate that the chosen location has the best possibility of solar and wind power generation, especially in the form of the hybrid energy system. Therefore, the chosen location is good for renewable energy generation.

7.2 Hourly Energy Generation Analysis

This case study analysis is mainly based on the 24 h based real time data of solar radiation (kW/m^2) and wind speed (m/s) of one day of February month, obtained from the website of NASA Prediction of Worldwide Energy Resources (NASA Prediction of Worldwide Energy Resources 2019). This data of February month mainly used to perform a 24 h analysis for understanding all the different situations that the hybrid system mainly faces. By utilizing this data in the simulation model of hybrid energy system, solar power, solar voltage, wind power and wind voltage are obtained as shown in Table 6.2. Table 6.2 indicates the hourly results of the hybrid energy system. To expand the use of renewable energy sources for controlling the dangerous impacts of climate change in Kavaratti, Lakshadweep Island, this analysis promotes the utilization of solar and wind power instead of non-renewable energy sources based power generation using the hybrid energy system concept. Figure 6.7 indicates the hourly scenario of solar power in kW. Figure 6.8 indicates the hourly scenario of wind power in kW. Figure 6.9 indicates the hourly scenario of excess power sending to the grid after fulfilling the load demand and Fig. 6.10

indicates the hourly scenario of required power taking from the grid for fulfilling the load demand in kW. The total energy generation and consumption scenario of hybrid energy system in Kavaratti, Lakshadweep, India is shown in Table 6.3.

From Figs. 6.5, 6.6 and Table 6.1, it is clear that the chosen location has the best possibility of solar and wind power generation, especially in the form of the hybrid energy system. From the Table 6.2, Figs. 6.6, 6.7, 6.8, 6.9, 6.10 and Table 6.3, it is clear that the hybrid system is profitable, feasible and environmentally friendly. Since this hybrid system utilizes solar and wind power sources so no greenhouse gases emitted, therefore the hybrid system is environmentally friendly. The total excess energy sending to the grid after fulfilling the load demand is more than the required energy taking from the grid for fulfilling the load demand, which indicates that the hybrid system is profitable. In the time duration of 0:00 to 5:00 h (12:00–5:00 AM), solar power is not available so the load demand is fulfilled by the

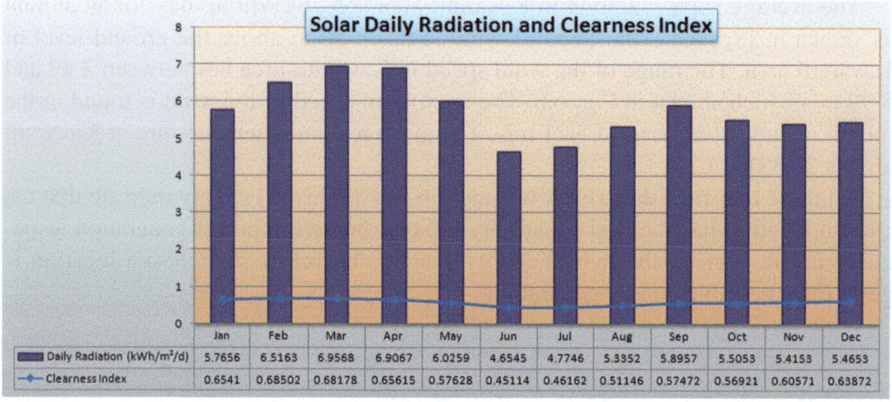

Fig. 6.5 The average solar radiation (kWh/m^2/d) and clearness index of Kavaratti, Lakshadweep, India

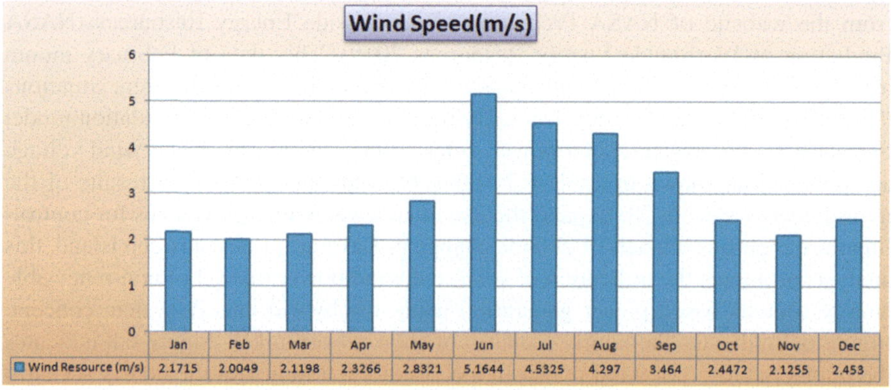

Fig. 6.6 The average wind speed (m/s) of Kavaratti, Lakshadweep, India

Table 6.1 The monthly average data of ambient temperature (°C)

Ambient temperature in °C					
Month	Minimum °C	Daily low °C	Average °C	Daily high °C	Maximum °C
January	27.1577	27.1577	27.1576	27.1577	27.1577
February	26.8477	26.8477	26.8477	26.8477	26.8477
March	27.1777	27.1777	27.1777	27.1777	27.1777
April	27.6977	27.6977	27.6975	27.6977	27.6977
May	28.1276	28.1276	28.1276	28.1276	28.1276
June	27.9976	27.9976	27.9979	27.9976	27.9976
July	27.2177	27.2177	27.2176	27.2177	27.2177
August	26.9177	26.9177	26.9179	26.9177	26.9177
September	26.8677	26.8677	26.8678	26.8677	26.8677
October	26.9077	26.9077	26.908	26.9077	26.9077
November	27.3777	27.3777	27.3777	27.3777	27.3777
December	27.6877	27.6877	27.6875	27.6877	27.6877
Annual	26.8477	27.3319	27.3345	27.3319	28.1276

Table 6.2 The hourly results of hybrid energy system

Hours of a day	Solar Radiation (kW/m²)	Wind Speed (m/s)	Solar Power (kW)	Solar Voltage (V)	Wind Power (kW)	Wind Voltage (V)
0	0	2.284543	0	0	3.12086	684.2
1	0	2.74815	0	0	9.857.71	1216
2	0	3.166489	0	0	22.98966	1857
3	0	3.42009	0	0	36.00651	2324
4	0	3.561609	0	0	45.48363	2612
5	0	4.052288	0	0	94.40113	3763
6	0	3.918661	0	0	78.24984	3426
7	0.0749713	3.699776	607.3	7793	56.53163	2912
8	0.2469629	3.181849	6359	25,220	23.63793	1883
9	0.4613641	3.202515	11,160	33,410	24.55041	1919
10	0.7416899	3.685511	12,870	35,880	55.296	2880
11	0.9480524	3.771765	13,570	36,840	63.07851	3076
12	1.015523	3.851352	13,750	37,080	70.98113	3263
13	0.9372602	4.18636	13,540	36,800	11.310774	4119
14	0.9209787	3.561713	13,490	36,730	45.51846	2613
15	0.6545269	3.318367	12,480	35,330	30.21761	2129
16	0.2806644	2.421205	7700	27,810	4.50785	822.3
17	0.1788829	2.479518	3448	18,570	5.23094	885.8
18	0.0133514	2.436653	19.24	1387	4.69168	838.9
19	0	2.497638	0	0	5.47345	906.1
20	0	2.775878	0	0	10.46673	1253
21	0	2.788477	0	0	10.7527	1270
22	0	2.073891	0	0	1.6667	500
23	0	1.290287	0	0	0.042273	79.63

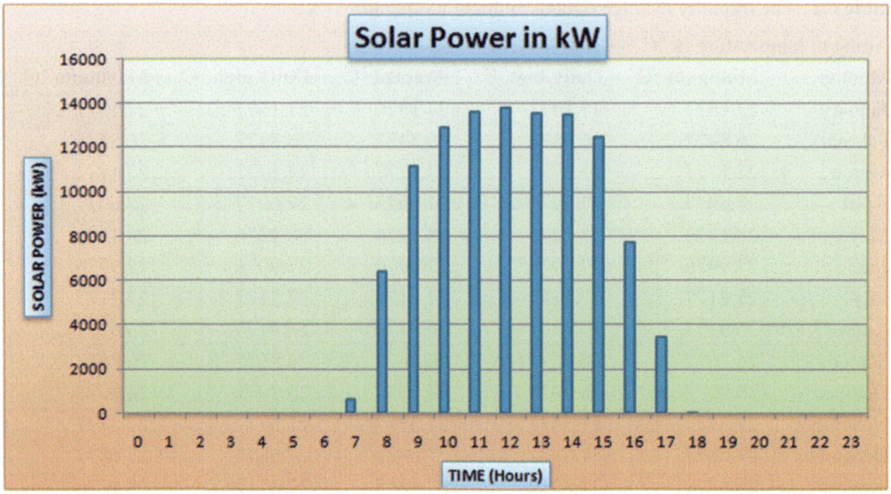

Fig. 6.7 The hourly scenario of solar power in kW

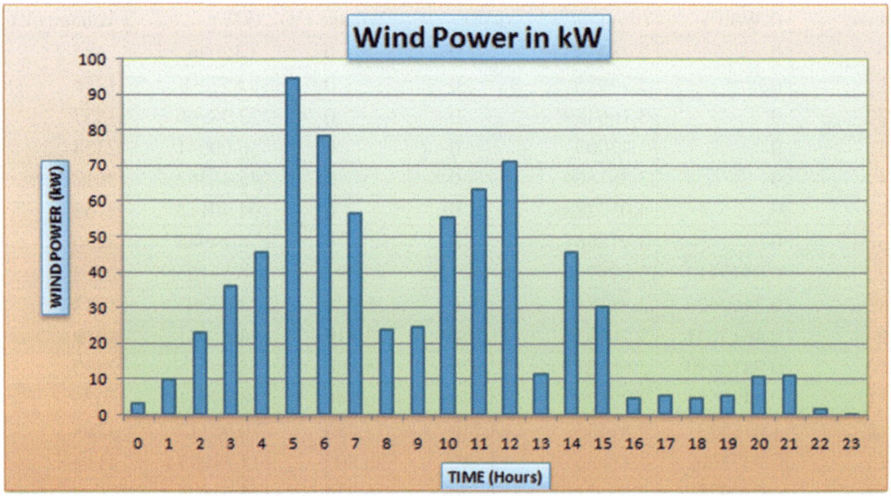

Fig. 6.8 The hourly scenario of wind power in kW

combined power of wind farm and the grid. In the time duration of 5:00–7:00 h (5:00–7:00 AM), solar power is not available so the load demand is fulfilled by wind power, and excess energy sending to the grid. In the time duration of 7:00–17:00 h (7:00–5:00 PM), load demand is fulfilled by the combined power of wind farm and PV array system, and excess energy sending to the grid. In the time duration of

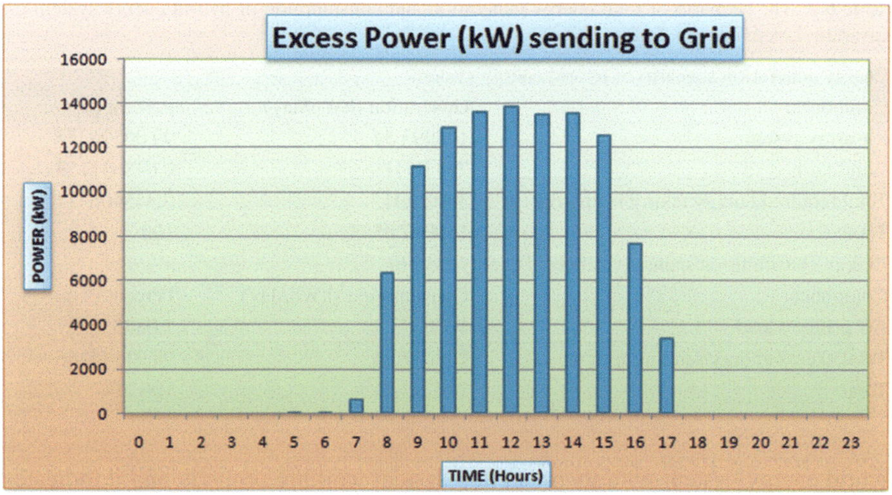

Fig. 6.9 The hourly scenario of excess power sending to grid after fulfilling the load demand in kW

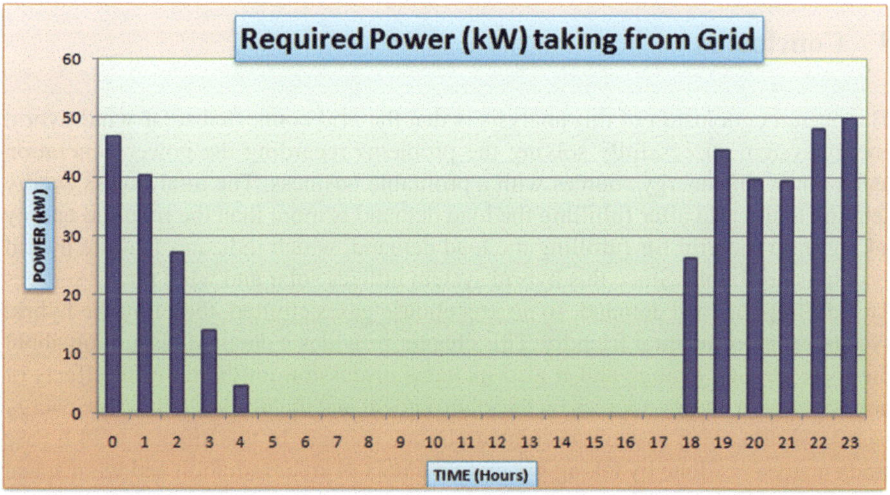

Fig. 6.10 The hourly scenario of required power taking from the grid for fulfilling the load demand in kW

17:00 to 23:00 h (5:00–11:00 PM), solar power is not available so the load demand is fulfilled by the combined power of wind farm and the grid. In the time duration of 23:00–0:00 h (11:00–2:00 PM), the combined voltage of the wind farm and PV array system is less than 440 V, so wind farm and PV array system disconnected from the GEL and load demand is fulfilled by grid. Therefore, it is clear that the

Table 6.3 The scenario of total energy generation and consumption of hybrid energy system on Kavaratti, Lakshadweep, India

Energy generation scenario of hybrid energy system		
Component	Generation (kWh/day)	Percent (%)
PV array system	108993.54	99.00596255
Wind farm	714.0641	0.648631153
Total required energy taking from grid	380.2504	0.345406313
Total	110087.8545	100
Energy consumption scenario of hybrid energy system		
Component	Consumption (kWh/day)	Percent (%)
AC primary load	1200	1.088555337
Total excess energy sending to grid	109037.85	98.91144466
Total	110237.85	100

hybrid energy system with its protection system working properly and it indicates that the hybrid system easily managing the fluctuation of energy from solar and wind energy sources.

8 Conclusion

The overall conclusion of this analysis is that the grid connected solar-wind hybrid energy system successfully solving the problems regarding the power generation using renewable energy sources with a profitable business. The total excess energy sending to the grid after fulfilling the load demand is more than the required energy taking from the grid for fulfilling the load demand, which indicates that the hybrid system is profitable. Since this hybrid system utilizes solar and wind power sources for fulfilling the load demand, so no greenhouse gases emitted, therefore the hybrid system is environmental friendly. This chapter provides a detailed discussion about the topic climate change and it also includes discussion on the harmful effects of non-renewable energy sources on the environment and the need of renewable energy based hybrid energy system to combat climate change. In this chapter a 24 h case study analysis is done by taking the real time data of solar radiation and wind speed of a selected location. The results of this analysis indicate that the hybrid system is profitable and environmentally friendly. This analysis simply gives an idea about to what extent there will be the generation of power and how much it will be helpful to this environment. In addition, this analysis also indicates that the hybrid energy system with its protection system working properly and also successfully managing the fluctuation of energy from solar and wind energy sources. This work can be further extended by connecting energy storage system or biogas power plant with the hybrid energy system for decreasing its dependency on the grid.

References

Abushnaf J, Rassau A (2018) Impact of energy management system on the sizing of a grid-connected PV/Battery system. Electr J 31(2):58–66. https://doi.org/10.1016/j.tej.2018.02.009

Advantages of the Hybrid System (2019) Power generators INMESOL. http://www.inmesol.com/hybrid-system/advantages-of-the-hybrid-system.asp. Accessed 12 July 2019

Aktas A, Erhan K, Özdemir S, Özdemir E (2018) Dynamic energy management for photovoltaic power system including hybrid energy storage in smart grid applications. Energy 162:72–82. https://doi.org/10.1016/j.energy.2018.08.016

Bhattacharjee S, Nandi C, Reang S (2018) Intelligent energy management controller for hybrid system. Paper presented at the 3rd international conference for Convergence in Technology (I2CT), IEEE, Pune, India, 6–8 April 2018. https://doi.org/10.1109/I2CT.2018.8529345

Bhikabhai Y (2005) Hybrid power systems and their potential in the Pacific islands, SOPAC Miscellaneous report 406. http://prdrse4all.spc.int/system/files/MR0406_0.pdf. Accessed 11 July 2019

Cause of Climate Change (2019) European commission. https://ec.europa.eu/clima/change/causes_en. Accessed 01 July 2019

Climate Change of Earth (2019) Wikipedia. https://en.wikipedia.org/wiki/Climate_change. Accessed 01 July 2019

Comodi G, Renzi M, Cioccolanti L, Caresana F, Pelagalli L (2015) Hybrid system with micro gas turbine and PV (photovoltaic) plant: guidelines for sizing and management strategies. Energy 89:226–235. https://doi.org/10.1016/j.energy.2015.07.072

Dave NR, Sinha M (2017) Simulation of solar and wind power plant using MATLAB for micro-grid. Kalpa Publ Eng 1:208–213

Effect of Non-Renewable Resources on the Environment (2019) Greentumble. https://greentumble.com/harmful-effects-of-non-renewable-resources-on-the-environment/. Accessed 02 July 2019

Gaur P, Singh S (2017) Investigations on issues in microgrids. J Clean Energ Technol 5(1):47–51. https://doi.org/10.18178/jocet.2017.5.1.342

Hosseini SA, Abyaneh HA, Sadeghi SHH, Razavi F, Nasiri A (2016) An overview of microgrid protection methods and the factors involved. Renew Sust Energ Rev 64:174–186. https://doi.org/10.1016/j.rser.2016.05.089

Hussain A, Kim H-M (2016) A hybrid framework for adaptive protection of microgrids based on IEC 61850. Int J Smart Home 10(5):285–296. https://doi.org/10.14257/ijsh.2016.10.5.26

Importance of Electricity – How It Changed People's Lives (2012) Articles factory. http://www.articlesfactory.com/articles/science/importance-of-electricity-how-it-changed-peoples-lives.html. Accessed 05 June 2018

Khorasani PG, Joorabian M, Seifosadat SG (2017) A new proposal for the design of hybrid AC/DC microgrids toward high power quality. Turkish J Electr Eng Comp Sci 25(5):4033–4049. https://doi.org/10.3906/elk-1609-74

Lai K, Illindala MS, Haj-ahmed MA (2015) Comprehensive protection strategy for an islanded microgrid using intelligent relays. IEEE Industry Applications Society Annual Meeting, IEEE, Addison, TX, USA, 18–22 October 2015. https://doi.org/10.1109/IAS.2015.7356952

Liu X, Wang P, Loh PC (2011) A hybrid AC/DC microgrid and its coordination control. IEEE Trans Smart Grid 2(2):278–286. https://doi.org/10.1109/TSG.2011.2116162

Marisarla C, Kumar KR (2013) A hybrid wind and solar energy system with battery energy storage for an isolated system. Int J Eng Innov Technol 3(3):99–104

Mirsaeidi S, Dong X, Shi S, Tzelepis D (2017a) Challenges, advances and future directions in protection of hybrid AC/DC microgrids. IET Renew Power Gener 11(12):1495–1502. https://doi.org/10.1049/iet-rpg.2017.0079

Mirsaeidi S, Dong X, Shi S, Tzelepis D (2017b) Challenges, advances and future directions in protection of hybrid AC/DC microgrids. IET Renew Power Gener 11(12):1495–1502. https://doi.org/10.1049/iet-rpg.2017.0079

Nandi C, Bhattacharjee S, Chakraborty S (2019) Climate change and energy dynamics with solutions: a case study in Egypt. In: Qudrat-Ullah H, Kayal A (eds) Climate change and energy dynamics in the Middle East, Understanding complex system. Springer, pp 225–257. https://doi.org/10.1007/978-3-030-11202-8_8

NASA Prediction of Worldwide Energy Resources (2019). https://power.larc.nasa.gov/. Accessed 15 Aug 2019

Nge CL, Ranaweera IU, Midtgård OM, Norum L (2019) A real-time energy management system for smart grid integrated photovoltaic generation with battery storage. Renew Energy 130:774–785. https://doi.org/10.1016/j.renene.2018.06.073

Paska J, Biczel P, Kłos M (2009) Hybrid power systems–an effective way of utilising primary energy sources. Renew Energy 34(11):2414–2421. https://doi.org/10.1016/j.renene.2009.02.018

Patil PG, Venkateshwarlu K, Patel MT (2015) Application of super capacitor energy storage in microgrid system. Int J Sci Eng Technol Res 4(3):589–594

Solar Cell Model (2019) MATLAB, MathWorks. https://www.mathworks.com/help/physmod/sps/ref/solarcell.html;jsessionid=504a9d98bf80d7f4d11916dfe528. Accessed 13 July 2019

Soon CC (2015) Development of a hybrid solar wind turbine for sustainable energy storage. Doctoral dissertation, Universiti Tun Hussein Onn Malaysia. https://core.ac.uk/download/pdf/42956060.pdf. Accessed 10 July 2019

The Impact of Electricity on Society (2019) Reference. https://www.reference.com/history/did-electricity-impact-society-3d108662bc468b61. Accessed 09 July 2019

There M, Suke S, Muley S (2011) Microgrid frequency stabilization with ultracapacitor based system. Int J Electr Electron Data Commun, pp 168–172. http://www.iraj.in/journal/journal_file/journal_pdf/1-150-1434614047168-172.pdf. Accessed 15 Aug 2019

Vuc G, Borlea I, Jigoria-Oprea D, Teslovan R (2013) Virtual power plant strategy for renewable resources aggregation. Paper presented at EUROCON 2013, IEEE, Zagreb, Croatia, 1–4 July 2013. https://doi.org/10.1109/EUROCON.2013.6625065

Chapter 7
Sustainable Solar Energy

George Loumakis

Abstract Solar energy is one of the most rapidly developing renewable energies in the world. The global solar PV market has been exceeding future market projections and giant untapped markets are only now waking up. For many years proposals have been made about using the Sahara desert as a place where a gigantic solar farm would be able to power the world. But is solar energy always the best solution out there? What can hold it back? What are some major drawbacks it has? Not all solar energy applications are created equal. Even amongst certain applications – for example PV- there are numerous technologies that are competing for greater efficiency, lower cost etc. In fact this can go even further, as apart from the more mainstream PV and solar thermal panels, other applications exist such as solar parabolic troughs, solar concentrators, solar desalination, solar ponds etc. This chapter will provide an overview of various solar technologies, explain their characteristics and discuss scenarios for best practice in regards to maximising energy yield, minimising environmental impacts and making sure that solar energy is used in the most sustainable way possible.

Keywords Solar energy · PV · Solar thermal · Non-conventional solar energy technologies · Solar towers · Passive solar

1 The Solar Resource

Sun deities have been worshiped across the world for millennia and a very brief search for their number online returns a very large number of results. Indicatively Merriam-Webster's encyclopedia of world religions (1999) alone lists hundreds of sun deities from all parts of the world amongst civilisations spanning millennia. That alone can be considered as proof of the importance that the sun has played throughout human evolution.

G. Loumakis (✉)
Glasgow Caledonian University, Glasgow, Scotland, UK
e-mail: George.Loumakis@gcu.ac.uk

H. Qudrat-Ullah, M. Asif (eds.), *Dynamics of Energy, Environment and Economy*, Lecture Notes in Energy 77, https://doi.org/10.1007/978-3-030-43578-3_7

Many reasons can be attributed to that, including the ability of the sun to bring an end to the perils that are found during nightfall in the form of predators, colder weather and losing the ability to see far away. On more scientific terms though, the sun is a literal life giver, as it is through photosynthesis that plants can grow. Plants are the basis of our ecosystem, so that alone shows the worth the sun has on life in general. Incidentally the word that usually describes the sun in the Greek language is ζωοδότης (pronounced zo-o-tho-tes) which when translated means lifegiver.

Our sun is a G-type main sequence star that makes up for 99.86% of the total mass of our solar system. It has a mean distance to the earth of 150,000,000 km, which is the equivalent of 1 astronomical unit (AU). It is composed of almost three quarters Hydrogen while the rest is predominantly Helium with trace of other elements by mass such as carbon, neon, iron and others. As with every star, the energy given off by the sun is due to nuclear fusion – more specifically hydrogen to helium which is a proton-proton reaction. Other very small amounts of energy from the carbon-oxygen-nitrogen cycle (CNO). This latter cycle is the predominant form of fusion in larger stars, typically at 1.3 times the mass of our sun.

All in all the sun gives us 174 PettaWatts (PW) of incoming solar radiation in the upper layers of our atmosphere, with almost 30% being reflected back and the rest being absorbed by our planet. When talking about solar energy it is useful to distinguish between 2 definitions that unfortunately end up being used almost interchangeably in real life.

- Insolation refers to the total amount of solar radiation energy received on a given surface and is measured in MJ/m^2. A very common unit is also the Wh/m^2 whereas an older unit that is rarely used nowadays is the Langley (equal to $1\ cal/cm^2$)
- Irradiance is the power (radiant flux) of light that is received over a surface and is usually measured in W/m^2.

Seeing the units that are used for insolation and irradiance, we can tell that insolation is a measurement of energy carried by light for a given surface area, whereas irradiance refers to the power of light over an area. When using the general term sunlight we talk about the electromagnetic radiation coming from the sun that has a spectrum spanning from infrared to UV. Sunlight can also be roughly broken down into categories, direct and diffuse light. Direct light is light coming to use via absence of cloud coverage, whereas diffuse light is light that comes through cloudy conditions.

The average value of extra-terrestrial solar radiation at the edge of our atmosphere is $1367\ W/m^2$ – known as the solar constant- with a variation of 3.3% higher in January to 3.3% lower in July. That variation can be explained by the actual Sun-Earth geometry. A rough approximation that even to this day is taught in elementary education is that the Earth revolves around the sun in a circle. This is an oversimplification, whereas the more accurate representation is that the earth moves around the sun in an elliptical orbit thus being closer in January and further away in June. This is showcased in Fig. 7.1 below.

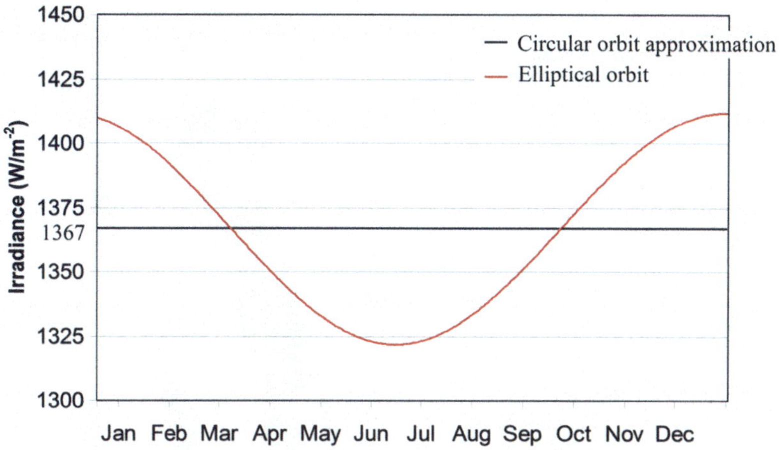

Fig. 7.1 Monthly extra-terrestrial irradiance for circular and elliptical orbits

At the Earth's surface at zenith, total direct radiation is 1050 W/m^2 with the total amount of direct, diffuse and reflected being 1120 W/m^2.These figures become extremely important in the context of solar energy because they are a very good indication of what we can expected as a best case scenario. Without taking any energy conversion whatsoever – something that will be discussed later on- it is handy to remember said values because it shows us the limitations of solar energy usage. If the question becomes "How much area would I need in order to have X amount of available energy" the answer can start being as simple as "No matter what happens you can never get more than 1120 W/m^2". This of'course is a gross oversimplification but can serve its purpose as a ball park figure. In regards to the solar energy spectrum, almost 52–55% of available irradiance is in the infrared (wavelength of above 700 nm) 42–43% visible (400–700 nm) with 3–5% UV (below 400 nm).

Light in a nutshell is electromagnetic radiation and its spectrum is a mixture of photos with different energies – the lower the wavelength the higher the energy they carry. As it is electromagnetic radiation it can be transmitted, reflected, absorbed, re-emitted and scattered. The absorption of radiation and re-emission of said radiation in the infrared spectrum especially is of great interest but as this is a chapter about solar energy and its applications and not the nature of light and electromagnetic radiation itself, there is no need to delve deeper into the above.

As solar radiation reaches the Earth, as mentioned above, some of it gets reflected by the atmosphere whereas the rest follows a complicated path of absorption, re emission, reflection etc. A breakdown of this can be seen below in Fig. 7.2.

Incidentally it is by seeing this that we can understand the effects that humans have on climate change. By releasing large amounts of carbon dioxide in the atmosphere the balance is tipped by increasing the amount of radiation absorbed by the atmosphere thus limiting the ability of our planet to cool down.

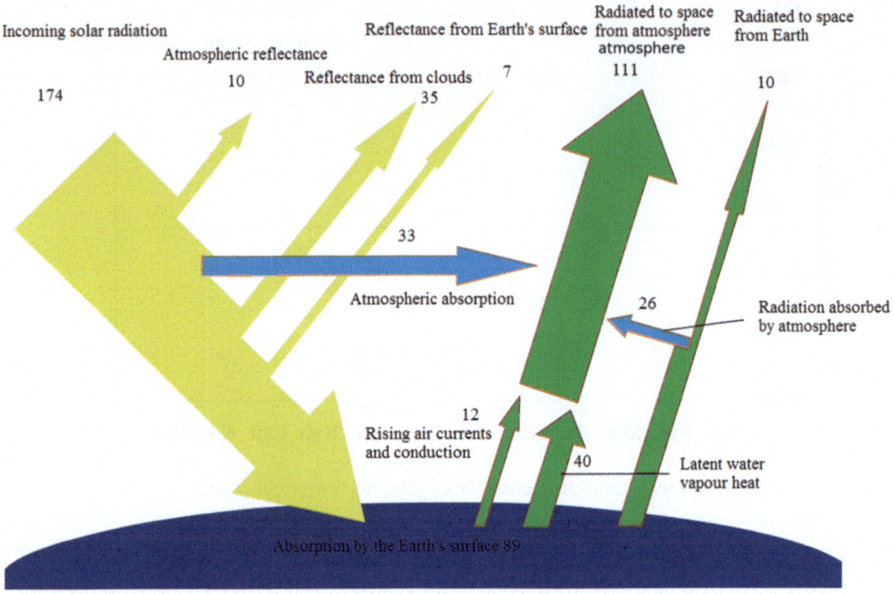

Fig. 7.2 Solar radiation interaction with the Earth - units are PW

Although it is a well-established fact that the Earth orbits the Sun, as also mentioned above, it is a useful tool to know the perceived movement of the Sun around the earth, especially for most calculations that have to do with solar energy. Hence we use the horizontal system which is centered on an observer. This system considers the Earth as the center of the universe. The Sun, planets, stars and other objects are all attached to the surface of the hypothetical celestial sphere. This very simplified model helps us predict the motion of the sun. The beauty of the horizontal system is that is acts as a natural reference of all observers as can be illustrated in the following Fig. 7.3.

With this system the position of Sun only requires knowledge of two angles, the elevation (usually found as h) and the azimuth angle g_s. The elevation determines the height angle of the sun (how high the sun is located in the horizon) and the azimuth shows the angle between due south and the position of the Sun (in the Northern hemisphere)-so effectively it can show us where the sun is located from east to west.

If it were to be described in very simple terms, then the sun has two simultaneous paths it follows. One is from east to west (sunrise to sundown, indicated by the azimuth angle) which happens on a daily basis and the other one determines the height the sun will be at (for the Northern hemisphere that would be lower for winter and higher for summer, indicated by the elevation). Between these two movements, the position of the sun can be determined for any given time and date. Finally we should remember that in the Northern hemisphere the sun will always be facing

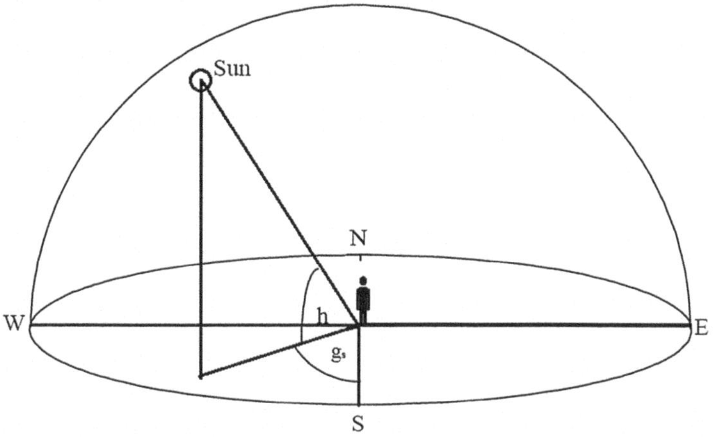

Fig. 7.3 The horizontal system

southwards, whereas it is the opposite in the southern hemisphere. As the math of calculating the perceived movement of the sun around the Earth gets a bit complicated, a very good online tool that helps us track the sun's path can be found at www.suncalc.org.

From a solar energy perspective all of that matters because we need to know how much solar energy is available to us but also at what angles the solar radiation will be incoming in order to find the ideal orientation of any of our solar devices. Each solar column in the below illustration effectively "contains" the same amount of energy. Due to different geometry though, it is dispersed at larger surfaces and thus would require either a larger installation to capture the same amount of energy, or requires to be angled- for that specific instance, as is illustrated in Fig. 7.4 below.

This of'course causes issues when trying to estimate how much available solar energy we have because that changes during the day. The sun rises, amount of available energy increases until it reaches a maximum value – solar noon, where the elevation is maximum- then decreases until sunset. This gets more complicated by two factors. The first one is that the elevation changes every day and the second one has to do with weather interference. More cloud coverage means more radiation is absorbed and scattered by the clouds, thus increasing the amount of diffuse radiation and lowering the amount of direct. Of'course under a very cloudy day the combined amount of diffuse and direct radiation would be lower than the amount we can get on a sunny day. It's no wonder that solar energy installations are more prevalent in places where the weather is typically less cloudy. Because of these complications with solar energy, assessing energy yield can be difficult so testing is done under the standard testing conditions (STC) which assume a temperature of 25°, an air mass of 1.5 (the sun being about 41° above the horizon) and an irradiance of 1000 W/m^2.

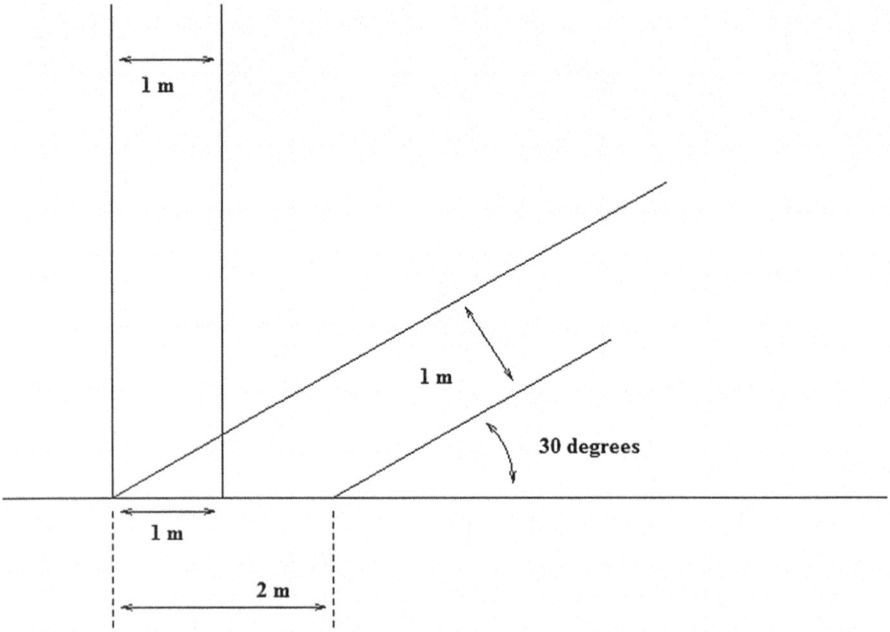

Fig. 7.4 Dispersion of light over surfaces

Finally some clarification need to be given between a common confusion because of two very similar sounding solar related units – the (peak) sun hour the sunlight hours. Sunlight hours show you the amount of time per day you have sunlight whereas the peak sun hour is the equivalent to the number of hours that the solar irradiance would be at a peak level of 1000 W/m^2. Simply put if a site receives a total of 3000 Wh/m^2 over a whole day, that would be the same as receiving 1000 W/m^2 for 3 h. This 1000 W/m^2 value is very close to the maximum possible available irradiance as discussed earlier.

1.1 Solar Photovoltaics

Solar power can be harnessed in a way that it provides electricity, which happens via using solar photovoltaics or PV as it is most well known. The word photovoltaic is a combination of two words from two different languages. Photo comes from the Greek word phos (φως) which means light, and voltaic is from the Italian pioneer of electricity Alessandro Volta. Solar photovoltaics (PV) comprise interconnected solar cells, electric devices that use the photoelectric effect in order to convert sunlight to DC electricity. The photoelectric effect, first noticed by Edmond Becquerel and subsequently explained by Albert Einstein is when some materials – semiconductors- emit electrons when light shines upon them. As such, semiconductors are the basis for every PV technology. When talking about solar panels, we are

Fig. 7.5 Small scale solar PV panel

effectively talking about many solar cells connected together. Figure 7.5 below shows a small scale PV panel in which the solar cells (in blue) are interconnected via the small metal contacts to the partly exposed bigger metal contact.

PV panels have a variety of uses that are, but not limited to, for electricity generation in buildings, small or large scale solar farms, building integrated PV where the PV panels act as part of the fabric of a building, floating PV installations, power sources for sensors, power sources for weather stations, power sources communications towers, power sources for road signs, PV assisted irrigation, solar lights, solar chargers etc. Effectively they act as small to large scale variable DC generators.

Silicon is the most common semiconductor material for making PV cells because it's fairly cheap and abundant. Other well-known semiconductors used for PV are Cadmium Telluride (CdTe) and Copper Indium Gallium Arsenide (CIGS). Silicon PV cells are broken down into three main categories – Monocrystalline Silicon PV cells, Polycrystalline Silicon PV cells (sometimes found as multicrystalline) and Amorphous Silicon PV cells.

Monocrystalline Silicon PV cells exhibit typically the highest efficiencies amongst the three with types with typical figures being between 13 and 17%. They are made from a single crystal of silicon that is grown from highly pure molten silicon using the Czochralski crystal growth process developed by Jan Czochralski. This process is completed by dipping a silicon seed crystal to molten silicon. This process is generally energy intensive and as such, monostrycalline tends to be more expensive overall. Physically it exhibits smooth textures and is typically thick. It has a very rigid structure and thus frame mounting is usually recommended.

Polycrystalline Silicon PV is also produced from highly pure molten silicon, but instead of a seeding, it is made using a casting process. It is effectively made from a slice cut from a block of silicon thus containing different crystal structures. Typical efficiencies are in the range of 11–15% making it slightly less efficient that monocrystalline but at a lower cost. A common characteristic of polycrystalline silicon PV is a blue appearance which is due to an anti-reflective layer.

Amorphous silicon is made from non-crystalline silicon and is typically very thin, with expected thicknesses of the semiconductor materials being only 0.5–2.0 μm thick. As such amorphous silicon can be flexible. The construction of said PV cells involves depositing a gaseous layer of amorphous silicon on a usually glass substrate. It exhibits the lowest efficiencies of all Silicon cell with expected ranges being between 6 and 9% and has a glossy dark to dark brown appearance. An interesting characteristic of amorphous silicon PV is that it has 20% better performance on hot environments and up to 12% better performance in low and diffuse light conditions when compared to equally rated crystalline modules. Because of these characteristics, amorphous silicon cells are used a lot in indoor environments for low energy applications such as calculators, sensors etc.

Apart from these three very popular types of solar cells newer ones include organic solar cells, polymer solar cells, quantum dot solar cells, dye-sensitized solar cells, copper zinc tin sulphide (CZTS) cells, perovskite cells and multijunction cells. A very brief rundown of some of these is shown below-note that this list is not exhaustive.

Dye sensitized cells	Typically low cost, thin film. Absorption of photons occurs in dye molecules at a porous structure of TiO_2 in an electrolyte. Lower efficiencies that typical thin film technologies but at a lower cost. The electrolyte has temperature stability problems that can potentially cause it to freeze at low temperature. Furthermore it contains various amounts of volatile organic compounds and as such it needs to sealed.
Organic solar cells and polymer solar cells.	Thin film technology, made from conducting polymers such as polyphenylene vinylene (PPV) and p-type copper phthalocyanine and n-type bisbenzimidazo[2,1-a:2′,1′-a']anthra[2,1,9-def:6,5,10-d'e'f']diisoquinoline-10,21-dione. Their main characteristic is that the analogous p-n junction isn't layered, but the donor and acceptor area is mixed together in an amorphous structure with different regions. They can be printed at a relatively low cost but currently their efficiencies are very low.
Perovskite solar cells	Made from $CaTiO_3$ (calcium titanate also known as Perovskite) and lead or tin halide based materials- cheap to make. This is a fast advancing technology almost quadrupling their efficiency in 7 years (from 3.6% to 21%). No commercial modules are available yet and they have issues with stability
Multi junction cells	This is a generic category of solar cells whose main characteristic is the usage of various semi conductors that absorb light at different frequencies in order to maximise energy capture.

When it comes to installing PV, the ideal scenario would be that we want the PV systems to be facing the sun at a right angle at all given times. That can be achieved via a tracking system which comes in two configurations, single axis and dual axis. A single axis tracking system tracks the sun from east to west, whereas a dual axis tracking system apart from east to west, also follows the elevation of the sun and makes sure the PV system is at any given time at a right angle. A well calibrated tracking system can lead to an increase of up to 40% in energy yield compared to a static system. This sounds very good in theory but the problem with tracking systems is that they cost energy to run (usually fed by the PV system itself) and make the system more complicated and this more costly. Whether the energy increase you get from the tracking system is enough to offset the energy cost for the tracking system to operate and the added complexity cost (leading to higher maintenance cost) depends greatly on the location, efficiency and status of the PV array, shading etc. As such a straight answer cannot be given as to whether tracking systems are worth it.

Carrying on from the above, for static installations there are certain rules of thumb we follow. These can be summed up very quickly with the following. Mind you though that this assumes an even distribution of solar energy throughout the year. For example, in cases in the Northern hemisphere where we have more solar energy during the summer, the ideal angle will be shallower (the sun is higher) whereas with increased solar energy during the winter months, the ideal angle is steeper (the sun is lower).

- For maximum efficiency throughout the year for a fixed installation, PV panels have to be facing south (for the northern hemisphere and vice versa for the southern)
- For latitudes below 25°, the optimum inclination angle for the PV is latitude∗0.87
- For latitudes between 25° and 50° the optimum angle is latitude∗0.87 + 3.1
- For latitudes above 50° we typically use the same angle as the latitude
- If we are looking for maximum energy capture before midday, the PV system should be facing east. For maximum energy capture after midday, the PV system is faced west.
- The available energy will always be equal to $E = H*A*\eta$ where H is your insolation (energy per surface area) at the angle of your panel, A is the effective area of the PV system and η the efficiency. The power is always $E = I*A*\eta$ where I is your irradiance (power per surface)

A very good database using GIS data regarding global available solar energy can be found from the Global Solar Atlas via this hyperlink https://globalsolaratlas.info/. Through that online tool we have at our disposal the available solar energy for most regions of the world gathered together using years of combined satellite and weather data.

A mention should be made about the rest of the components that make up a PV system. These come under the umbrella term "Balance of System (BOS) components" and they can include cables, fuses, earthing and lightning protection, batteries, charge regulators, low voltage disconnects, inverters, tracking systems, sensors, frames, fittings etc. An interesting fact that shows how much the price of PV panels has dropped dramatically over the years has to do with how much money the BOS contributes to the system itself. In 2008, 68% of the cost of a PV system was the panels themselves, whereas in 2015 that percentage had dropped 32% according to industry's average figures.

When it comes to how PV systems are connected there are again different ways. The most popular method in places where there is adequate infrastructure, the PV system is grid connected. If the PV system is installed on a building, the generated electricity is exported and sold to the grid, whereas the building buys electricity from the grid. The prices that electricity is sold and bought are negotiated and it isn't uncommon to have feed in tariffs that make the export of electricity a profitable affair. In the absence of a grid then the PV system is a stand alone system, usually coupled with a back up system in the form of batteries. There are more specialized (and somewhat rare) installations where a building is powered both by PV and the grid at the same time. That requires separate electrical lines and as such isn't commonly used.

Thus, most connection regimes will fall under the following four categories

1. PV electricity directly to the building – requires careful planning, sizing and a back up system.
2. PV electricity and grid electricity used directly in the building.
3. PV electricity exported to the grid and grid electricity to the building – very common in places with existing electrical infrastructure.
4. PV electricity used in the building and the excess sold to the grid – very uncommon.

1.2 Energy Payback Period and Issues

PV panels – as with every technology – require energy to be made. That is mostly attributed to the melting of Silicon, as discussed earlier, but large amounts of energy are used up in mining the materials, making the electrical components, assembling the panels etc. The Fraunhofer institute, a major research center for PV technologies has done some work on calculating energy payback periods for PV and depending on the place in the world, the type of installation and the technology used, by assuming a global irradiance of 1000 kWh/m^2 annually energy payback is from 3.5 years and less. Having said that, a PV installation that is badly maintained is shaded and has corroded electrical connections – to name a few issues- will be grossly underperforming and as such above values wouldn't apply.

The disposal of PV panels can be an issue as well as currently there are only very few companies that recycle PV panel materials. Finally something that should be mentioned is the expected loss of efficiency of a PV panel. Typical figures are around the 0.5% loss of efficiency per year. That is a combination of various issues such as the glazing/encapsulant becoming less transparent due to the effects of solar radiation, material defects, electronics deterioration etc.

A final issue about PV might seem a bit counter intuitive. This has to do with the fact that they tend to exhibit lower performance with increasing temperatures-which is what tends to happen with increased insolation. As such, PV panels can benefit from passive and/or active cooling. The PVT panel tends to address this issue. PVT stands for Photovoltaic Thermal and is a hybrid PV/solar thermal panel. The front of the device is a normal solar PV panel and on the back of it tubing is circulating a thermal fluid using the same operating principle as a solar thermal collector- discussed in more detail below. This way the PV panel is actively cooled down thus having its performance increased and at the same time we also have heat production via the thermal panel part. These kinds of devices tend to be a bit more complex to manufacture, install, operate and maintain. Other active efforts to cool PV panels utilize Peltier devices (thermoelectric generators). The typical passive way to cool a PV panel down is to raise it slightly – in a roof installation- so air is circulated below it. This has an interest prospect in the case of BIPV where the waste heat could be used for the building itself.

1.3 Solar Thermal Collectors

In contrast to solar PV that produces electricity, solar thermal collectors – also known plainly as solar collectors – produce heat. There are many different types and classifications of solar collector, but most commonly they can be regarded as specialized types of heat exchanger that absorb solar radiation from the sun in the form of heat and then transfer that heat into a circulating fluid. This fluid is usually either air and used to heat a living or other indoor space directly, or a liquid and used to generate hot water stored in a tank. In these types of collectors, that heat can be distributed along a network of tubing via a medium called solar fluid and is usually a mixture of water and a type of antifreeze in varying concentrations.

In terms of how the fluid or air is circulated, solar collectors can be either passive or active systems. Passive solar collectors use the natural buoyancy of water or air in order to ensure circulation through the collector. Since hot water is less dense than cold water, the cold water entering at the inlet is heated up and then starts circulating naturally. This effect is called thermosiphoning and solar water heaters of this kind are very common in the Mediterranean countries (Mauthner et al. 2016). Passive and active systems are showcased below in Fig. 7.6.

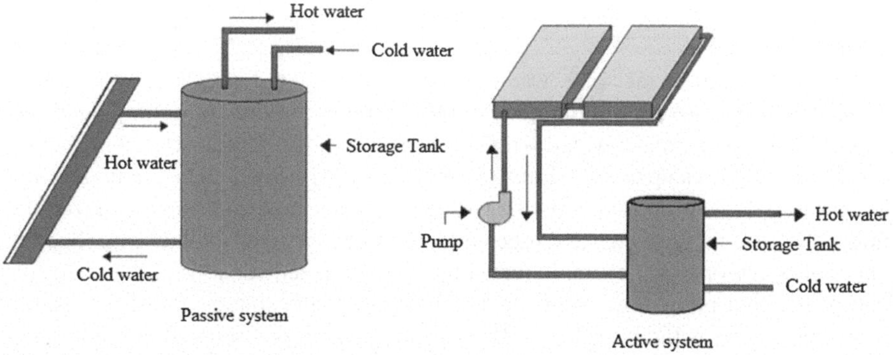

Fig. 7.6 Passive and active solar systems

Active solar collectors for water heating use pumps to circulate the water or working fluid. Although pumps need energy, the main advantages of active solar collectors, compared with passive systems, is that the flow can be regulated according to requirements and there are fewer restrictions of the configuration of the system. In the case of air collectors, instead of pumps, fans are predominantly used.

1.4 History of Solar Collectors

The first known use for solar energy can be found at the seventh century BC where concave glass pieces were used to focus sunlight in order to start a fire. The ancient Greeks and Romans have been recorded to use elements of passive solar techniques in houses in order to utilise heat gain from the sun. In fact Priene, Πριήνη in ancient Greek, which was an ancient Greek city located at the base of mount Mycale, in the west cost of central Anatolia, modern Turkey, was one of the first examples of using wide city planning and optimised passive solar orientation (Perlin 2013).

The first recorded scientific study for the fundamentals behind solar thermal collectors was made by a Swiss naturalist named Horace Bénédict de Saussure in the 1760s in Geneva, Switzerland. De Saussure is credited, amongst others, as the founder of alpinism and modern meteorology. According to his observations, the air temperature was significantly higher when the sun rays passed through glass surfaces into a glass covered structure such as a conservatory, a coach or a room with windows.

What Horace Bénédict de Saussure was witnessing was in fact a manifestation similar to the greenhouse effect. In order to test out his hypothesis in 1767 he built an insulated box whose bottom was painted black in order to absorb as much sunlight as possible and the top was covered with two sheets of glass. According to his observations, when the box was placed orientated towards the sun, the inside temperature would far exceed the temperature needed to boil water. (Solarcooking.org 2013) This insulated box formed a basis for the design of modern solar thermal air heaters and demonstrated the greenhouse effect. A very similar design to this box was the solar oven used by Sir John Herschel during his South African expedition in the 1830s in order to cook food (Turner 2013). A rendition of De Saussure's hot box can be seen below.

The first commercial solar water heater was patented in 1891 by Clarence M. Kemp. It was a very simple design with cylindrical tanks fitted in pine boxes lined with felt paper and covered with single pane glazing. It was offered in various sizes from 121 litres up to 2650 litres (Bainbridge 1981). Since then, many different designs have been proposed, with the two most common types being the glazed flat plate collector and the evacuated tube collector. Other designs include air collectors and pool collectors – which are in fact simplified unglazed versions of flat plate collectors. Another popular design is the batch or integrated storage collector – sometimes referred to as the breadbox solar collector- which is discussed in more detail later on.

1.5 Flat Plate Collectors

Flat plate collectors consist of a dark flat-plate absorber, a transparent glazing that allows solar energy to pass through but reduces heat losses and a heat transfer fluid that is usually water mixed with glycol to act as antifreeze, to remove heat from the absorber and insulation. The absorber consists of a thin sheet usually of aluminium or copper to which black or selective absorber paint is applied. The working fluid absorbs heat from the absorber and is then circulated through a heat exchanger to an insulated water tank and therefore it is a closed loop system (Duffie and Beckman 2013). A simplified outline of a flat plate collector can be seen in Fig. 7.7 below.

Typical materials for flat plate collectors include aluminum for the absorber plate, copper for the tubing and glass for the glazing. Unconventional materials such as polymers have been used as well with varying results when it comes to efficiencies and costs (Loumakis 2018).

1.6 Evacuated Tube Collectors

Evacuated tube collectors, also called cylindrical absorber arrays or heat pipe evacuated tube collectors consist of a metal heat pipe, made usually from copper, to which an absorber plate is attached, inside a vacuum-sealed solar tube. The heat pipe is hollow and the space inside contains alcohol such as glycol or purified water at reduced pressure. When solar radiation is absorbed by the surface of the absorber, the liquid evaporates and rises to the top of the pipe which is sealed and inserted into a manifold. Water, or a water –glycol mixture, flows through the manifold and picks up the heat, while the fluid in the heat pipe after releasing the captured heat, condenses and flows back down the tube for the process to be repeated. (Duffie and Beckman 2013)

An advantage of heat pipes over direct-flow evacuated-tubes is the "dry" connection between the absorber and the header or manifold which makes installation

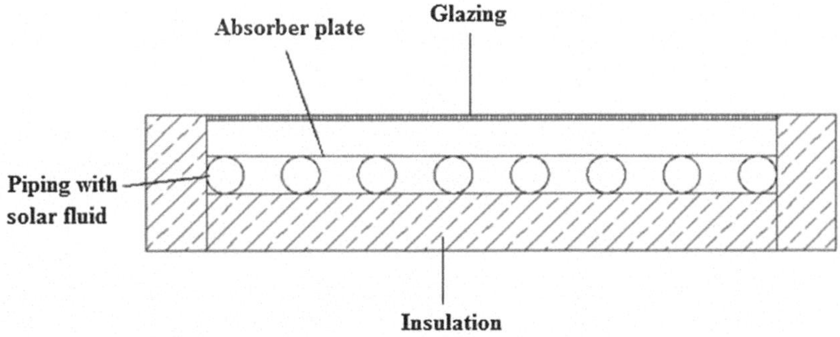

Fig. 7.7 Cut away of a flat plate collector

easier and also means that individual tubes can be exchanged without emptying the entire system of its working fluid. A drawback of heat pipe collectors is that they must be mounted with a minimum tilt angle of around 25° in order to allow the internal fluid in the heat pipe to circulate. The vacuum that surrounds the outside of the tube greatly reduces convection and conduction heat loss to the outside, therefore achieving greater efficiency than flat-plate collectors, especially in colder conditions. This advantage is largely lost in warmer climates, except in those cases where very hot water is desirable, for example commercial process water. The high temperatures that can occur may require special system design to avoid or mitigate overheating conditions (Heliodyne 2011). A design for an evacuated tube collector is shown below in Fig. 7.8.

1.7 Batch Collectors

Smyth et al. (2006) investigated the Integrated Collector Storage Solar Water Heater (ICSSWH), sometimes referred to as a "breadbox collector", which was developed from very early systems comprising of a simple black tank placed facing the sun. The ICSSWH, by its combined collection and storage function suffers substantial heat losses to ambient, especially at night-time and non-collection periods. They mention that in order to be viable economically, advancements incorporating glazing systems, insulation, configurations with reflectors, internal and external baffles, use of evacuation and phase change materials have been made. Their advantages over flat plate collectors and evacuated tube collectors are that they resist freezing

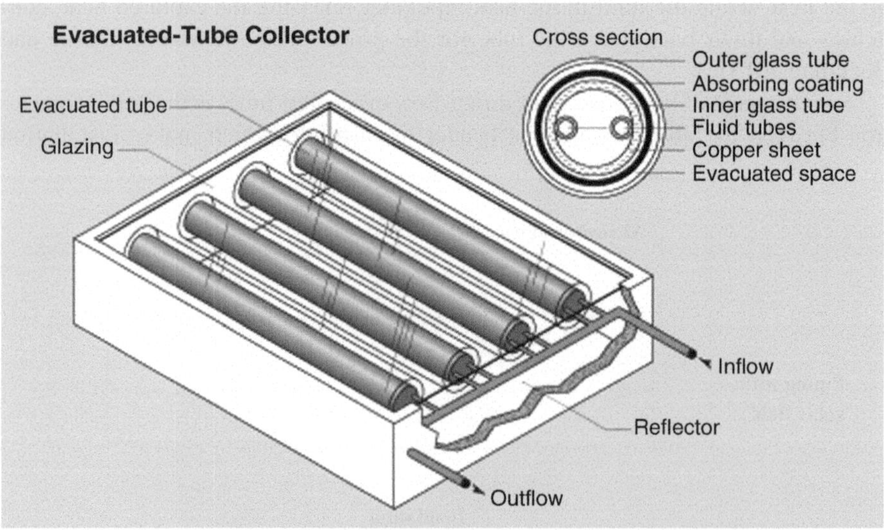

Fig. 7.8 Evacuated tube collector (Public domain image – US Department of Energy)

in most climates and have generally lower costs. However they remain much less common than the aforementioned types of collectors. A novel design of integrated collector storage solar water heater by Souliotis et al showed that the thermal behaviour of the system was comparable to that of a flat plate thermosiphon collector with a thermal loss coefficient during the night similar for both collectors (Souliotis et al. 2011). Helal et al. (2011) designed an integrated collector storage solar water heater using parabolic reflectors in order to cover the needs for a family composed of four individuals. The comparison between their system -both theoretical and experimental- and conventional integrated collector storage solar water heaters with symmetrical and asymmetrical compound parabolic concentrating reflectors shows better performance, despite the much simpler design. A design for a batch collector can be seen below in Fig. 7.9.

1.8 Potential for Solar Collectors

Flat plate collectors and evacuated tube collectors make up 93.2% of total installed capacity and most of the remaining 6.8% are unglazed solar collectors, used mainly to heat swimming pools, or air-heating collectors used for space heating. Evacuated tube collectors have become the most prevalent type with usage of 71.1% in total compared to 22.1% for plat plate collectors, attributed mostly to the huge growth of the Chinese market in the last few years, where evacuated tube collectors are the

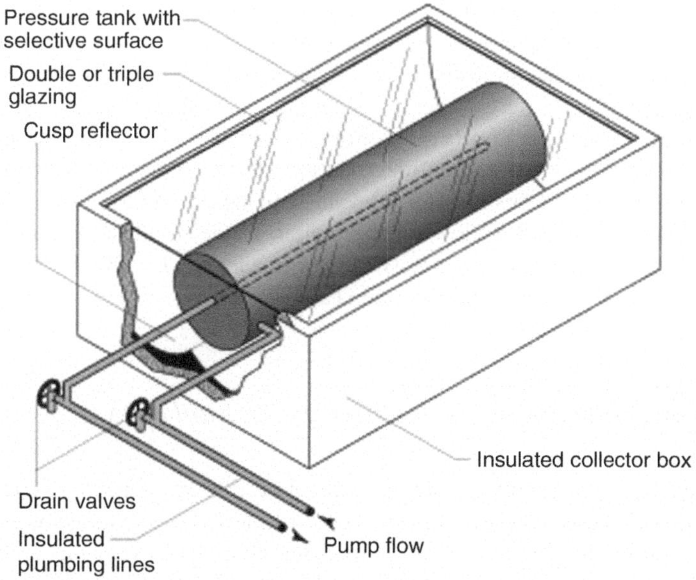

Fig. 7.9 Batch collector (Public domain image – US Department of Energy)

preferred technology. The countries with the largest capacity (per head of population) are Israel, Cyprus, Austria, Greece and Barbados which all boast a very high solar resource. The total capacity in operation at the end of 2016 was 410.2 GW, corresponding to a total of 586 million m^2 of collector area. In total numbers, China is by far the biggest user of solar thermal technologies, showing a total capacity of 289.5 GW and Europe being the second user with a capacity of 47.5 GW. This figure accounts both for water and air collectors (Mauthner et al. 2016). Another distinction that can be made is between active systems and passive systems. In the United States, Europe and Australia pumped systems are predominant, whereas in other large markets, namely in Japan, Brazil and China, thermosiphon systems are more common. Thermosiphon systems have better efficiencies in warmer climates, whereas pumped systems with heat exchangers are better suited for colder climates (Mauthner et al. 2016).

1.9 Solar Energy Markets

At the time of writing solar energy is providing 3% of the world's total electricity demand (Enerdata 2019). This figure might seem disheartening at first, but keep in mind that the solar energy markets are constantly evolving and growing. As an energy type, electricity is usually considered as being more useful than heat since it can very easily be transformed to other types of energy, is very easily transported over long distances and is used in a multitude of applications. As such, solar PV and not solar thermal is usually the energy technology we think of when talking about solar energy. Furthermore, solar PV has a more modern appeal whereas solar thermal seems old and antiquated – not true from a technical perspective, but public perception is a major driving force for the markets work. Strangely enough according to the International Energy Agency's estimations, both the solar thermal market and the solar PV market are fairly similar in size with solar PV experiencing a much larger rate of growth. As aforementioned though, the markets are constantly evolving and in order to keep this chapter relevant you should be monitoring the energy markets to have an updated view on the matter. Some good publications on that matter – updated annually include BP's world energy (BP 2019), IEA's World Energy Outlook (IEA 2018) and Enerdata (2019).

1.10 Non-conventional Solar Energy Technologies

PV systems and solar thermal systems dominate the solar energy market. As mentioned earlier PV systems can in various sizes, from extremely small such as pocket calculators, power for sensors, solar lights, solar chargers to very big such as solar farms. Solar thermal systems are mostly used in domestic and industrial settings but are fairly rare for very large scale installation.

Apart from these two above mentioned solar energy applications though, some other less known ones exist and are beginning to gain momentum with some already having a proved track records, whereas others have some pilot installations. Whereas with PV and solar thermal systems there is a distinction between electricity and heat generation (with some exceptions coming from PVT systems), with non-conventional solar energy technologies, this distinction stops being very apparent. Many non-conventional systems can be used both for heating and electricity and they are typically more versatile. Moreover some of this applications use solar energy for other purposes than to give useful energy back. So although they cannot be classified as energy generating applications, they are still applications of solar energy. The following paragraphs will provide a brief overview of some of these technologies and applications.

1.11 Large Scale Applications

1.11.1 Solar PV

Although more well known for their small scale and domestic applications, some large scale industrial applications of PV exist. Currently the largest PV plant in the world, the Solar Star plant in Rosamond, California went live in June 2015. It is rated at 579 MW using approximately 1,720,000 panels and taking up a space of 1300 hectares (SEIA 2015; Sunpower 2014).

1.11.2 Solar Parabolic Troughs

Solar parabolic troughs take advantage of the ability of a parabolic mirror to focus the sun's rays at a specific focal point. At that focal point there is a tube, usually a Dewar tube, which contains the solar fluid that is to be heated. A Dewar tube is a double walled tube with an evacuated space between the walls and is used to maintain a contained substance at high temperature (National Renewable Energy Laboratory 2019). The solar fluid is then circulated and used in high heat applications. Another way that solar parabolic troughs can be used is to heat water in order to create steam that is then used in a steam turbine for electricity production, with the rest of the heat being used for low grade heat applications such as domestic heating. A simplified illustration of this is shown in Fig. 7.10 below.

1.11.3 Operating Solar Parabolic Trough Plants

Two famous examples of large scale solar parabolic trough plants can be found in USA, where this technology was pioneered. The Solana plant, located in Arizona, is currently the largest parabolic trough plant in the world. Furthermore, this plant is

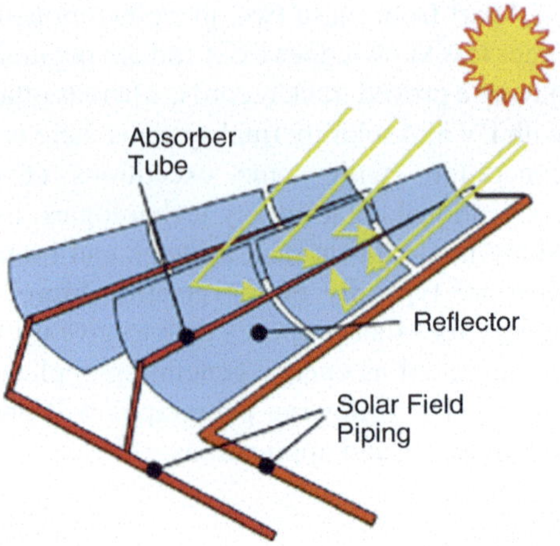

Fig. 7.10 Solar parabolic trough, creative commons license-design by Andrew Buck

using molten salt for heat storage. Once the sun sets and there isn't enough solar radiation to continue producing heat, the molten salt releases the heat stored, thus being able to keep generating energy for hours after sundown. It is rated at 280 MW and takes up a space of 7.8 km^2.

A more traditional solar parabolic trough plant is the SEGS plant located in California. The SEGS plant is currently the second largest solar thermal facility in the world while being the largest parabolic trough installation. It is located in the Mojave desert and is rated at 354 MW, being able to provide 662 GWh annually. It takes up a space of 6.5 km^2 and has 936,384 mirrors in total. To provide a sense of scale, if all the mirrors used for the project were lined up, then the total length would be 370 km.

Solar parabolic troughs can operate at a smaller scale as well, but due to their complexity, size and very precise geometry they are usually more common in large scale applications.

1.11.4 Solar Towers

The concept of solar towers is another solar heating project. An array of mirrors using heliostats is tracking the sun and their focal point is located in a tower. Power plants of this type were featured in the opening scenes of the Blade Runner 2049 film, but in terms of commercial applications, the first plant of its kind is the PS10 solar power plant in Seville, Spain. That specific plant can produce steam at 275 °C, that in its turn is diverted to a steam turbine that is able to provide 11 MW of electricity. An alternative route would be to use the heat directly, but with the limitations of heat transfer over large distances, the electricity generation seems like a more viable route.

A second plant located next to the PS10 plant is also operational. It is called the PS20 solar power plant using the same abovementioned technology and is rated at

20 MW of electricity. It is effectively a bigger version of the PS10 plant. Up until 2014 it was the largest plant of this type, when the controversial Ivanpah plant surpassed it.

The Ivanpah solar tower plant – shown below in Fig. 7.11- is located in the Mojave desert, California, US. It consists of 173,500 mirrors taking up an area of 14 km^2. It is rated at 392 MW- a massive upscale compared to the much smaller PS10 and PS20 plants. It has not avoided controversy though because of its need to use natural gas as a start up fuel in order to avoid losing steam over cloudy periods (Breaking Energy 2015). Due to non ideal weather conditions, the plant currently uses 60% more natural gas than originally planned, leading to some doubts whether it should continue to be classified as a renewable energy project.

If anything this shows that this technology has a very big potential but it has to overcome some burdens. Using external energy sources to kick start a renewable energy installation isn't unheard of – many vertical axis wind turbines require help with their start up. Nonetheless, this need for an external source will provide some challenges for the future of this technology. Another limiting factor can be the massive scale that these plants come from, making them potentially unsuitable for smaller sites.

Fig. 7.11 Ivanpah solar power plant, creative commons license-photo by Craig Butz

Fig. 7.12 Solar furnace, Odeillo, France, Creative commons license – photo by H. Zell

1.11.5 Solar Furnaces

Solar furnaces use a very similar technology to Solar towers, that can be used both in small and large scale. In smaller scales this application is more commonly found as a solar cooker and will be covered later on in a bit more detail. Solar furnaces use parabolic mirrors to focus sunlight onto a specific focal point. Water can then be circulated at the focal point, providing steam that can then be used for a steam turbine. The first solar furnace in the world was developed by Professor Félix Trombe in Mont Louis in 1949 and has been reported to develop temperatures between 2000 °C up to 3500 °C.

The current largest solar furnace project is located in Odeillo, France. It consists of 10,000 mirrors that redirect sunlight to a large concave mirror. This area that the sunlight is focused is roughly the same as a cooking pot (P-O Life 2019) (Fig. 7.12).

1.11.6 Solar Ponds

A solar pond is pool that is filled with stratified salty water of varying salinities with the saltiest layer being brine (20% salt concentration or above). Because of the density gradient that is caused by the different concentrations of salt, the bottom of the lake accumulates water with greater salinity than the top of the lake. The water is transparent while the bottom of the lake is painted black thus having a high absorption coefficient. That causes the bottom of the lake to heat up, and in its turn to heat up the very salty layer. Typically warm water would be expected to rise but the increased density of the layer prevents the heated water to rise through convection, leading to greatly increased temperatures than wouldn't be anticipated normally. The heat trapped in the bottom layer of the solar pond can be used in various heating applications. The working principle is shown in Fig. 7.13.

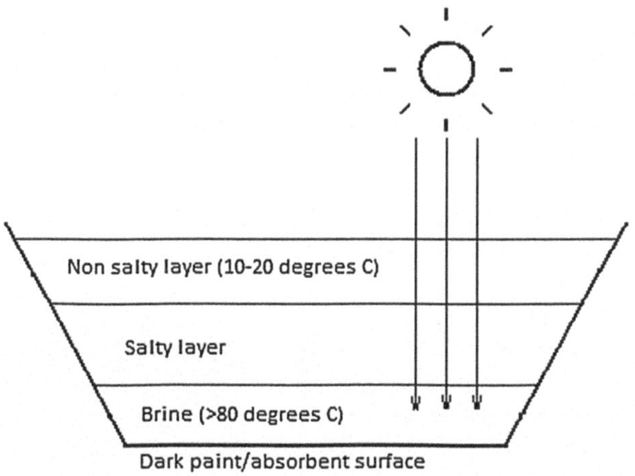

Fig. 7.13 Solar pond

1.11.7 Solar Desalination

Solar energy can also be used to desalinate water. There are two methods that can be used to desalinate water, the direct and indirect method.

In the indirect method, the energy needed to drive a conventional desalination plant, whether it is by reverse osmosis, multiple stage flash distillation etc. is provided by a PV system. Complementary energy sources such as wind are also often used, depending on where the installation might be located. An example of this set up can be seen in the Utrik Atoll reverse osmosis plant in the Marshall Islands (The Marshall Islands Journal 2014).

The direct method of solar desalination uses the principle of the solar still and is mostly used for small scale applications. Various designs exist, but the working principle in all of them is that solar energy is used to evaporate freshwater from brackish or seawater. The saline water is placed in a container and the container is covered with a sloping clear glass or plastic covering. Solar radiation is transmitted through the transparent layer and heats up the container thus causing the evaporation of the saline water. Water vapour then condenses on the interior surface of the covering. It is used mostly in arid regions and smaller scale solar stills are often used in lifeboats or survival kits. The Texas AgriLife Extension Service in College Station, Texas tested four different solar still designs and it was shown that as little as 0.7 m^2 of surface can produce enough fresh water for the survival of one person. (Sustainable Sanitation and Water Management 2017). Another direct method is to use the heat generated from solar thermal applications directly in order to evaporate large amounts of sea water as is seen in Fig. 7.14.

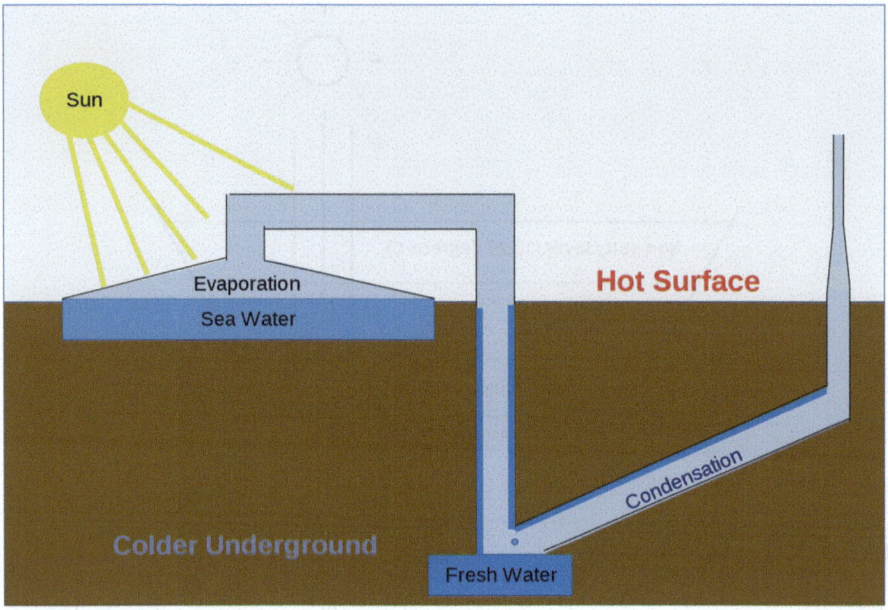

Fig. 7.14 Solar desalination, by Engelbert Niehaus – LibreOffice Draw Images exported to SVG, CC BY-SA 4.0, https://en.wikipedia.org/w/index.php?curid=55237502

1.11.8 Solar Disinfection

Solar disinfection (SODIS) is a method of disinfection of fresh water using solar energy and as such it's a very different application of solar energy. The method uses clear Polyethylene terephthalate (PET) bottles that are filled with the water and set out in the sun for at least 6 h. The UV-A rays in sunlight kill germs such as viruses, bacteria and parasites. The method also works when air and water temperatures are low, but higher temperatures speed up the process. Bacteria, viruses and parasites all have different sensitivities to UV-A radiation, with bacteria being easy to kill and viruses being killed in the 6 h recommended by SODIS. Parasites might take up to 10 h to kill, especially cryptosporidia cysts. SODIS is recommended by the World Health Organisation (WHO), UNICEF, and the Red Cross as a way to treat drinking water in developing countries. (SODIS 2019)

1.11.9 Passive Solar Design

Passive solar design is a method of designing a house in order to achieve heating and/or cooling naturally using the sun. Typically, passive solar heating involves the collection of solar energy through well oriented windows. The correct orientation depends on which part of the world the building is located and the kind of application that is needed. For all around heating of a house in the Northern hemisphere, south facing windows are used, since they maximise the energy capture throughout the day. The energy that is absorbed during the day is stored in a thermal mass that comprises of building materials that have high heat capacities. These can be Trombe walls, floor tiles, heavy brick walls etc. The stored solar energy can be distributed naturally back the surroundings through the naturally occurring mechanisms of convection and radiation showcased below in Fig. 7.15 (McMullan 2007).

Passive building design can be used in order to cool a space as well, as the heated elements can cause a natural air circulation if the heated air is directed away from the building, thus drawing cooler air in. In a similar concept evaporation of water assisted by solar energy can help as is evident in the case of draft tower as can be seen in the picture below – Fig. 7.16 from the Zion National Park.

Water is allowed to evaporate at the top of the tower – typically by using cooling pads. The evaporation subsequently cools the incoming air causing it to travel downwards and into the building that brings the temperature down. This is the same working principles that is utilized in traditional Iranian architecture with windcatchers (bâdgir in Persian) shown in Fig. 7.17 below.

1.11.10 Small Scale Applications

A mention should be made about the small scale applications that solar energy has a well. As mentioned before, solar stills are very important and crucial additions to lifeboats as well as survival packages, as they can provide a good amount of fresh water using minimal resources. Smaller working versions of solar furnaces can be found in solar cookers. Solar cookers –both commercial and DIY designs are very popular choices for campers and during outdoor excursions and have applications in rural communities in developing countries as well. Various designs exist with even foldable versions being sold as survival gear. The operating principle is the same in all of these applications. The cooking vessel -typically a lightweight metal pot- is placed in the focal point of the mirrors and thus absorbs a large amount of energy-shown below in Fig. 7.18.

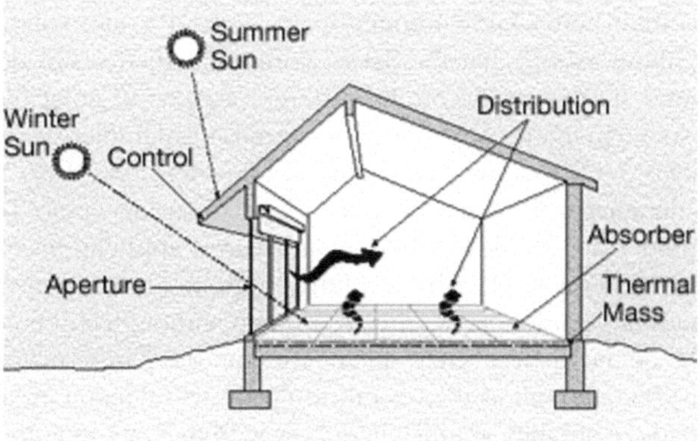

Fig. 7.15 Passive solar design, Public domain image, Office of Energy Efficiency and Renewable Energy

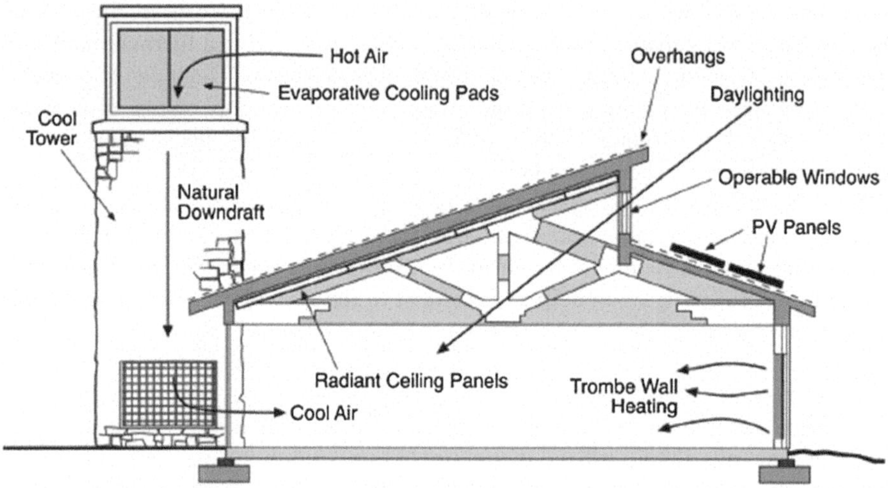

Fig. 7.16 Solar cooling Public domain image P. Torcellini, R. Judkoff, and S. Hayter, National Renewable Energy Laboratory

Solar showers are also a popular choice amongst campers. The simplest design consists of a dark vinyl bag that contains water. The bag is left facing the sun during the day and that provides enough hot water for a shower, or other applications. Different designs, borrowing from solar thermal collectors exist as well including more complicated designs such as spiral wound tubing encased in a transparent cover.

1.12 Recap of Non-conventional Solar Technologies

This was only a brief mention of some solar technologies that are not as well known as their better-known counterparts, solar PV and solar thermal collectors. It goes without saying that the list is neither exhaustive nor definitive but will serve as a good starting guide on the different aspects of solar technology for anyone interested. No matter how complex (or not) a solar energy application is though, there is always a limitation that should not be forgotten. What is the maximum amount of solar energy we can ever have on 1 m^2 of surface area? This should always be on our mind when thinking about solar energy applications- whether that it the mo PV installations tend to have a fairly high environmental footprint due to the need to melt large amounts of silicon. As such, although still better than fossil fuels the tend the be the dirtiest renewable energy out there in terms of CO_2 emissions. This is on average though as the overall footprint will depend greatly on how the panels were made, were they are installed, how well they are maintained etc. It's always good to

Fig. 7.17 Bâdgir Creative commons license by Diego Delso

be careful when discussing environmental footprints as they tend to be very case specific in the case of solar energy – and other renewable energy sources in general. They also tend to have a high material usage per unit of energy produced. That can be attributed to the fairly low output they have thus bringing their material usage per energy cost up.

1.13 Worked Examples for Solar Energy

Here follow two examples of how we can estimate solar yield in the case of PV. For solar thermal applications this becomes a more complex issue that is dictated by how much hot water we want, at what temperature and for what tank size. Having said that, the approach is very similar with the main difference being that the

Fig. 7.18 Example of a solar cooker Creative commons license By Erik Burton from Kapaa

efficiency of solar thermal collectors isn't constant and is affect by the temperature difference between the collector and the ambient temperature. Calculations tend to get a bit tedious and as such they are left out from this general chapter on solar energy. Duffie and Beckman's (2013) seminal work contains excellent examples and is an invaluable source for people that want to go deeper into the workings of solar thermal collectors.

1.13.1 Good vs Bad Site

Using data from the Global Solar Atlas we can identify what the best and the worst places in the world are when it comes to solar potential. These appear to be Antofagasta in Chile (most available solar energy) and Yamane Shi, Sichuan in China. The website allows you to download the GIS data for similar exercises like these.

Assuming the optimal angle and orientation, the first site experiences -on average – 8.06 kWh/m^2 per day whereas the second experiences only 2.41 kWh/m^2 per day. We are not including areas that aren't covered by the Global solar atlas as typically in these regions the amount of solar availability is even less and in extreme cases can be zero for months – in Antarctica for example.

Currently one of the most efficient commercial PV panels in the world (at the moment of writing) is the SunPower X22 360 W Panel boasting a 22.2% efficiency -under STC (Sunpower 2019). As we have no real access to the sites and for the sake of keeping things even, we can assume that the conditions laid out by STC are indeed met. That panel has an area of 1.63m^2 and remembering that available energy from a PV panel is always $E = \eta * A * H$ then we have

$$E_A = 0.222*1.63*8.06 = 2.92\,\text{kWh per day for Antofagasta}$$

$$E_Y = 0.222*1.63*2.41 = 0.87\,\text{kWh per day for Ya'an}$$

That is the energy available from the panel, not the energy we are able to get finally, because we have to account for BOS losses via the performance ratio. Very well maintained and calibrated systems can exhibit a PR of >90% so assuming that, the respective energies now become $E_A = 2.63$ kWh per day and $E_Y = 0.78$ kWh per day. In order to be more accurate we would have to apply temperature corrections, correct for shading and use the net area of the panel instead of the gross.

Plus we shouldn't forget that this answer is only about maximum possible energy yield – it doesn't take into account engineering works, infrastructure etc. For example these specific sites might be completely unsuitable for solar energy installations. It is just a very rough estimation of what we can in theory expect as a best and worst scenario.

1.13.2 Sahara Desert Example

Some proponents of solar energy have proposed a hypothetical gigantic solar park in the Sahara desert that would act as a global energy solution. Instead of theorising about such a park we can actually – very roughly- work out what would be needed. This will give us a basis that we can use for future discussions.

World electricity consumption for 2018 was 22,964 TWh (Enerdata 2019). The Sahara desert is a vast place but by looking at available GIS data we can see that an average of 2500 kWh/m^2 per year is expected for tilted insolation at the optimum PV tilt. We will use this as a our basis for our calculations.

Assuming the same PV panel as earlier and the same performance ratio we can deduce that each panel would be able to give us, annually:

$$E = 0.222 * 2500 * 1.63 * 0.90 = 814\,\text{kWh}.$$

As such in order to cover our total energy needs as humanity, we would need 28,211,302,211 panels- coming from the consumption of electricity divided by the electricity per panel. Since we know that each panel is 1.63 m^2, we can easily deduce that the total area needed to be covered by PV would be equal to 45,984 km^2. The good news is that the Sahara desert is 9,200,000 so in theory there is enough space for our gigantic solar park. By no means is this an accurate estimation and it can only serve as a very rough ballpark figure to show us how monumental of an undertaking that would be.

Of' course it isn't as simple as that. Some of the things we have to factor in are-and not limited to

- Transport of that electricity – as a ballpark figure in the US almost 6% of electricity is lost via transmission and distribution. These losses would become much greater over long distances.

- Technical issues of installing PV in a place with poor infrastructure
- Dust that can shade the PV panels and thus reduce their efficiency
- Loss of efficiency due to high temperatures
- Thermal stress of the materials
- Electricity storage issues – and losses due to it. Storing energy would become paramount since the world uses electricity at different rates and at different times. That is on top of the fact that the Sun also sets.
- Amount of materials needed – both for the PV panels and the infrastructure.
- Border issues- as with every big energy projects that spans across borders the issue now becomes, who does what.
- Maintenance
- The above figure shows the size of the panels and not the installation itself. The panels should be spaced apart to increase cooling and allow for maintenance – especially crucial for such and example
- Financing such a task

If this task is so unrealistic then why mention it here? It's always nice to put things under perspective and understand how much of something we need. We have the technical constraints that are very well known but we also have to content with our energy hungry nature. Despite all efforts for energy conservation, our energy demand keeps increasing. To add insult to injury, we should also remember that electricity is only one form of energy. If we need electricity to help with transport – via electrifying cars for example- then the demand and consumption will get even higher. Despite all that, the future of solar energy seems bright and it is only one of our weapons against climate change.

2 Conclusions

This chapter presented a brief overview of various solar energy technologies and their uses. Is solar energy a perfect energy source? Of'course it's not – but no other energy is. Wider adoption of solar energy presents challenges, some of which are technical. It can be argued that solar energy will be one of the major drivers for increased energy storage – molten salt storage is already used in large scale solar thermal installations and batteries for PV systems are very regularly used. The environmental cost of solar energy cannot be dismissed though as with the increased usage of PV we will inevitably have an increased need to dispose of non working panels, panels that reached the end of their useful life, panels that malfunctioned, broke etc. Despite that though, the potential of solar energy is massive and if used in sustainable ways by taking into account good locations, technologies that reflect our needs and managing our energy usage patterns it will play a very major role in our energy future.

References

Bainbridge D (1981) The integral passive solar water heater book. In Solar Usage Now. Retrieved from http://www.mangus.ro/pdf/Manualincalziresolarapasiva.pdf

BP (2019) BP energy outlook

Breaking Energy (2015) Ivanpah 170% increase in output. Retrieved from http://breakingenergy.com/2015/06/17/ivanpah-solar-production-up-170-in-2015/

Duffie JA, Beckman WA (2013) Solar engineering of thermal processes. In: Solar engineering of thermal processes, 4th edn. Wiley, Hoboken

Enerdata (2019) Wind & solar share in electricity production data | Enerdata. Retrieved August 26, 2019, from 2019 website: https://yearbook.enerdata.net/renewables/wind-solar-share-electricity-production.html

Fraunhofer ISE (2019) Photovoltaics Report. Retrieved from www.ise.fraunhofer.de

Helal O, Chaouachi B, Gabsi S (2011) Design and thermal performance of an ICS solar water heater based on three parabolic sections. Sol Energy 85(10):2421–2432. https://doi.org/10.1016/j.solener.2011.06.021

Heliodyne (2011) Solar flat plate vs evacuated tube collectors.pdf (p 2). Retrieved from http://www.heliodyne.com/

International Energy Agency (2018) WEO 2018. Retrieved August 26, 2019, from https://www.iea.org/weo2018/

Loumakis G (2018) Development and testing of a low cost solar thermal collector. Glasgow Caledonian University, UK

Mauthner F, Weiss W, Spork-Dur. (2016) Solar heat worldwide 2016

McMullan R (2007) Environmental science in building, 6th edn. Palgrave Macmillan

National Renewable Energy Laboratory (2019) Solar parabolic troughs. Retrieved from http://www.nrel.gov/csp/troughnet/solar_field.html

Perlin J (2013, September 10) Let it shine: the 6,000-year story of solar energy (First). New World Library, Novato; Revised edition

P-O Life (2019) Solar furnaces. Retrieved from http://www.anglophone-direct.com/Mont-Louis-Font-Romeu-Odeillo-Via

SEIA (2015) Major solar projects in the US. Retrieved from http://www.seia.org/sites/default/files/resources/PUBLICMPL 9-29-2015.pdf

Smyth M, Eames PC, Norton B (2006) Integrated collector storage solar water heaters. Renew Sustain Energy 10(6):503–538. https://doi.org/10.1016/j.rser.2004.11.001

SODIS (2019) Solar disinfection. Retrieved from http://www.sodis.ch/methode/index_EN

Solarcooking.org (2013) Horace de saussure and his hot boxes of the 1700's. Retrieved from http://solarcooking.org/saussure.htm

Souliotis M, Quinlan P, Smyth M, Tripanagnostopoulos Y, Zacharopoulos A, Ramirez M, Yianoulis P (2011) Heat retaining integrated collector storage solar water heater with asymmetric CPC reflector. Sol Energy 85(10):2474–2487. https://doi.org/10.1016/j.solener.2011.07.005

Sunpower (2014) Solar star projects. Retrieved from http://us.sunpower.com/sites/sunpower/files/media-library/fact-sheets/fs-solar-star-projects-factsheet.pdf

Sunpower (2019) Sunpower-x-series-x22-360-residential-solar-panels.pdf | SunPower. Retrieved August 26, 2019, from https://us.sunpower.com/sunpower-x-series-x22-360-residential-solar-panels

Sustainable Sanitation and Water Management (2017) Desalination. Retrieved from http://www.sswm.info/content/desalination

The Marshall Islands Journal (2014) Utrik RO plant. Retrieved from http://www.marshallislands-journal.com/Archive1-17-14-page.html

Turner S (2013) History of solar energy | SolarEnergyBase.com. Retrieved from http://www.solar-energybase.com/history-of-solar-energy/

Chapter 8
Coupling Behavior-Based Intervention with Pro-Environmentalism. The Dynamics of Energy Usage, Crisis and Its Conservation

Tharindu Prabatha, Kh Md Nahiduzzaman, Hirushie Karunathilake, Kasun Hewage, Rehan Sadiq, Shahria Alam, and Pamela Shaw

Abstract Higher cost and crisis of energy vis-à-vis increasing consumption have been a twin but contested paradox. The rapidly growing energy demand has prompted many countries, including Canada, to undertake manifold energy-saving initiatives. However, these are predominantly technology driven and no apparent measures are taken yet to address and modify the end users' behavior at the residential sector. In order to reduce the rate of growth of the residential energy consumption, it is critical to engage the end users through better education and awareness while using the inherent pro-environmental behavior (PEB). This is even more critical for groups, such as new immigrants and Canadian indigenous communities. Given this background, this chapter essentially presents (i) the importance of behavior-based, non-technical interventions on end users' perceptions of energy conservation; and (ii) its impact on the nature of consumption at the household level. Empirical findings from the East and current practices from the West are drawn to investigate these phenomena. The other part sheds light on the prospect and need for behavior based interventions toward reduced energy consumption. While the time of use (ToU) is in effect, some forms of PEB exists among the residential users in

---

T. Prabatha · K. Md Nahiduzzaman (✉) · K. Hewage · R. Sadiq · S. Alam
School of Engineering, The University of British Columbia (UBC) Okanagan, Kelowna, BC, Canada
e-mail: Tharindu.Prabatha@alumni.ubc.ca; Kh.Nahiduzzaman@ubc.ca; Kasun.Hewage@ubc.ca; Rehan.Sadiq@ubc.ca; Shahria.Alam@ubc.ca

H. Karunathilake
Department of Mechanical Engineering, University of Moratuwa, Katubedda, Moratuwa, Sri Lanka
e-mail: hirushiek@uom.lk

P. Shaw
Community Planning, Vancouver Island University (VIU), Nanaimo, BC, Canada
e-mail: Pam.Shaw@viu.ca

H. Qudrat-Ullah, M. Asif (eds.), *Dynamics of Energy, Environment and Economy*, Lecture Notes in Energy 77, https://doi.org/10.1007/978-3-030-43578-3_8

Ontario. With this taken into account, the paper calls for a renewed policy insights on 'investment' and 'curtailment' behavior approaches to assess the 'longevity' effects on energy consumption. This, in turn, stems the foundational need for 'collaborative' think-tanks of multi-disciplinary professionals, including engineers, urban developers, environmentalists, planners, sociologists, economists, and psychologists.

Keywords Pro-environmental behavior · Behavioral interventions · Energy crisis · Energy conservation behavior · Energy efficiency

Glossary

BC	British Columbia
DSM	Demand-side management
ECB	Energy conscious behavior
GHG	Greenhouse gas
MURB	Multi-unit residential buildings
NECB	National Energy Code of Canada for Buildings
NG	Natural gas
PEB	Pro-environmental behavior
SDG	Sustainable development goals
ToU	Time-of-use

1 Introduction

Forty percent of global energy consumption and 12% of the global green house gas (GHG) emissions are accounted under the building sector (Environment and Climate Change Canada 2016; Friess and Rakhshan 2017). Moreover, the increasing energy prices create considerable economic pressures on the building sector. These economic and environmental pressures have made building energy efficiency a priority theme in Canada. Therefore, new policies and standards developed in parallel with the climate change mitigation agenda have increased the emphasis on reducing energy use and GHG emissions associated with the building sector (Environment Canada 2015; Karunathilake et al. 2018). These measures provide multiple benefits to the building owners and communities in the form of reduced energy bills, improved energy security, and socio-economic growth in addition to the environmental benefits (Natural Resources Canada 2017). Therefore, improving the energy performance is essential for the environmental and economic sustainability of buildings.

A building that can operate with a lower energy consumption compared to an average building that operates under similar conditions of climate and use without compromising the occupant satisfaction can be identified as an energy efficient building. In net-zero buildings renewable energy sources are used to cater the energy demand, which is already reduced by the use of energy efficiency measures (Torcellini et al. 2006). Over 70% of energy use, environmental foot print, and human health impacts associated with buildings occur during the operational phase of the buildings (Buyle et al. 2012; Deru et al. 2011). Therefore, the operational phase has received a lot of attention in recent energy performance enhancement initiatives such as BC Energy Step Code. Operational energy performance can be improved by two main strategies including technical measures and behavioral changes (Karunathilake et al. 2018). Retrofitting, using energy efficient appliances, and implementing energy conservation technologies are some of the popular technological energy performance enhancement approaches used in building sector (Roulet 2006; Meier and Lamberts 2002). Building energy codes across the world mandate these energy efficient technological advancements, and these approaches have shown significant energy savings. Being one of the most innovative building energy codes currently available, the Energy STEP Code developed by the province of British Columbia, Canada, aims to make all the new constructions energy efficient and net-zero ready by 2032 (Natural Resources Canada 2018). However, the BC Energy STEP Code has room for improvement by prescribing how to achieve these energy efficiency goals through robust investment decisions. Moreover, there lies a question whether Canada has put enough emphasis to develop net-zero ready energy conscious citizens who are compatible with net-zero energy buildings.

Demand-side management (DSM) programs that can decrease energy use at building level can primarily be classified into three routes; (I) replacing the existing building stock with low energy and passive designs, (II) develop and increase the penetration of low energy (efficient) appliances and equipment, (III) promote and instill energy conscious-behaviours among users (Wood and Newborough 2003). Despite the proven benefits, technological energy efficiency achievements including low energy buildings and efficient appliances have their own limitations in practice due to high costs. Even so, the investments made on energy efficiency in a given building result in benefits localized to that particular building alone, and the impact on the energy performance of other buildings is minimal. On the other hand, behavioral changes require much lower investments to implement. Nevertheless, once implemented, the energy efficiency culture developed among the occupants in a given building propagates around the community, creating a widespread impact on many buildings in contrast to technological measures. Therefore, developing policies to sufficiently promote net-zero ready occupants in parallel to the technological measures can significantly increase the effectiveness of building energy efficiency improvement efforts.

Literature suggests that significant energy savings can be achieved through behavioral pattern changes created by the use of feedback mechanisms (Xu et al. 2018). However, predicting the level of behavioral pattern changes and its impacts on the energy consumption is not a very straightforward process. Moreover, while behavioural modifications can deliver significant benefits once put in place, it is more challenging to change existing patterns of energy use behaviour than it is to simply instigate a technological intervention. Therefore, it is evident that both behavioral and technological energy efficiency solutions have direct and indirect environmental, economic, and social impacts. Hence, careful evaluation of the costs, benefits, and deployment challenges associated with these approaches is needed to identify the best solutions for a given context. Moreover, studies need to be conducted to quantify the effectiveness of non-technical measures in improving the building energy performance, and to identify the best pathways for promoting energy-consciousness. On this backdrop, this chapter specifically focuses on studying the importance of behavior-based interventions on end users' perceptions of energy conservation and their impacts on the energy use trends in households. Moreover, empirical results from Saudi Arabia and a Canadian case-study are used in investigating the aforementioned phenomena. This article will inform the energy planners about the effectiveness of behavioral changes in achieving energy and cost savings. Moreover, policy makers will benefit from the integrated knowledge in improving energy efficiency policies to reach intended emissions targets.

2 Energy Sources and End Uses

Canada mainly relies on five fuels including renewables, natural gas, crude oil, coal, and uranium for the energy needs of the country (National Energy Board of Canada 2019). These energy sources are converted into secondary energy uses such as electricity, motor gasoline, etc. The trend changes of different secondary energy uses over the time is elaborated in Fig. 8.1 (Natural Resources Canada 2019a).

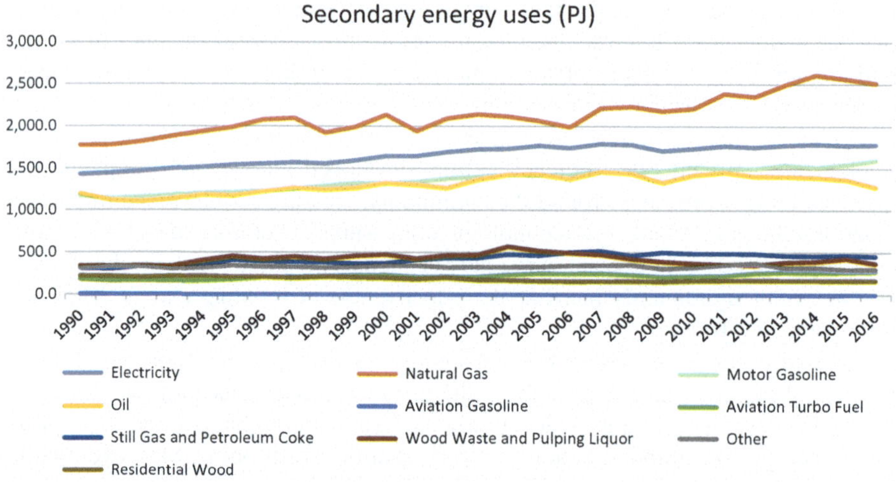

Fig. 8.1 Secondary energy uses (Natural Resources Canada 2019a)

According to Fig. 8.1, natural gas is the main secondary energy source that is closely followed by the electricity. Both these categories have shown increasing trends over the time while NG has shown a relatively higher growth. Motor gasoline (including ethanol) have shown a slight growth over the years following a similar gradient to the electricity. Use of oils (including diesel fuel oil, light fuel oil, kerosene and heavy fuel oil) is similar to motor gasoline except for the reduction indicated after 2011. Other secondary energy uses have not shown significant variations over the years. Even though some energy sources are being used less compared to others, they have higher emissions per unit energy delivered. Therefore, it is important to compare the emissions associated with different secondary energy categories over the years. The historical trends are indicated in Fig. 8.2 (Natural Resources Canada 2019a).

Emissions intensity of NG has shown a slight increment over the years according to Fig. 8.2. However, the impacts associated with this change cannot be neglected as NG is the most used secondary energy source in Canada. It is important to note that emissions associated with unit energy of electricity has reduced by one-third from 2000 to 2016. This greatly helps off-set residential emissions resulted from energy use. The emissions and the energy prices associated with NG and electricity have considerable impacts on residential energy use trends as they are the most commonly used secondary energy sources in Canadian residential sector. The emissions associated with other secondary energy sources have not changed much over the time except for still gas and petroleum coke. However, the contribution of these energy sources to the national GHG emissions through various means including transportation and industries should not be missed out.

The above-discussed energy sources are used for many end-use sectors, including residential, commercial, industrial, transportation, and agriculture (Natural Resources Canada 2011). Out of these sectors, transportation is identified as the least efficient sector with 75% losses while the residential sector converts 75% of the energy supply to useful work (National Energy Board of Canada 2019). However, when considering all the energy end-use sectors in Canada, only 33% of the

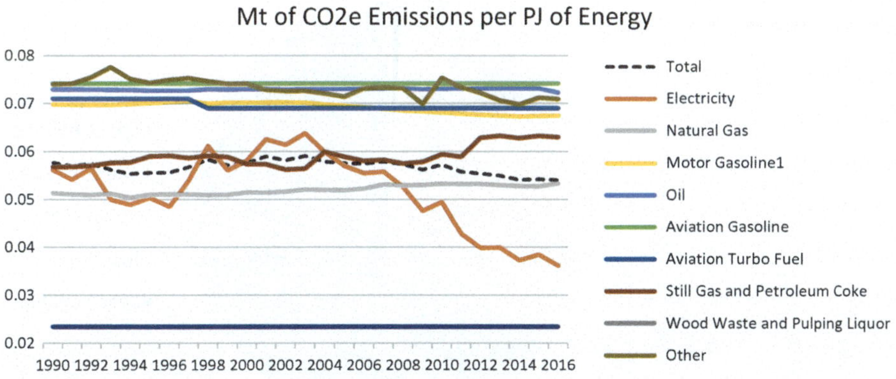

Fig. 8.2 Emissions per unit energy (Natural Resources Canada 2019a)

supplied energy is converted to useful work (National Energy Board of Canada 2019). Therefore, energy conservation practices and efficiency improvements can greatly contribute in reaching Canadian emission targets. Therefore, paying enough attention towards minimising energy losses and energy conservation practices are as important as striving to increase the renewable energy penetration in the country. With this understanding, improving the energy performance of building sector has received much attention.

Residential buildings contain multiple energy consuming components including space heating and cooling, water heating, lighting, and appliances. Figure 8.3 shows the percentage contribution of different energy end-uses for the overall energy consumption of Canadian residential building sector (Natural Resources Canada 2011, 2015a).

Space heating accounts for the highest energy consumption of Canadian residential buildings. Approximately 50% and 25% of the residential heating needs are catered by natural gas and electricity respectively. The heating performance of houses is significantly improved with higher penetration of energy efficient furnaces and improved building envelops (Natural Resources Canada 2019b; Karunathilake et al. 2018). Water heating is the second highest energy consumer of residential buildings. At present water heating is mainly done by natural gas and electricity with percentage contributions of 68% and 29% respectively (Natural Resources Canada 2019b). It is important to notice that use of heating oil, wood, and other fuels are significantly decreased over time. Efficiency of all major appliances such as refrigerators, freezers, dishwashers, clothes washers, and cloths driers have improved from 1990 to 2015. On the other hand, the use of minor appliances including electronics have increased over the time. Therefore, the energy savings generated by efficiency improvement of major appliances have been off-set by the minor appliances (Natural Resources Canada 2019b). Lighting energy requirements of a household has shown a huge reduction (18%) from 1990 to 2015. Natural Resources Canada (NRCan) attributes this efficiency improvement to multiple

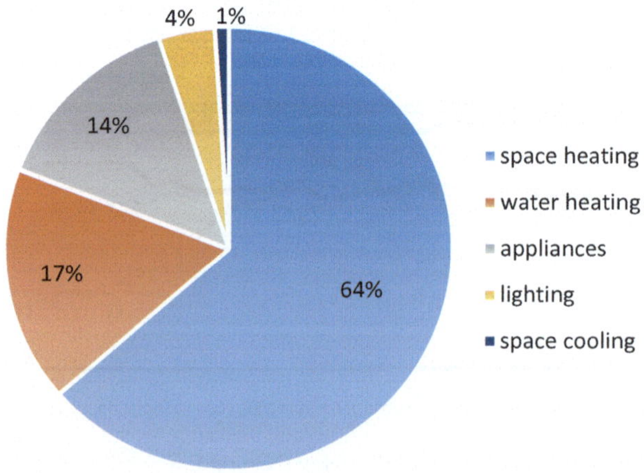

Fig. 8.3 Building energy end-uses (Natural Resources Canada 2011, 2015a)

factors including the market penetration of LEDs, motion detectors, timers, task lighting, light dimmers, and behavioral practices such as turning-off unnecessary lamps (Natural Resources Canada 2019b; Karunathilake et al. 2018).

Over the past three decades, significant changes occurred in Canadian residential building sector. The number of households in Canada increased from 9.9 million to 14.1 million from 1990 to 2015 (Natural Resources Canada 2019b). The other noticeable change is the increase in the cooled area percentage with respect to occupied spaces (Natural Resources Canada 2011, 2019b). The percentage differences of the changes occurred from 1990 to 2015 in residential buildings are indicated in Fig. 8.4 below. These information are derived from statistics provided by NRCan (Natural Resources Canada 2019b).

Energy use varies significantly with weather, occupancy patterns, and the energy efficiency of the equipment being used. NRCan attributes residential energy consumption changes over the said time period to five factors including activity, structure, service level, weather, and energy efficiency attempts (Natural Resources Canada 2019b; National Energy Board of Canada 2016). Except the structure and energy efficiency efforts, all the other factors have resulted significant increases in total energy consumption. If the energy efficiency efforts were not present, then the total energy consumption of the residential building sector would have increased by 54% by 2015 compared to 1990. Percentage increment of energy use was limited to 8% due to energy efficiency interventions including introduction of CFLs and LEDs, insulation upgrades, roof and window modifications, and use of energy efficient appliances (Natural Resources Canada 2019b). The said measures were able to avoid 27.8Mt of GHGs in 2015. Nevertheless, the GHG emissions associated with residential buildings are predicted to be increased by 15% compared to 2012 levels due to growing energy demand (Natural Resources Canada 2019b; Environment Canada 2014). Therefore, the energy efficiency improvements have to be a continuously improving process in order to bring the residential energy demand and related GHG emissions to a plateau or to a decreasing trend despite of the growing number of households.

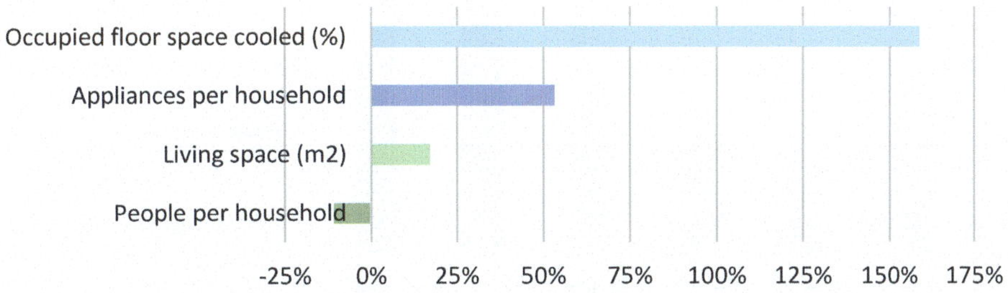

Fig. 8.4 Percentage differences of the changes occurred from 1990 to 2015

3 Energy Efficiency Strategies: A Bird's Eye View

Introduction of technological interventions in the form of retrofits is popular in Canada for achieving energy efficiency in buildings. Energy retrofits can be classified into three categories, namely minor energy retrofits, major energy retrofits, and deep building retrofits. Minor retrofits are defined as focused energy saving measures that can achieve energy savings up to 15% from the original consumption. Similarly major and deep retrofits are expected to produce energy savings from 15 to 40% and greater than 40% respectively (Natural Resources Canada 2016).

Energy efficiency measures are popular in residential building sectors of BC and many other provinces of Canada. Statistics show that at 82% of the residential buildings have adopted one or more energy conservation measures and 37% of the households have implemented energy retrofits (Statistics Canada 2011). Energy efficiency of new residential buildings is expected to be improved by 27% compared to 1997 with the introduction of new building codes such as NECB and BC Energy STEP Code (Natural Resources Canada 2015b; Government of Canada 2016). Moreover, 0.7% annual reduction in energy use per unit indoor area is expected from 2016 to 2040 with the planned energy efficiency efforts (National Energy Board of Canada 2016).

A case study in the Major Retrofit Guidelines (MRG) report prepared by NRCan shows some insightful cost benefit figures for energy retrofits. Costing information including investment cost, savings, and payback period from the mentioned case study is summarised in Fig. 8.5 (Natural Resources Canada 2016).

Cost information in Fig. 8.5 indicate that higher energy savings can be achieved with higher initial investments. Moreover, it shows that the payback period for other retrofits except for AHU motor considered in the case study generally increases with the increasing investments. It can be seen that if higher energy savings in the range of C$5000 are anticipated, then the investments are in the range of C$70,000–100,000 leading to major and deep retrofits. However, this level of

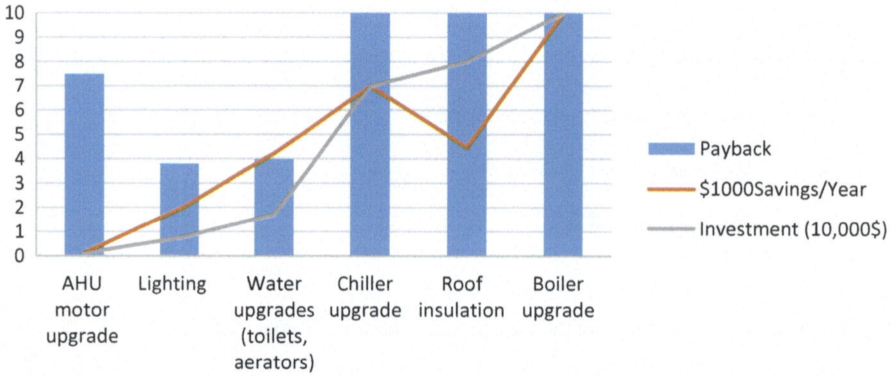

Fig. 8.5 Cost information of retrofits (Natural Resources Canada 2016)

investments is not always possible for residential building owners spanning from single family detached houses to multi unit residential buildings. When it comes to rentals, the tenants get to pay the utility bills while the owner is responsible for the retrofits. In situations such as MURBs, the owners of the apartment units cannot do any major or deep retrofits by themselves without the consent of strata management. MRG highlights that aligning these deep retrofits with scheduled maintenance activities produces better results. Therefore, the interventions have to be waited until a suitable execution point reaches in the building energy system to reap the maximum economic benefits. Essentially, there are visible challenges hindering the implementation of technical energy efficiency interventions due to financial constraints, irregular division of costs and benefits, and management and scheduling complications.

In contrary to retrofitting interventions, behavioral changes can produce energy savings at minimal or zero costs. Both responsibility and benefits of energy efficient behaviors are held by the building occupants and not by the owners. Behavioral changes can be initiated at any point of time without waiting for major energy system renovations. These characteristics of behavioral interventions reduce the complications in cost benefit divisions, scheduling issues, and financial constraints. Therefore, it can be argued that the behavioral changes are an effective tool that can take the message of Canadian energy efficiency practices to the buildings that have difficulties in implementing technical measures. On the other hand, building automation systems and the sensor-based technologies have their limitations in response times. For an example, lights and heating elements that can operate automatically based on occupant presence and absence take some time before turning off after the occupants leave the space, leaving some room for further improvement. Unarguably, this technology is important in avoiding any prolonged energy losses due to mistakes of occupants. However, if this technology can be combined with energy conscious behavior, energy wastage can be minimized further. Moreover, there is a potential that people pay relatively less attention to energy conservation when energy efficiency technologies and building automation techniques are implemented without proper awareness regarding occupant responsibility. This issue can be easily resolved by energy conscious behavior and awareness programs that continuously remind occupants about their role in energy efficiency. A recent study reveals that buildings occupied by educated occupants have 1.3% higher energy efficiency compared to the other buildings. Authors of the study attribute this observation to energy conscious technology choices and behavior (Kavousian et al. 2015). This strengthens the fact that the implementation and usage of the energy efficient technologies can be improved by fostering the energy conscious behavior and awareness of the occupants. Therefore, it is important to investigate methods to educate the occupants about energy conscious behavior in parallel to energy efficient technologies in order to reach the Canadian emission targets as expected.

Occupant behavior is identified as one of the main uncertainties faced by the energy planners in system designing and operation scheduling. Literature suggest that the energy wastage occurring due to the uncertainties introduced to planning by the occupancy patterns and occupant behavior can reach up to 30% from the intended energy use (Jia et al. 2017; Wang and Shao 2017). Codes and standards such as NECB (National Energy Code for Buildings), BC Energy STEP Code, and ASHRAE are widely used to help energy designers to overcome these challenges. However, the unpredictable energy behavior of the occupants still causes losses in the energy system operation due to the mismatch in user expectations and government policies, because the energy standards look at the building energy performance in an energy efficiency lens while the building occupants are more focused on the economic performance of their energy system. This priority mismatches in decisions further poses challenges to energy system operation schedules due to the discrepancies caused by unexpected behavioral patterns (Wang and Shao 2017). If the occupant energy behavior can be made consistent by awareness programs leading to energy conscious behavior, then money, time, and effort in energy system designing and operation scheduling can be reduced significantly while minimizing the energy losses in the practical operation. Many technological interventions such as sub-metering techniques, internet of things (IoT) based techniques, non-intrusive load monitoring techniques are being implemented with the aim of reducing energy wastage by optimizing the energy system operation with real-time monitoring (Anderson et al. 2012; Martín-Garín et al. 2018). However, these approaches need expert knowledge and high initial investments. Moreover, it is difficult to implement some of these technical measures in buildings at a later stage after construction. Therefore, promoting the consistent adoption of energy conscious behaviors can be easier compared to the implementation of technical solutions proposed.

Energy conscious behavior (ECB) can be defined as the process of eliminating non-essential energy uses through actions initiated by the occupants. These behaviors can be in many forms including turning-off unnecessary lights and unplugging unused appliances, selecting appropriate room temperatures to match the needs and occupancy levels, wearing suitable cloths to avoid unnecessary increments in heating and cooling loads, and avoiding the instances of running the washers and driers at partial loads (Mills and Schleich 2012; U.S. Department of Energy 2016). The idle energy consumption of devices are found to be in the range of 7.3–10.7% of total household energy use according to the studies done in Australia, China, and Japan (Kelly 2012; Ouyang and Hokao 2009). For an example, the energy used by a microwave oven for heating is lesser than the energy consumed to operate its' digital display throughout the day (Province of British Columbia 2008). This indicates that simple attitude changes can produce significant energy changes without any additional investments. ECB is an effective tool that is strongly interlinked with energy conservation measures as discussed above. However, implementation of behavioral interventions has its own challenges if effective ways to bring it to the people were not found. Literature highlights that consumer behavior is largely influenced by the costs, required effort, and the convenience (Steg 2008). Public awareness programs are an effective way to promote conservation techniques that don't incur significant costs, require less user effort, and cause minimal inconveniences. Effectiveness of the behavioral interventions can be further enhanced with feedback mechanisms to inform the users about their energy performance (Steg 2008). Research shows that 5–15% energy savings can be achieved by providing detailed real-time energy use information through smart meters, mobile phone applications (Darby 2006). A mobile phone application to report the residential energy use is successfully practiced by Ontario residents.

Financial incentives are identified as another effective way for promoting energy conscious behavior within all the social groups including Canadian citizens, indigenous people, and immigrants. These incentives can be introduced in the form of taxes and subsidies (Lindén et al. 2006). Moreover, energy prices can be used as a direct influence to change the energy use trends. Time-of-use (ToU) pricing technique have proven to provide greater leverage to the utilities to direct the less urgent energy use applications such as cloths washing towards off-peak hours (Strbac 2008). This approach is been proven to produce mutual benefits to both utilities and occupants according to a research conducted in Sweden (Campillo et al. 2016). The users can shift their less priority energy needs to off-peak times and benefit from the low energy bills while the utilities can manage the peak demand with minimal installed capacity as the load profile is becoming flatter when the loads are shifted to off-peak times. In some provinces including British Columbia, ToU approach is only applied to commercial and industrial sectors while Ontario (ON) has employed this for residential buildings too. ON uses three time periods including off-peak, mid-peak, and peak to apply ToU pricing (Ontario Energy Board 2019). However, public awareness and acceptance of the ToU schemes are still at a questionable state according to the expert opinion. Nevertheless, outcomes of previous studies indicate that maintaining peak power rates around four times higher compared to off-peak rates produce positive results (Pittis 2019). Block tariff schemes are another commonly used billing strategy by the utilities to control the cumulative electrical energy usage of buildings. In this method, users have to pay incrementally higher prices after exceeding pre-set energy use thresholds. However, this approach is not directly supporting the load curve management. Looking at the energy use from a different perspective, governments are introducing the Carbon taxing policies in the view of reducing GHG emissions. These taxes apply on fossil fuel production, distribution, and use in relation to the associated CO_2 emissions. However, the effectiveness of carbon tax in regions where the energy cost is significantly low is arguable (National Energy Board of Canada 2019). Considering all these factors, literature suggest the use of appropriately high energy prices that can create significant cost increments and reductions depending on the energy consumption practices to foster energy conscious behavior in communities (Loi and Loo 2016).

The above discussion emphasizes the need for combining technical and social factors with economic forces to guide residential buildings towards greater energy efficiency. Multidisciplinary studies are needed to strengthen the residential energy demand management programs (Steg 2008). Upcoming sections of this article discusses the lessons learned from an empirical study on behavioral changes conducted in Saudi Arabia to discuss the practical challenges and opportunities in implementing energy conscious behavior.

4 Lessons Learned: An Oxymoron from the East and West

Climate change mitigation efforts and human behavior can greatly vary from one country to another with the differences in socio-economic dimensions. However, lessons learned through studying the best and contemporary practices along with the underlying causes can be globally applicable. Therefore, this section compares energy use and pro-environmental behavior (PEB) trends with reference to Saudi Arabia and Canada.

4.1 Saudi Arabia in Focus

Saudi Arabia is amongst the most energy intensive countries when comes to per capita consumption (Alshehry and Belloumi 2015). The conventional perspective about the daily energy behavior of the Saudi population could be characterized as 'not-so-thoughtful' (Nahiduzzaman et al. 2018). Therefore, people seem to demonstrate rather lavish energy consumption behavior. Understandably, this is partially attributed to the fact of relatively cheaper energy prices resulting from substantial governmental subsidies and higher per capita income. However, a recent study by Nahiduzzaman et al. (2018) suggests that some degree of energy conservation practices exists among the families, which could be understood as partial form of PEB, away from the ideal behavior that refers to practices consciously seeking to minimize the negative impact of one's actions through reduced resource and energy consumption, and waste generation (Kollmuss and Agyeman 2002). However, this seems to imply that if people from countries with higher energy availability per capita implicitly possess the sense PEB with a record of its partial demonstration in daily consumption, essence of PEB could be easily adapted in countries like Canada where cost of energy is relatively higher. Furthermore, it is going to potentially encourage individual users to reflect that ideology into regular practices, if context-specific non-technical behavior based interventions are designed to implement. Due to positive directives emanating from the 'unorthodox' result from an energy-rich context, the study by Nahiduzzaman et al. (2018) is particularly critical. If the inherent PEB embedded in energy-rich individuals could be unleashed into an energy-responsive behavior, the outcome might be more promising where cost of energy is much higher. Bearing that in mind, the key lessons of the study are worth a deliberation. The study suggests that there is always a pressing need for behavioral (non-technical) intervention to establish the norm of energy conservation at the individual levels. The contextualized interventions with a multi-pronged approach (e.g. a composition of video based education, stickers based messages, weekly meetings, etc.) could potentially save more energy while establishing a pressing sense for conservation. The study further amends for a positive relation between the adults members in the family and their level of education with the total energy consumption while number of female members in family, household size, building area demonstrate a negative association attributing to 'economies of scale' and existence of some degree of PEB. Hoisting the importance of 'investment', aiming to enhance the energy efficiency standard of a dwelling, that study emphasizes on the 'curtailment' behavior, which seeks to achieve energy saving by altering behavior (Han et al. 2013; Testa et al. 2016) in order to establish such norm for conservation. The essence stem further questions that whether (i) Canadian families is ready to be potentially intervened through non-technical behavioral tools; and (ii) if the inherent PEB among the Canadians could be capitalized into a responsible energy saving behavior.

4.2 Energy Pricing and PEB in Canada

While the energy demand at the residential sector is on the rise, whether the families have been really efficient in energy usage as opposed to the potential savings remains to be answered. Following are amongst the most notable energy efficient measurement tools that government found to be effective in downsizing the burgeoning demand for energy in Canada: fostering innovation and competitiveness, development and enforcement of regulations, standards and codes, the administration of voluntary certification, benchmarking and information-based programs, and domestic and international partnerships (Ipsos Reid 2015). The 'innovation' and 'standards and code' parts entail hard core technology driven measures that are equally vital to address the current mounting energy demand while achieving higher efficiency in usage. However, they have been conceptualized and practiced with rather a narrow insight. They largely exclude the potential role of the individual (*end-*) users that would have been manifested through everyday behavior in order to contribute to achieving higher energy efficiency (Nahiduzzaman et al. 2018; Chelleri et al. 2015, 2016). Therefore, 'end-user' centric intervention targeting individual's behavior to curtail the level of consumption and gain higher efficiency of usage resulting in the reduction of overall energy demand is somewhat absent in research axioms and think-tanks. As a result, the concept of and need for such non-technological intervention in energy research and practices are pervasively missing across the Canadian provinces: (i) addressing consumption behavior of the individuals; (ii) potential for better efficiency after a detail profile of end users' consumption behavior.

Understandably, ToU is a regulatory tool for behavioral intervention. It is designed to control the volume of energy usage during off-peak, mid-peak and on-peak hours through differentiated price rates. The resulting outcomes suggests that each household tends to use two-third of their electricity during off-peak hours while the remainder being split between mid-peak and on-peak (Ipsos Reid 2015). Governed by the prices, the residential users tend to operate in certain ways to manage the monthly energy bills in line with their financial affordability. This seems to advocate that the temporal dimension of pricing has partially unleashed the inherent PEB of the residential users. This is a 'sporadic' and weaker demonstration of PEB that is not consciously addressed to unearth its full potentials. Such situation-specific demonstration could be perceived as *pro*^{2} (pro- pro)-*environmental behavioral* (P^2EB), which has not been attempted to unleash and capitalize into energy saving actions with scientifically designed non-technical interventions.

Canada being a diversified immigrant-driven society demands a thorough and thoughtful approach for behavior based intervention that encourages better conservation practices. In order to achieve this, a multipronged approach should be adopted combining videos and stickers based conservation messages, and regular dissemination of information through social and TV media, community based dialogues, workshops and meetings, among others. A future study considering the diverse societal landscape should substantiate the context appropriate modes for interventions. However, it needs to be carefully designed after a thorough identification and characterization of the families through the lens of cultural, socio-economic, education, demographic, linguistic, and religious diversities (Nahiduzzaman et al. 2018). These features might be critical and even sensitive, especially for the new immigrants, families with low-income and education, and first nation Canadians (Ng 1998; Wee and Choong 2019). This is more so when many families do not even aware about ToU and its intended benefits (Ipsos Reid 2015; Shi et al. 2019). While they might be at constant struggle to survive than living, ToU alone would barely be able to transform the PEB into a desirable conservative actions.

5 Conclusion: A Way Forward

Unarguably, there is a continued need for technological interventions to regulate the current and future energy consumption and sequent crisis. However, this alone could only do a little without having a marital tie with non-technical interventions that fosters PEB among the clusters of families with various demographic, social, economic, cultural, religious, and personal traits. Therefore, it is compelling to embrace the significant importance of twin-ventured intervention where individual's behavior and smart technologies to combine in order to reduce the surge of residential energy demand. The growing need to establish a sustainable practice of conservation at the households is also critical to meet the sustainable development goals (SDGs) and envision 2030 that Canada pledged for (UN 2018; Gupta and Vegelin 2016).

As hinted out earlier, the current academic think-tank associated with energy research in Canada is somewhat lacking the perspectives of interdisciplinary focus despite of the efforts made by knowledge networks, such as Pacific Institute for Climate Solutions (PICS). Human interaction and behavioral demonstration in the built environment demands for greater unions of scholars to fill the vacuum of trans-disciplinary knowledge and perspectives. On that front, energy research should be conducted while taking all pertinent fields of knowledge, including engineering, environment, urban planning and design, architecture, psychology, economics and sociology into account. Failing to do so, the outcomes of the scholarly endeavors would merely be partial to deal with the ongoing energy crisis and stemming adversities. Furthermore, contemporary endeavors to achieve energy security and sustainability would be questionable while leaving the future with uncertainties. Therefore, this paper calls for a 'collaborative' think-tank for energy research where 'humane' attributes are to be equally accounted along with other essential parameters. In order to do that, the strategic thought-led policies need to undergo certain changes where the 'collaborative' think-tank would inform the executive actions in different forms of interventions, leading to an energy efficiency and 'net-zero' development. There is also a need for renewed policy insights and resulting actions on integrating 'investment' and 'curtailment' behavior approaches into research and practices. This will help assess the 'longevity' effects of twin interventions in actualizing P^2EB into an energy responsive behavior.

Acknowledgments The authors gratefully acknowledge the funding and support provided by FortisBC Inc., Green Construction Research and Training Centre (GCRTC) at the University of British Columbia (UBC), and Mitacs Canada to conduct this research.

References

Alshehry AS, Belloumi M (2015) Energy consumption, carbon dioxide emissions and economic growth: the case of Saudi Arabia. Renew Sust Energ Rev 41:237–247

Anderson K, Ocneanu AF, Benitez D, Carlson D, Rowe A, Bergés M (2012) BLUED: a fully labeled public dataset for event-based non-intrusive load monitoring research. In: Proceedings of the 2nd KDD Workshop on Data Mining Applications and Sustainable, October 2011, pp 1–5

Buyle M, Braet J, Audenaert A (2012) LCA in the construction industry: a review. Int J Energy Environ 6(4):397–405

Campillo J, Dahlquist E, Wallin F, Vassileva I (2016) Is real-time electricity pricing suitable for residential users without demand-side management? Energy 109:310–325

Chelleri L, Kua HW, Rodrigues JP, Thondhlana G, Nahiduzzaman KM, Abdullatif A (2015) 'Exploring the user-driven implications in building urban sustainability and resilience: lessons from OURS CITIES global network study cases', in Proceedings of 8th Conference of the International Forum on Urbanism (IFoU). Basel, Switzerland: MDPI, 58–69. doi: 10.3390/ifou-A006

Chelleri L, Kua HW, Rodrigues JP, Nahiduzzaman KM, Thondhlana G (2016) Are people responsive to a more sustainable, decentralized, and user-driven management of urban metabolism? Sustainability 8(3):1–12

Darby, S. (2006) The Effectiveness of Feedback on Energy Consumption a Review for Defra of the Literature on Metering, Billing and direct Displays. Oxford.

Deru M et al (2011) U.S. Department of Energy commercial reference building models of the national building stock. Publ., no. February 2011, pp 1–118

Environment and Climate Change Canada (2016) Overview of 2014 Reported Emissions. Gatineau QC.

Environment Canada (2014) Canada's emissions trends. Gatineau QC

Environment Canada (2015) The 2015 progress report of the federal sustainable development strategy. Gatineau QC.

Friess WA, Rakhshan K (2017) A review of passive envelope measures for improved building energy efficiency in the UAE. Renew Sustain Energy Rev 72:485–496. January 2016

Government of Canada (no date) New National Energy Code for Buildings provides Canadians with energy-efficient solutions, 2016. Available at: http://www.marketwired.com/press-release/new-national-energy-code-buildings-provides-canadians-with-energy-efficient-solutions-2105505.htm (Accessed: 18 April 2016).
Gupta J, Vegelin C (2016) Sustainable development goals and inclusive development, vol 16. Int Environ Agreements Polit Law Econ, pp 433–448
Han Q, Nieuwenhijsen I, de Vries B, Blokhuis E, Schaefer W (2013) Intervention strategy to stimulate energy-saving behavior of local residents, vol 52. Energy Policy, pp 706–715
Ipsos Reid (2015) Time of Use and Electricity Bill Research: Residential and Business Survey. Toronto.
Jia M, Srinivasan RS, Raheem AA (2017) From occupancy to occupant behavior: an analytical survey of data acquisition technologies, modeling methodologies and simulation coupling mechanisms for building energy efficiency. Renew Sustain Energy Rev 68:525–540. (October 2016)
Karunathilake, H., Hewage, K. and Sadiq, R. (2018) 'Opportunities and challenges in energy demand reduction for Canadian residential sector: A review', Renewable and Sustainable Energy Reviews. 82:2005–2016. https://doi.org/10.1016/j.rser.2017.07.021.
Kavousian A, Rajagopal R, Fischer M (2015) Ranking appliance energy efficiency in households: utilizing smart meter data and energy efficiency frontiers to estimate and identify the determinants of appliance energy efficiency in residential buildings. Energ Buildings 99:220–230
Kelly G (2012) Sustainability at home: policy measures for energy-efficient appliances. Renew Sust Energ Rev 16(9):6851–6860
Kollmuss A, Agyeman J (2002) Mind the gap: why do people act environmentally and what are the barriers to pro-environmental behavior? Environ Educ Res 8(3):239–260
Lindén AL, Carlsson-Kanyama A, Eriksson B (2006) Efficient and inefficient aspects of residential energy behaviour: what are the policy instruments for change? Energy Policy 34(14):1918–1927
Loi TSA, Loo SL (2016) The impact of Singapore's residential electricity conservation efforts and the way forward. Insights from the bounds testing approach. Energy Policy 98:735–743
A. Martín-Garín, J. A. Millán-García, A. Baïri, J. Millán-Medel, and J. M. Sala-Lizarraga, "Environmental monitoring system based on an Open Source Platform and the Internet of Things for a building energy retrofit," Autom Constr, vol. 87, no. January, pp. 201–214, 2018
Meier, A., Olofsson, T. and Lamberts, R. (2002) 'What Is an Energy-Efficient Building ?', in IX Encontro Nacional de Tecnologia do Ambiente Construido. Parana, Brazil, pp. 3–12.
Mills B, Schleich J (2012) Residential energy-efficient technology adoption, energy conservation, knowledge, and attitudes: an analysis of European countries. Energy Policy 49:616–628
Nahiduzzaman KM, Aldosary AS, Abdallah AS, Asif M, Kua HW, Alqadhib AM (2018) Households energy conservation in Saudi Arabia: lessons learnt from change-agents driven interventions program, Journal of Cleaner Production, 185:998–1014. https://doi.org/10.1016/j.jclepro.2018.03.052.
National Energy Board of Canada (2016) Canada's Energy Future 2016: Energy Supply and Demand Projections to 2040. Calgary.
National Energy Board of Canada (2019) Canada's energy transition - An energy market assessment. Ottawa.
Natural Resources Canada (2011) Energy Efficiency Trends in Canada - 1990 to 2009. Ottawa.
Natural Resources Canada (2015a) Energy Fact Book 2015–2016. Ottawa.
Natural Resources Canada (2015b) Canada's energy code. Available at: https://www.nrcan.gc.ca/energy/efficiency/buildings/20675 (Accessed: 5 March 2016).
Natural Resources Canada (2016) Major Energy Retrofit Guidelines: for Commercial and Institutional Buildings. Ottawa.
Natural Resources Canada (2017) Build smart Canada's buildings strategy: a key driver of the Pan-Canadian framework on clean growth and climate change. New Brunswick
Natural Resources Canada (2018) Waste Heat Recovery and Upgrading. Available at: http://www.nrcan.gc.ca/node/5601 (Accessed: 21 August 2018).

Natural Resources Canada Canada's secondary energy use (Final Demand) and GHG emissions by energy source I Natural Resources Canada. [Online]. Available: http://oee.nrcan.gc.ca/corporate/statistics/neud/dpa/showTable.cfm?type=HB&sector=aaa&juris=ca-&rn=1&page=2&CFID=5635943&CFTOKEN=822c32f40807e2c8-8FF6CC55-ECE5-8695-2D7A518A7E20132D. Accessed: 29 July 2019a

Natural Resources Canada Energy use in Canada: trends publications I Natural Resources Canada. [Online]. Available: http://oee.nrcan.gc.ca/publications/statistics/trends/2015/residential.cfm. Accessed 11 July 2019b

Ng CF (1998) Canada as a new place: the immigrant's experience. J Environ Psychol 18(1):55–67

Ontario Energy Board Managing costs with time-of-use rates I Ontario Energy Board. [Online]. Available: https://www.oeb.ca/rates-and-your-bill/electricity-rates/managing-costs-time-use-rates. Accessed 15 Jul 2019

Ouyang J, Hokao K (2009) Energy-saving potential by improving occupants' behavior in urban residential sector in Hangzhou City, China. Energ Buildings 41(7):711–720

Pittis, D. (2017) The high cost of cheap electricity: Don Pittis I CBC News, CBC News. Available at: https://www.cbc.ca/news/business/energy-electricity-cost-1.4428048 (Accessed: 15 July 2019).

Province of British Columbia (2008) Energy Efficient Buildings Strategy: More Action, Less Energy. Victoria.

Roulet CA (ed) (2006) Indoor air quality and energy performance of buildings. In: Proceedings of healthy buildings 2006, pp 37–47

Shi D, Wang L, Wang Z (2019) What affects individual energy conservation behavior: personal habits, external conditions or values? An empirical study based on a survey of college students, vol 128. Energy Policy, pp 150–161

Statistics Canada (2011) Households and the Environment: Energy Use. Ottawa. https://doi.org/ISSN1920-7522.

Steg L (2008) Promoting household energy conservation. Energy Policy 36(12):4449–4453

Strbac G (2008) Demand side management: benefits and challenges. Energy Policy 36(12):4419–4426

Testa F, Cosic A, Iraldo F (2016) Determining factors of curtailment and purchasing energy related behaviours, vol 112. J Clean Prod, pp 3810–3819

Torcellini, P. et al. (2006) 'Zero Energy Buildings: A Critical Look at the Definition', in 2006 ACEEE Summer Study on Energy Efficiency in Buildings. Pacific Grove, CA. Available at: http://www.nrel.gov/docs/fy06osti/39833.pdf.

U.S. Department of Energy (2016) Tips: Home Office and Electronics. Available at: https://www.energy.gov/energysaver/appliances-and-electronics/energy-efficient-computers-home-office-equipment-and (Accessed: 3 March 2016).

United Nations (2018) Sustainable Development Goals Report 2018, United Nations Department of Economic and Social Affairs. New York.

Wang Y, Shao L (2017) Understanding occupancy pattern and improving building energy efficiency through Wi-Fi based indoor positioning. Build Environ 114:106–117

Wee SC, Choong WW (2019) Gamification: predicting the effectiveness of variety game design elements to intrinsically motivate users' energy conservation behaviour, vol 233. J Environ Manag, pp 97–106

Wood G, Newborough M (2003) Dynamic energy-consumption indicators for domestic appliances: environment, behaviour and design. Energ Buildings 35(8):821–841

Xu F et al (2018) Classifier economics of semi-intrusive load monitoring. Int J Electr Power Energy Syst 103:224–232

Chapter 9
Modelling the Wind Supply Chain to Reduce Emissions: How Could Affect Transmission Congestion?

Milton M. Herrera, Isaac Dyner, and Mauricio Uriona Maldonado

Abstract Delays in new transmission line construction produce drawbacks that can lead to a precarious energy planning. As there is a delay between the licensing process and complete lines, it could create uncertainty as well as transmission congestion. In this context, it is essential to estimate the impact the transmission congestion might have on the reduction of emissions. To assess the effects of power grid congestion on emissions; first, it was necessary to design a simulation model, then some simulation scenarios were provided for analyzing the dynamic behavior of the wind-power supply chain. This chapter presents the findings of the case study conducted in Brazil. Our analysis shows that options to reduce emissions through the current energy policy of Brazil might is affected by delays in transmission construction.

Keywords Simulation · Wind power · Supply chain · Asynchronous policy · Electricity transmission · System dynamics modeling · Emissions

M. M. Herrera (✉)
Economic Sciences Research Centre, Universidad Militar Nueva Granada, Bogotá, Colombia
e-mail: milton.herrera@unimilitar.edu.co

I. Dyner
Faculty of Natural Science and Engineering, Universidad Jorge Tadeo Lozano, Bogotá, Colombia
e-mail: isaac.dynerr@utadeo.edu.co

M. Uriona Maldonado
Department of Industrial and Systems Engineering and the Sustainable and Renewable Energy Innovation Group (Sinergia), Federal University of Santa Catarina, Campus UFSC, Trindade, Florianopolis, Brazil
e-mail: m.uriona@ufsc.br

H. Qudrat-Ullah, M. Asif (eds.), *Dynamics of Energy, Environment and Economy*, Lecture Notes in Energy 77, https://doi.org/10.1007/978-3-030-43578-3_9

1 Introduction

Rapid industrialization and urbanization lead to an increase in energy use, mainly in developing countries. Given that the energy industry is one of the main contributors to CO_2 emissions worldwide, developing countries face an urgent need to reduce energy consumption and improve environmental conditions (Coelho and Goldemberg 2013; Franco et al. 2015; Espinosa et al. 2019). Governments have implemented several strategies and policies for mitigating emissions. Previous studies have analyzed policy alternatives focusing on the emission reduction and the promotion of clean technologies (Haselip et al. 2005; Liu et al. 2014; Qudrat-Ullah 2014; Blumberga et al. 2015; Cardenas et al. 2016), although rarely associated to power grid congestion. Thus, assessing the effects of power grid congestion on clean technologies is a relevant topic for research and policy studies (Herrera et al. 2019b).

Energy policy implementation may bring undesirable behavior affecting the growth of renewable energy supply chains (Wee et al. 2012; Cucchiella and D'Adamo 2013; Herrera et al. 2018). In the case of developing countries such as Brazil, the growth of wind industry has increased wind power generation in the Brazilian grid, because of the adopted policy, however insufficient transmission capacity has generated a gap between generation and transmission. Although Brazil government has been effective in promoting wind power, the targets of emissions reduction might compromise by the insufficient transmission lines in the long term.

As renewable energy relies heavily on climate conditions, Brazil has been affected by climate variation due to its dependence on hydropower sources for electricity supply (De Lucena et al. 2009). Consequently, the government has actively encouraged wind industry expansion mostly in northeast Brazil (Agencia Nacional de Energía Eléctrica-ANEEL 2017; Rego and de Oliveira Ribeiro 2018). Thus, the potential for wind power generation in this region has been fundamental to guarantee energy security in the context of the current environmental issues (Lima et al. 2015; Ribeiro et al. 2016; De Jong et al. 2017).

Although Brazil promotes the generation of electricity using wind sources to ensure energy security, there are serious problems that affect its deployment. On the one side, the dispatch of wind power could be affected due to deficient transmission capacity from the northeastern region. Depending on different conditions, the built a power grid might take around 10 years, which disturbs connections and dispatches of power from wind farms. On the other side, the fragmentation in energy policy design and implementation may result in poor resource allocation and delays in the growth of wind power (Herrera et al. 2019a). The lack of wind energy policy coordination might impact on the performance and management of strategic resources supporting the supply chain.

Under these conditions, it is essential to evaluate the impact that the transmission congestion might have on the reduction of emissions and the growth of the wind industry. Whilst the reduction of emissions is almost a consensual policy issue, this chapter seeks to assess energy policy alternatives for the wind power supply chain, incorporating transmission congestion concerns. Thus, the chapter explicitly addresses the following research questions regarding transmission congestion issue and its effects on the emission reduction: What are the effects of the insufficient transmission capacity on emission reduction? How might Brazil achieve its emission target with limited transmission capacity? What can the wind power supply chain do to face transmission congestion?

The assessment of emissions on the energy system, which is the measure adopted in Brazil for the wind industry, is crucial because it guides the future investment and consequently, the composition of the electricity generation portfolio. Thus, energy planning becomes an essential issue for decision making on the dynamics of clean technologies diffusion and energy policy planning (Naill 1992). Recently, an increasing number of research on energy issues have led to developing models to facilitate energy planning process (Ahmad et al. 2016). These models derive from different modeling approaches such as operational research (Bazmi and Zahedi 2011; Henao et al. 2012; Viteri et al. 2019) and economics (Fortes et al. 2014; Cabalu et al. 2015), but some cases do not consider uncertainties regarding delays and nonlinearities among the variables that describe the system (Cardenas et al. 2016; Morcillo et al. 2017). Over the years, several researchers have used system dynamics to model problems in the electricity sector (Ahmad et al. 2016). System dynamics is useful to analyze social and physical interactions in the system – that many time has uncertainties in the long term – which make difficult the search for solutions (Sterman 1981; Ford 1997; Dyner and Larsen 2001). Consequently, system dynamics may offer an alternative approach to policy assessment and give a response to the above questions.

This chapter is organized as follows. Section 2 discusses transmission system concerns and barriers to wind-power supply chain expansion. The modeling approach and the scenarios used in the study is presented in Sect. 3. Section 4 exhibits the results for climate policy scenarios, including transmission system barriers. Finally, this chapter presents conclusions and policy implications in Sect. 5.

2 Transmission System Issues

The Brazil government has a low-carbon policy in place that promotes the production of electricity using renewable energy sources to mitigate climate change and improve energy security (Santos et al. 2018). Since the early 2000s, the growth of the wind industry in Brazil has progressively added power to the Brazilian grid, primarily driven by the auction-based power expansion plan in its regulated market. Currently, approximately 15 GW of wind capacity has been installed across the country, and almost 4 GW of capacity is contracted to be added to the regulated market by 2020 (ABEEólica 2019). However, the rapid growth of the wind industry has faced severe problems due to the construction delays of electricity transmission (De Jong et al. 2016; Global Transmission Report 2016; Herrera et al. 2017; Miranda et al. 2017). Thus, the transmission issue could cause drawbacks that affect the increase of the wind industry and the emissions reduction targets.

The Transmission System Operator (ONS) is responsible for defining the minimum interconnection requirements for all power plant requests, including those using renewable energy (Feltes et al. 2012). Whilst the requirements for evaluating the impact of the generation of new wind farms on the interconnected system must be presented by the developer and approved by the ONS, there are delays for awarding licensing which influence on the connection of the finished wind farms to power grid (Cardoso Júnior et al. 2014; Bayer et al. 2018; Herrera et al. 2019b). At present, over 30% of wind power projects with expired implementation deadlines still do not have a grid connection (Bayer 2018).

Interconnected power system planning is considered as a complex issue due to its interaction with other aspects such as social, technical and environmental. One impact both transmission infrastructures and wind farms are the immobilized area overtime followed by disturbing of land use (Moreira et al. 2015; Ribeiro et al. 2016; Brannstrom et al. 2017). Considering that, the Brazilian authorities implement processes to verify the project's viability regarding its environmental impacts for stablishing the main terms and conditions of its implementation (Bayer et al. 2018). Nevertheless, some regions environments have had altered by wind farms (Brannstrom et al. 2017; Gorayeb et al. 2018), and consequently also by transmission infrastructures.

Since 2013, the risk of delays in the transmission grid extension was transferred to the wind project developers, namely if the transmission grid is delayed, then wind developers are obliged to supply energy (Bayer et al. 2018; Global Transmission Report 2018). This situation makes wind farms a higher risk investment than what it usually be. Despite the additional transmission capacity is planned at the beginning of the auctions, it does not reduce the transmission connection risk because the planning process also includes transmission capacity that is still in the construction process or delayed.

Interactions between supply, demand and the long delays associated with transmission construction have a significant influence on the electricity market, mainly in the price. When the power grid is constrained, it may distort electricity prices (Herrera et al. 2019b). However, too much capacity results in very high costs (Ochoa and van Ackere 2015). Thus, a balance between capacities of supply – power generation and power grid – including demand may regulate the electricity prices.

For wind power, the increasing of the transmission capacity is of great importance. To start a construction project of transmission infrastructure, developers must process environmental licensing. Delay in processing this license can lead to a delay in the wind power generation inside or outside a region. Thus, limited transmission capacity can significantly reduce wind power capacities as well as disturbing emission reduction targets. The next section proposes a modeling approach which contributes to these issues.

3 Modeling Approach

As discussed above, Brazil faces some concerns about transmission capacity along the northeastern region. For overcoming the challenges of the effects of transmission congestion on emission reductions targets of Brazil, this section adapted the simulation model developed by Herrera et al. (2018). As the system dynamics approach takes into account feedbacks process and delays to represent and understand the dynamic behavior in the long run (Sterman 2000; Qudrat-Ullah 2016), this research adopted it. Additional to this, the system dynamics approach is useful to design and asses the effects of policy on a complex system characterized by uncertainty about long-term (Forrester 1961). The system dynamics approach has been used to model behavior patterns associated with the energy system for more than 30 years (Sterman 1981; Naill and Belanger 1989; Ford 1997; Dyner 2000).

For facilitating the assessment of policy alternatives of the case study, the following steps were applied:

(i) the problem statement with dynamic behavior hypothesis was represented through a causal loop diagram,
(ii) a stock-and-flow diagram was designed to represent the energy system of Brazil, this step included parameter estimation, model validation, and policy design and finally,
(iii) the limitations of the model were declared.

3.1 *Dynamic Behavior Hypothesis*

The dynamics and structural complexities of the Brazilian electricity market are captured from the dynamic behavior hypothesis. The dynamic behavior hypothesis is represented with two causal loop diagram, including the interactions among variables of the energy system. The first diagram exhibits transmission system concerns, while the second diagram showed a contribution to this issue. These diagrams comprise three dynamics – capacity, market, and emissions – which characterize the behavior of the wind power supply chain modeled.

Given that the building capacity for the power grid in Brazil is more slowly, the response capacity of the wind power supply chain for attended the electricity demand could be disturbed. This situation is created due to the lack of synchronization between policies aims to generation and transmission. Figure 9.1 represents the dynamic behavior hypothesis which incorporates unsynchronized energy policy. The interaction between demand and supply are represented in the balancing loop B1 and B2. On the one side, when electricity price increases, demand decreases in the long term, which influences on capacity margin – calculated as the difference between capacity generation and electricity demand – which closes the balancing loop B1. On the other side, as price increases, incentives for renewable energy also

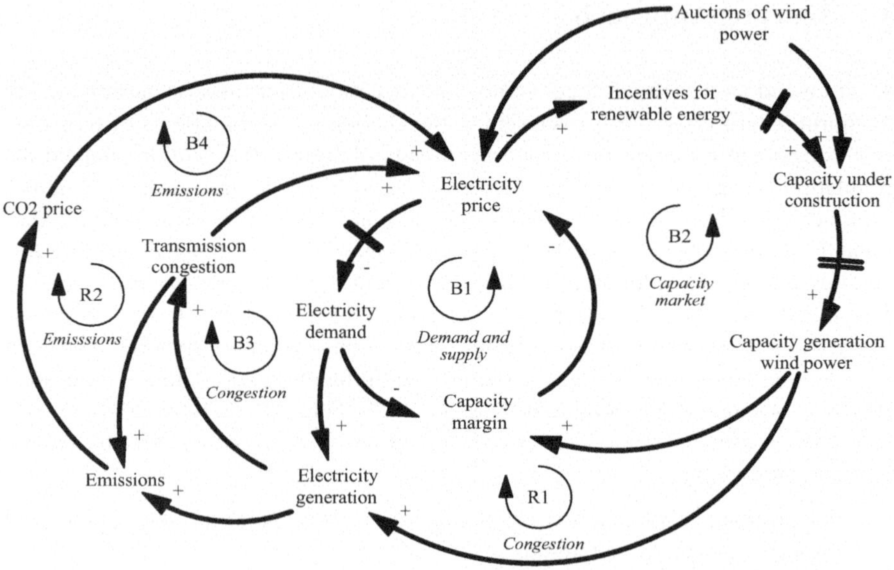

Fig. 9.1 Dynamic behavior hypothesis for the carbon market and capacity market with unsynchronized policy

increases, then sustained higher price have to influence on capacity generation in the long term, closing the balancing loop B2.

The balancing loop B3 describes how electricity generation influences transmission capacity. As electricity generation increases and transmission capacity is insufficient, generating congestion. Transmission congestion depends on the electricity that must be transmitted and significantly affect electricity price (Herrera et al. 2019b). Given that the price increases, capacity generation increases, and congestion increases, the reinforcing loop R1 is generated.

The balancing loop B4 and reinforcing loop R2 represent the relationships of the carbon market – emission and CO_2 price – with electricity demand and capacity generation. The balancing loop B4 shows the internalization of environmental cost proposed by Franco et al. (2015). As emission and CO_2 price increase, that in the mid-term influences electricity price, creating the reinforcing loop R2.

Whilst the installed capacity for wind power generation plays an essential role in the CO_2 emission reduction, syncing up with transmission capacity is indispensable. Figure 9.2 shows the dynamic behavior hypothesis proposed to synchronize generation capacity with the transmission. The future capacity under construction depends on the projected capacity auctions, both wind generation and transmission capacity. When the projected capacity auctions are synchronized, the electricity price decreases, forming the balancing loop B3. Consequently, the wind power generation increases and emissions decrease, creating the reinforcing loop R4. While the transmission capacity is available, the emission generation will reduce.

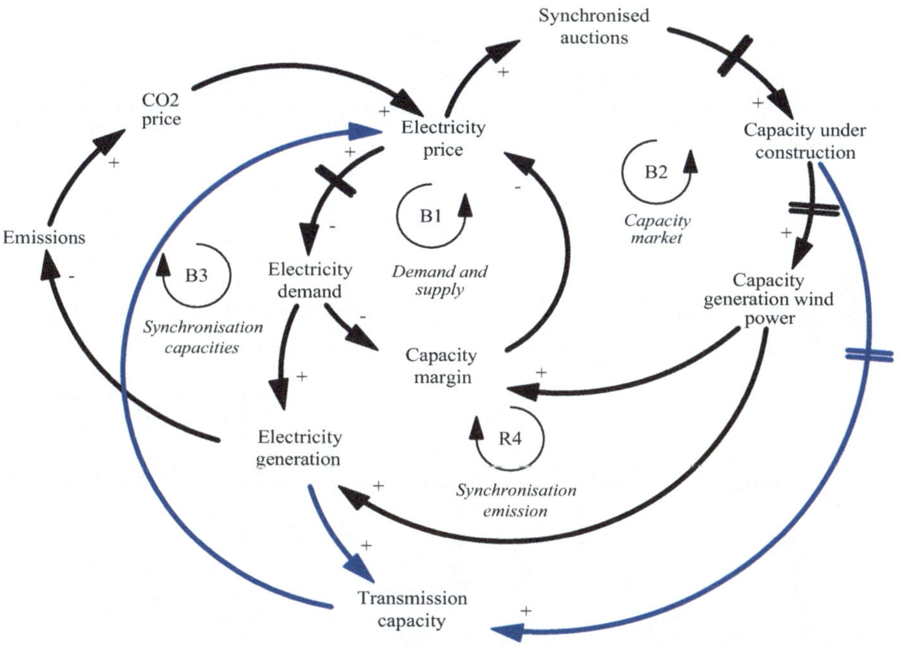

Fig. 9.2 Dynamic behavior hypothesis for the carbon market and capacity market with synchronized policy

3.2 *Model Description*

The dynamics and structure of the carbon market and capacity market are represented through a stock and flow diagram, as illustrated in Fig. 9.3. This diagram shows strategy resources of the supply chain – suppliers, industry and wind farm – within stocks, flows which represents decision making of the energy system (e.g. starting production) and drivers of the wind industry – production cycle time and construction time. The stock is calculated by a first-order integral equation, while the flow is a set of differential equations. Stocks integrate flows into and out of stock, and the net flow into the stock is the rate of change of the stock (Bala et al. 2017). Also, a structure for the simulation model is explained in Appendix.

The model considers two policy scenarios for the assessment of the questions proclaimed and the dynamic behavior hypothesis presented in Sects. 1 and 3.1, respectively: (i) unsynchronized policy, this scenario presents a resource allocation for power generation and transmission not synchronized and (ii) synchronized policy, this scenario considers alternatives to synchronizing the building of the wind power generation and transmission capacity. For both policy alternatives, the most relevant equations are presented in Table 9.1. Whilst the scenario 1 depends on incoming and outgoing flows, as can been seen below, scenario 2 contemplates the standard deviation both lead time of transmission projects and the installed

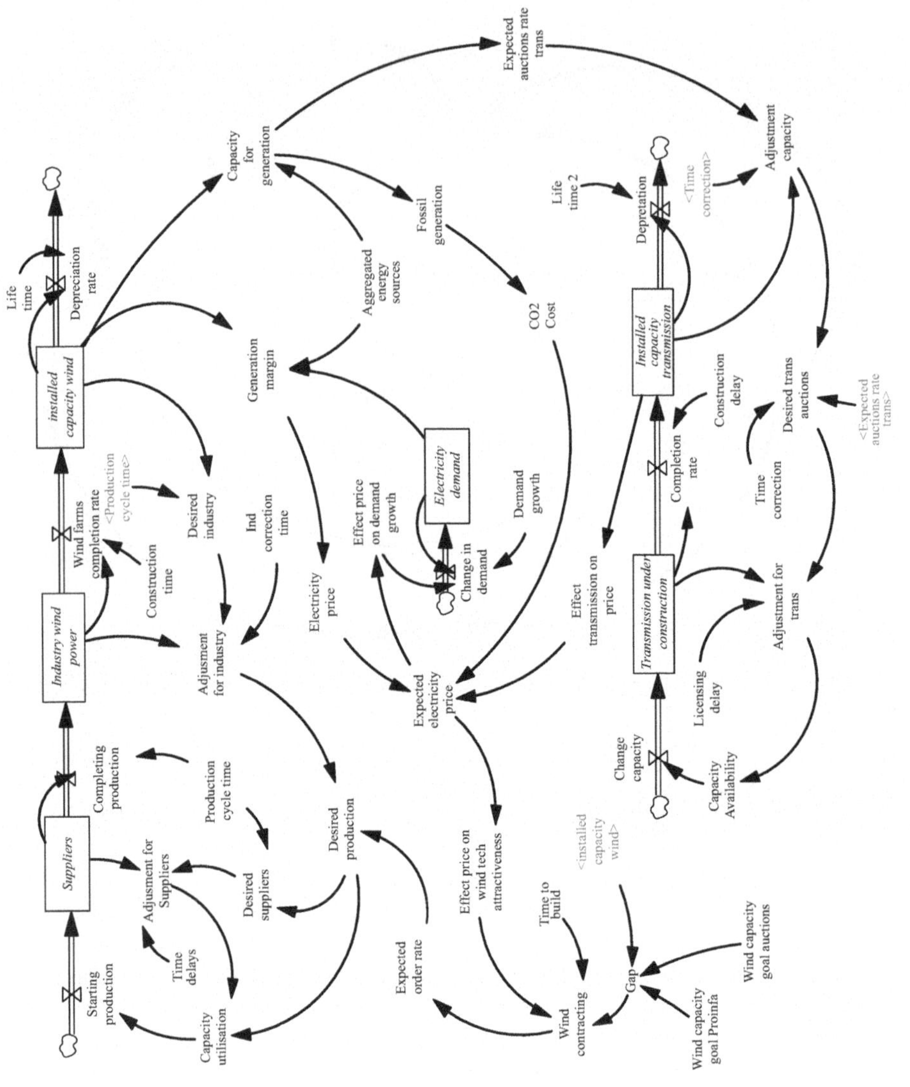

Fig. 9.3 Stock and flow diagram for the carbon market and capacity market

generation capacity to synchronize the building of transmission system and power generation.

A synthesis of the major assumptions of the model and input data incorporated are presented.

- The simulation model used the database of the Brazilian energy agency ANEEL, which publishes the auctions rounds for the contracted capacity of wind power that took place in the period between December 2009 and December 2018.

Table 9.1 Definition of proposed policy

Scenario	Definition	Policy
Scenario 1	This scenario considers delays in transmission lines construction that affect the dynamic of the supply chain.	$K(t)=K(t-dt)+\int_{T}^{t=0}\frac{\text{CR}}{\beta(t)}-\frac{D}{L(t)}d(t)$
Scenario 2	This scenario deems a stylized function to synchronize the transmission lines construction with the increase of wind power.	$\text{EK}=\frac{\text{K}}{K+\frac{1}{(\Delta IC*S_{LT}+LT*S_{IC})}}$

K(t): Installed capacity of transmission
CR: Completion rate
D: Depreciation rate
$\beta(t)$: Construction delay
$L(t)$: Life time
EK: Effect of transmission on the expected electricity price
ΔIC: Increase of the installed generation capacity
LT: Lead time
S_{LT}: Standard deviation of lead-time
S_{IC}: Standard deviation of the installed generation capacity

- To validate and evaluate the dynamic behavior, the model employed time series of the installed capacity projection of wind power obtained by (ABEEolica 2017).
- The simulation model takes into account the values of average bids concerning the variation of electricity demand.
- Other generation technologies are considered within the model. The initial data on the installed capacity of each technology corresponds to the year 2018, according to Brazil's energy matrix reported by (ANEEL 2018). This assumption is taken into account to calculate the market share expansion of the Brazilian wind power.
- Over 31% of the wind projects that established a power purchase agreement in 2010 had been affected by network delays by the time the implementation deadline was research in 2013 (Bayer 2018). Thus, one assumption made in the simulation model is that transmission congestion reached was 30% per year, including grid load loss.

The simulation model was extensively validated through several tests for consistency and conceptual validation (Sterman 1984; Barlas 1996; Oliva 2003; Qudrat-Ullah 2016). The main propose of behavior validity is to compare the model-generated behavior to the observed behavior of the real system (Qudrat-Ullah and Seong 2010). Figure 9.4 illustrates a behavior reproduction test to assess the model's ability to reproduction dynamic behavior.

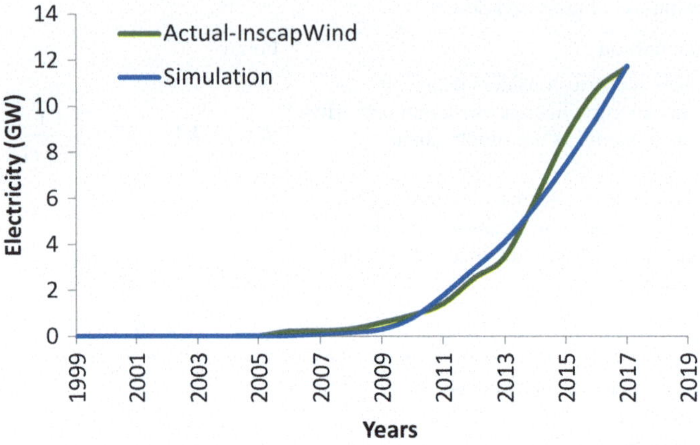

Fig. 9.4 Model testing for the installed capacity of the wind power supply chain

3.3 Limitations

The simulation model was developed to assess policy alternatives and their effects on the emission reduction in the case of Brazil. To this end, this chapter does not consider two aspects: on the one hand, though the interconnections among regions of Brazil are relevant, these were not contemplating due to the aggregated level used to develop the simulation model. On the other hand, technical aspects such as the hourly net load and spilled energy were not considered due to these have been addressed by other authors (De Jong et al. 2016, 2017).

4 Results

Despite the installed wind power in Brazil, principally in the northeastern region, grow to almost 16 GW by 2020 (De Jong et al. 2017), this growth might is affected due to the limited transmission. This section discusses how Brazil might synchronize its energy policy to overcome this issue. Also, an analysis of the impacts of transmission congestion on the emission reduction is provided.

4.1 What Are the Effects of Insufficient Transmission Capacity on Emission Reduction?

From the simulations, the dynamic behavior of installed capacity was calculated for both scenarios of policy. Figure 9.5 shows a steady increase in the installed capacity of wind power. If there enough transmission capacity, it is also assumed that the

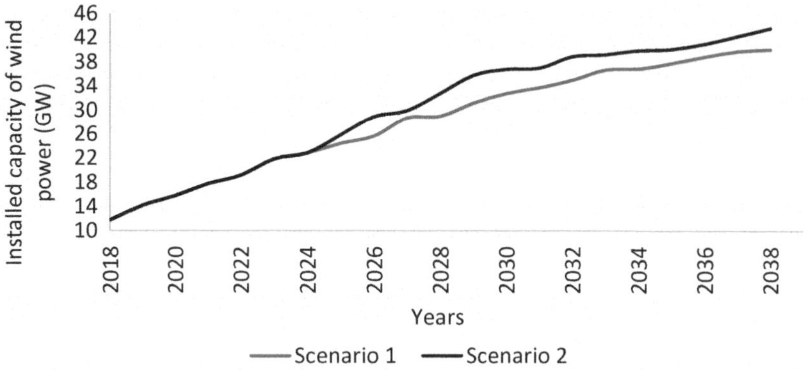

Fig. 9.5 Dynamic of the behavior from an installed capacity of wind power

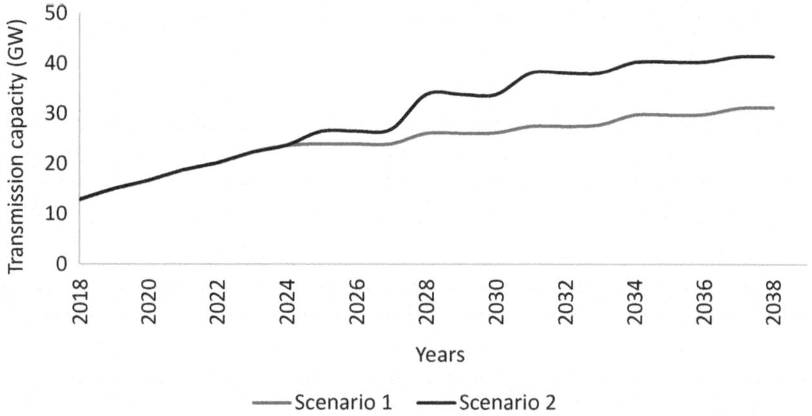

Fig. 9.6 Dynamic of the behavior from transmission capacity

wind power generation increases, for instance in Scenario 2; however, the delays for building new transmission capacity could influence the diffusion of the Brazilian wind industry in the long run (e.g. Scenario 1). Previous studies that assess the diffusion of wind power in Brazil coincide with these results (De Jong et al. 2017; Herrera et al. 2019b).

The extension of the transmission system is of great importance for the development of the wind industry (Miranda et al. 2017; Bayer et al. 2018; Herrera et al. 2019a). Despite that the wind projects are located much nearer electricity load centers (De Jong et al. 2015), the accumulated delays of the transmission projects in the last year have involved challenges and opportunities for the wind industry in Brazil (Global Transmission Report 2018). Figure 9.6 illustrates the problem of the accumulated delays of the transmission projects. Though Brazil has a long history of using auctions to increase its generation capacity, the transmission capacity has slow growth due to several adjust of the policy (Aquila et al. 2017; Bradshaw 2017; Bayer 2018).

As the carbon emission can be obtained for power generation in terms of kg of CO2eq emitted per MWh produced, the power grid plays an important role to guarantee the emission reduction. Widespread delays may explain the high share of the wind projects to transmission grid projects in Brazil (Aquila et al. 2016, 2017; Bayer et al. 2018; Gorayeb et al. 2018; Herrera et al. 2019b). Figure 9.7 shows the effects of insufficient transmission capacity on emission reduction. Scenario 1 presents how the transmission grid delays significantly could affect the emission reduction in the long run, while scenario 2 synchronizes the lead time of the transmission projects to reduce this issue. Although increased CO_2 emissions begin in the year 2024, the additional wind power capacity might start operational, after the licensing process and construction delays associated with the power grid.

Brazil promised to reduce its CO_2 emissions by 37% and 43% below 2005 levels in 2025 and 2030, respectively. Although the Brazilian government has promoted actively implementation of renewable sources to reduce CO_2 emissions, insufficient electricity transmission has been one of the main drawbacks by a large amount of stranded renewable generation. These results show that a lower generation of electricity, especially renewable energy, leads to higher CO2 emissions related to the electricity sector as well as higher prices.

A lack of appropriate long-term financing is a barrier in the market consolidation of renewable sources (Silva et al. 2013). In this way, the electricity price plays an essential role in incentive the investment of new wind farms as well as the power grid expansion. Figure 9.8 shows the dynamic behavior of the proposed scenarios. Scenario 1 shows higher electricity price, which continually increases, while scenario 2 induces a lower price due to the synchronization – between power generation and power grid – as from saving of the response time. Scenario 1 represents a significant risk may lead to reduced public acceptance of clean technologies. In general, scenario 2 highlights the need for transmission grid extension to contribute to the reduction of the electricity market prices.

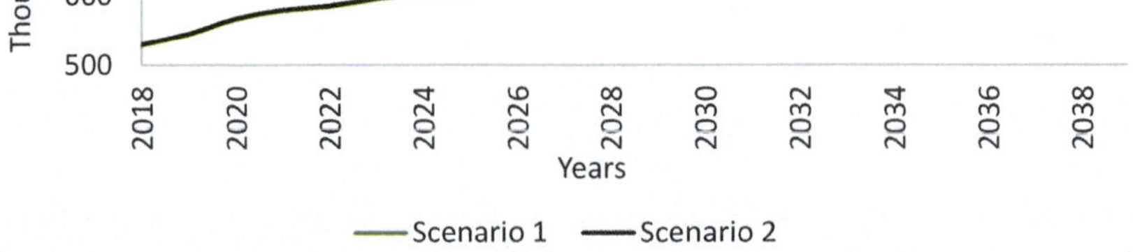

Fig. 9.7 Simulated carbon emissions from the electricity market of Brazil

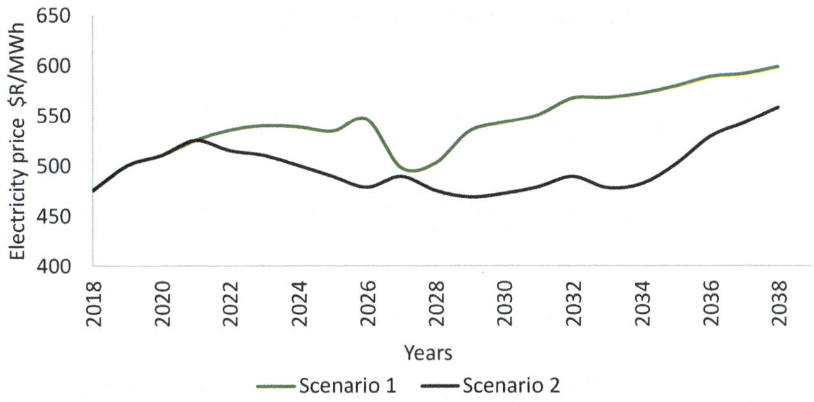

Fig. 9.8 Simulation of the electricity price under two scenarios

4.2 How Might Brazil Achieve Its Emission Target with Limited Transmission Capacity?

At first, planning of the energy system in Brazil based on transmission implementation being faster than generation operated well. According to simulation results, there is sufficient wind power capacity available to reduce CO_2 emissions; however, the power grid is insufficient to guarantee the energy supply. Results show that even if the construction delays of the power grid capacity are identified, the gap between power generation and electricity dispatch would still not alleviate. This situation suggests that the planning of auctions should mediate coordination between the expansion of the wind power and transmission system.

Previous studies have shown that the energy policy mechanism must be agreed among stakeholders to define liabilities and penalties when the required grid expansion is not completed on time (Bayer et al. 2018; Herrera et al. 2019a). Additional to this, the planning process should contemplate smaller distances from the power grid to wind farms to reduce transmission cost and electricity losses. Currently, the financial risk for transmission grid delays was implemented in some generation auctions and transferred to the wind project developer (ANEEL 2017). However, the risk of revenues in the starting of the project lifetime significantly could be disturbed the cash flows. This situation could lead to some investors to drop out of wind auctions. In this respect, the planning of wind auctions should take into account the following aspects:

(i) defining the maximum generation capacity that can be connected to the power grid,
(ii) the auction prices should be based on the availability of transmission capacity where will be connected the wind project,
(iii) the planning process should be synchronized with licensing processes to mitigate delays in granting.

4.3 What Can the Wind Power Supply Chain Do to Face Transmission Congestion?

The importance of a systematic analysis of transmission congestion focuses on feedback arguments that may jointly accelerate the effects nonlinearly on the wind power supply chain, unfavorably impacting the CO_2 emissions reduction. On the one side, Fig. 9.9 shows the dynamic behavior of the wind power supply chain without synchronized policy. The average stock for generation capacity of suppliers is 33.2 GW, while wind power industry reaches 26.8 GW throughout the simulation period. This scenario supposes significantly gap between suppliers and industry, which could generate backlog or surplus of the resources in the long term. Previous studies also have shown the effects of backlog or surplus on the electricity prices for the renewable energy supply chain (Herrera et al. 2018).

On the other side, the synchronization of the wind power supply chain contributes to reducing the difference of the generation capacity between suppliers and the wind industry. Figure 9.10 shows how the saving of lead time for the transmission construction supports to mitigate the difference among players of the supply chain. In this case, the average stock for the generation capacity of suppliers is 29 GW, while wind power industry reaches 28.1 GW throughout the simulation period. This scenario involves that the generation capacity has a minor impact along the supply chain, reducing the backlog and surplus of the resources.

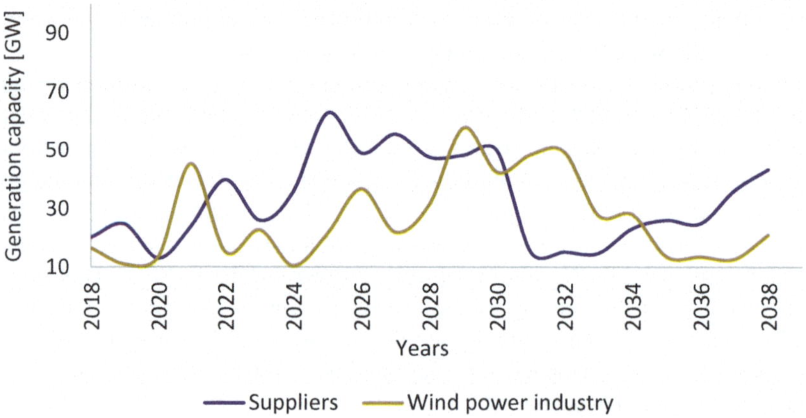

Fig. 9.9 Simulation of the dynamic behavior of the wind power supply chain without synchronized policy

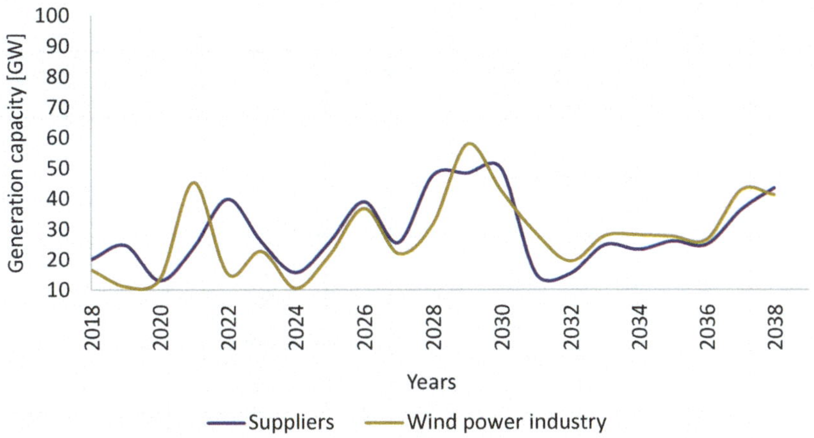

Fig. 9.10 Simulation of the dynamic behavior of the wind power supply chain with synchronized policy

5 Conclusions and Policy Implications

This chapter provides a simulation model for the analysis of the dynamic behavior for the wind power supply chain in terms of the emission reduction and synchronization of the energy policy. Results from the simulation model show that the synchronization of the energy policy has a more significant impact on both emission reductions and the diffusion of the wind industry. These results coincide with other studies obtained by Cardenas et al. (2016). Whilst the effect-in tariff policy focuses on expanding the capacity of clean technologies without affecting fossil fuel technologies as much (Cardenas et al. 2016), this chapter assesses other alternatives which include the transmission congestion issue.

Although Brazil is a country with a robust planning process of the energy system, two factors contribute to the deceleration of the diffusion of wind power, with undesirable effects on emission reduction:

(i) the construction delays for the power grid influence on the dynamic behavior of the supply chain, which reduces the generation capacity in the long term,
(ii) the lack of synchronization of auctions policy to align power generation and the power grid.

Based on the results of this chapter, a few suggestions for energy and climate policy in Brazil are proposed. Given that the conditions for the renewable generation in Brazil are adequate to mitigate CO_2 emissions, the country's energy system vulnerability, as well as use technologies not yet technically mature, should be taken into account.

By utilizing our proposed simulation model, future research can investigate other related issues. For instance, according to Dranka and Ferreira (2018), the coal power plants and gas power plants in Brazil indirectly foot the externality bill for damage caused by emissions; therefore, future research should include testing carbon tax from different electricity generation technologies and its effects in the long run.

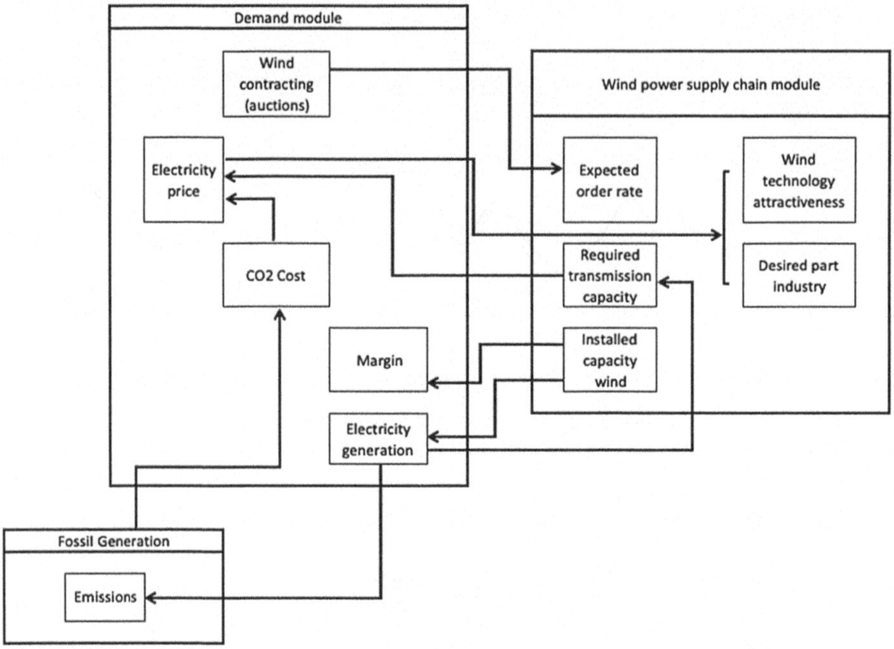

Fig. 9.11 Model structure – wind power supply chain, demand and fossil generation

Appendix: Model Design

The simulation model comprehends three modules: demand, wind power supply chain (including transmission capacity) and fossil generation, as illustrated in Fig. 9.11. The demand module is comprising wind auctions which regulate the electricity market; this influences an expected order rate on the wind-power supply chain module. As the electricity price influence demand, wind technology is more attractiveness, electricity generation increases, and desired transmission increases. The fossil generation module includes emissions which affect the behavior of the carbon market.

Acknowledgments The authors are grateful to Economic Sciences Research Centre at the Universidad Militar Nueva Granada, Faculty of Natural Sciences and Engineering at the Universidad Jorge Tadeo Lozano and the Sustainable and Renewable Energy Innovation Group (Sinergia) at the Federal University of Santa Catarina for providing the necessary financial support.

References

ABEEolica (2017) Dados Mensais-Novembro de 2017. Brazil
ABEEólica (2019) Dados ABEEólica. http://www.abeeolica.org.br. Accessed 29 Mar 2018
Agencia Nacional de Energía Eléctrica-ANEEL (2017) Leiloes de Geraçao. http://www.aneel.gov.br/geracao4. Accessed 12 Oct 2017

Ahmad S, Mat Tahar R, Muhammad-Sukki F et al (2016) Application of system dynamics approach in electricity sector modeling: a review. Renew Sust Energ Rev 56:29–37. https://doi.org/10.1016/j.rser.2015.11.034

ANEEL (2017) Leilões de Transmissão. In: http://www.aneel.gov.br/transmissao4

ANEEL (2018) Informações Técnicas. http://relatorios.aneel.gov.br/_layouts/xlviewer.aspx?id=/RelatoriosSAS/RelSampRegCC.xlsx&Source=http://relatorios.aneel.gov.br/RelatoriosSAS/Forms/AllItems.aspx&DefaultItemOpen=1

Aquila G, Rocha LCS, Rotela Junior P et al (2016) Wind power generation: an impact analysis of incentive strategies for cleaner energy provision in Brazil. J Clean Prod 137:1100–1108. https://doi.org/10.1016/j.jclepro.2016.07.207

Aquila G, de Oliveira Pamplona E, de Queiroz AR et al (2017) An overview of incentive policies for the expansion of renewable energy generation in electricity power systems and the Brazilian experience. Renew Sust Energ Rev 70:1090–1098. https://doi.org/10.1016/j.rser.2016.12.013

Bala BK, Arshad FM, Noh KM (2017) System dynamics: modelling and simulation. Springer, Singapore

Barlas Y (1996) Formal aspects of model validity and validation in system dynamics. Syst Dyn Rev 12:183–210. https://doi.org/10.1002/(SICI)1099-1727(199623)12:3<183::AID SDR103>3.0.CO;2 4

Bayer B (2018) Experience with auctions for wind power in Brazil. Renew Sust Energ Rev 81:2644–2658. https://doi.org/10.1016/j.rser.2017.06.070

Bayer B, Berthold L, Moreno Rodrigo de Freitas B (2018) The Brazilian experience with auctions for wind power: an assessment of project delays and potential mitigation measures. Energy Policy 122:97–117. https://doi.org/10.1016/j.enpol.2018.07.004

Bazmi AA, Zahedi G (2011) Sustainable energy systems: role of optimization modeling techniques in power generation and supply – a review. Renew Sust Energ Rev 15:3480–3500. https://doi.org/10.1016/j.rser.2011.05.003

Blumberga D, Blumberga A, Barisa A et al (2015) Modelling the Latvian power market to evaluate its environmental long-term performance. Appl Energy 162:1593–1600. https://doi.org/10.1016/j.apenergy.2015.06.016

Bradshaw A (2017) Regulatory change and innovation in Latin America: the case of renewable energy in Brazil. Util Policy 49:156–164. https://doi.org/10.1016/j.jup.2017.01.006

Brannstrom C, Gorayeb A, de Sousa MJ et al (2017) Is Brazilian wind power development sustainable? Insights from a review of conflicts in Ceará state. Renew Sust Energ Rev 67:62–71. https://doi.org/10.1016/j.rser.2016.08.047

Cabalu H, Koshy P, Corong E et al (2015) Modelling the impact of energy policies on the Philippine economy: carbon tax, energy efficiency, and changes in the energy mix. Econ Anal Policy 48:222–237. https://doi.org/10.1016/j.eap.2015.11.014

Cardenas LM, Franco CJ, Dyner I (2016) Assessing emissions-mitigation energy policy under integrated supply and demand analysis: the Colombian case. J Clean Prod 112:3759–3773. https://doi.org/10.1016/j.jclepro.2015.08.089

Cardoso Júnior RAF, Magrini A, da Hora AF (2014) Environmental licensing process of power transmission in Brazil update analysis: case study of the Madeira transmission system. Energy Policy 67:281–289. https://doi.org/10.1016/j.enpol.2013.12.040

Coelho ST, Goldemberg J (2013) Energy access: lessons learned in Brazil and perspectives for replication in other developing countries. Energy Policy 61:1088–1096. https://doi.org/10.1016/j.enpol.2013.05.062

Cucchiella F, D'Adamo I (2013) Issue on supply chain of renewable energy. Energy Convers Manag 76:774–780. https://doi.org/10.1016/j.enconman.2013.07.081

Da Silva NF, Rosa LP, Freitas MAV, Pereira MG (2013) Wind energy in Brazil: from the power sector's expansion crisis model to the favorable environment. Renew Sust Energ Rev 22:686–697. https://doi.org/10.1016/j.rser.2012.12.054

De Jong P, Kiperstok A, Torres EA (2015) Economic and environmental analysis of electricity generation technologies in Brazil. Renew Sust Energ Rev 52:725–739. https://doi.org/10.1016/j.rser.2015.06.064

De Jong P, Kiperstok A, Sánchez AS et al (2016) Integrating large scale wind power into the electricity grid in the northeast of Brazil. Energy 100:401–415. https://doi.org/10.1016/j.energy.2015.12.026

De Jong P, Dargaville R, Silver J et al (2017) Forecasting high proportions of wind energy supplying the Brazilian northeast electricity grid. Appl Energy 195:538–555. https://doi.org/10.1016/j.apenergy.2017.03.058

De Lucena AFP, Szklo AS, Schaeffer R et al (2009) The vulnerability of renewable energy to climate change in Brazil. Energy Policy 37:879–889. https://doi.org/10.1016/j.enpol.2008.10.029

Dranka GG, Ferreira P (2018) Planning for a renewable future in the Brazilian power system. Energy 164:496–511. https://doi.org/10.1016/j.energy.2018.08.164

Dyner I (2000) Energy modelling platforms for policy and strategy support. J Oper Res Soc 51:136–144

Dyner I, Larsen ER (2001) From planning to strategy in the electricity industry. Energy Policy 29:1145–1154. https://doi.org/10.1016/S0301-4215(01)00040-4

Espinosa M, Cadena ÁI, Behrentz E (2019) Challenges in greenhouse gas mitigation in developing countries: a case study of the Colombian transport sector. Energy Policy 124:111–122. https://doi.org/10.1016/j.enpol.2018.09.039

Feltes JW, Fernandes BS, Portugal PMM, Leite AP (2012) Transmission expansion in Brazil for renewables. IEEE Power Energy Soc Gen Meet:1–7. https://doi.org/10.1109/PESGM.2012.6345660

Ford A (1997) System dynamics and the electric power industry. Syst Dyn Rev 13:57–85. https://doi.org/10.1002/(SICI)1099-1727(199721)13:1<57::AID-SDR117>3.0.CO;2-B

Forrester JW (1961) Industrial Dynamics, MIT Press, Cambridge, MA p 197

Fortes P, Pereira R, Pereira A, Seixas J (2014) Integrated technological-economic modeling platform for energy and climate policy analysis. Energy 73:716–730. https://doi.org/10.1016/j.energy.2014.06.075

Franco CJ, Castaneda M, Dyner I (2015) Simulating the new British electricity-market reform. Eur J Oper Res 245:273–285. https://doi.org/10.1016/j.ejor.2015.02.040

Global Transmission Report (2016) Brazil's economic downturn: recession and lack of funds affect grid development. In: https://www.globaltransmission.info/archive.php?id=26657

Global Transmission Report (2018) Update on Brazilian power sector: key highlights of 2017. In: https://www.globaltransmission.info/archive.php?id=32000

Gorayeb A, Brannstrom C, de Andrade Meireles AJ, de Sousa MJ (2018) Wind power gone bad: critiquing wind power planning processes in northeastern Brazil. Energy Res Soc Sci 40:82–88. https://doi.org/10.1016/j.erss.2017.11.027

Haselip J, Dyner I, Cherni J (2005) Electricity market reform in Argentina: assessing the impact for the poor in Buenos Aires. Util Policy 13:1–14. https://doi.org/10.1016/j.jup.2004.03.001

Henao F, Cherni JA, Jaramillo P, Dyner I (2012) A multicriteria approach to sustainable energy supply for the rural poor. Eur J Oper Res 218:801–809. https://doi.org/10.1016/j.ejor.2011.11.033

Herrera MM, Dyner I, Cosenz F (2017) Effects of the penetration of wind power in the Brazilian electricity market. Rev Ing Ind 15:309–319

Herrera MM, Dyner I, Cosenz F (2018) Alternative energy policy for mitigating the asynchrony of the wind-power industry's supply chain in Brazil. In: Qudrat-Ullah H (ed) Innovative solutions for sustainable supply chains. Springer, Cham, pp 199–221

Herrera MM, Cosenz F, Dyner I (2019a) How to support energy policy coordination? Findings from the Brazilian wind industry. Electr J 32:106636. https://doi.org/10.1016/j.tej.2019.106636

Herrera MM, Dyner I, Cosenz F (2019b) Assessing the effect of transmission constraints on wind power expansion in Northeast Brazil. Util Policy 59:100924. https://doi.org/10.1016/j.jup.2019.05.010

Lima DKS, Leão RPS, dos Santos ACS et al (2015) Estimating the offshore wind resources of the State of Ceará in Brazil. Renew Energy 83:203–221. https://doi.org/10.1016/j.renene.2015.04.025

Liu X, Mao G, Ren J et al (2014) How might China achieve its 2020 emissions target? A scenario analysis of energy consumption and CO2 emissions using the system dynamics model. J Clean Prod 103:401–410. https://doi.org/10.1016/j.jclepro.2014.12.080

Miranda R, Soria R, Schaeffer R et al (2017) Contributions to the analysis of "integrating large scale wind power into the electricity grid in the northeast of Brazil" [Energy 100 (2016) 401–415]. Energy 118:1198–1209. https://doi.org/10.1016/j.energy.2016.10.138

Morcillo JD, Franco CJ, Angulo F (2017) Delays in electricity market models. Energy Strat Rev 16:24–32. https://doi.org/10.1016/j.esr.2017.02.004

Moreira JML, Cesaretti MA, Carajilescov P, Maiorino JR (2015) Sustainability deterioration of electricity generation in Brazil. Energy Policy 87:334–346. https://doi.org/10.1016/j.enpol.2015.09.021

Naill RF (1992) A system dynamics model for national energy policy planning. Syst Dyn Rev 8:1–19. https://doi.org/10.1002/sdr.4260080102

Naill RF, Belanger SD (1989) A system dynamics model for national energy policy planning. Comput Manag Complex Syst 8:423–433

Ochoa C, van Ackere A (2015) Winners and losers of market coupling. Energy 80:522–534. https://doi.org/10.1016/j.energy.2014.11.088

Oliva R (2003) Model calibration as a testing strategy for system dynamics models. Eur J Oper Res 151:552–568. https://doi.org/10.1016/S0377-2217(02)00622-7

Qudrat-Ullah H (2014) Green power in Ontario: a dynamic model-based analysis. Energy 77:859–870. https://doi.org/10.1016/j.energy.2014.09.072

Qudrat-Ullah H (2016) The physics of stocks and flows of energy systems applications in energy policy. Springer, London

Qudrat-Ullah H, Seong BS (2010) How to do structural validity of a system dynamics type simulation model: the case of an energy policy model. Energy Policy 38:2216–2224. https://doi.org/10.1016/j.enpol.2009.12.009

Rego EE, de Oliveira RC (2018) Successful Brazilian experience for promoting wind energy generation. Electr J 31:13–17. https://doi.org/10.1016/j.tej.2018.02.003

Ribeiro AED, Arouca MC, Coelho DM (2016) Electric energy generation from small-scale solar and wind power in Brazil: the influence of location, area and shape. Renew Energy 85:554–563. https://doi.org/10.1016/j.renene.2015.06.071

Santos MJ, Ferreira P, Araújo M et al (2018) Scenarios for the future Brazilian power sector based on a multi-criteria assessment. J Clean Prod 167:938–950. https://doi.org/10.1016/j.jclepro.2017.03.145

Sterman J (1981) The energy transition and the economy: a system dynamics approach. Massachusetts Institute of Technology

Sterman JD (1984) Appropriate summary statistics for evaluating the historical fit of system dynamics models. Dynamica 10:51–66

Sterman JD (2000) Business dynamics: systems thinking and modeling for a complex world. McGraw-Hill, New Delhi

Viteri JP, Henao F, Cherni J, Dyner I (2019) Optimizing the insertion of renewable energy in the off-grid regions of Colombia. J Clean Prod 235:535–548. https://doi.org/10.1016/j.jclepro.2019.06.327

Wee H-M, Yang W-H, Chou C-W, Padilan MV (2012) Renewable energy supply chains, performance, application barriers, and strategies for further development. Renew Sust Energ Rev 16:5451–5465. https://doi.org/10.1016/j.rser.2012.06.006

Part IV
Dynamics of Energy Policy and Climate Change

Chapter 10
Energy Policy and Climate Change

Kankana Dubey

Abstract Is economic growth and development a zero-sum game vis-à-vis the environment: i.e. must economic welfare always come at the cost of environmental damage? This is one of the most fundamental questions facing humanity today – whether there is a way for both the economy and the environment to flourish, or if we must always make trade-offs between what's best for each. Avoiding knee-jerk reactions wholly in favor of either side is important for society, as evidenced by the disorder experienced in several countries around the world at the hands of self-declared guardians of the environment. The fact is that climate change is not limited to a national boundary but is tied to international actions and reactions. While the price an economy pays for a sustainable future is purely a local price, the social costs resulting from environmental damage is shared globally. Hence, even if a government completely capitulated to the environmental extremists' demands, there would still be environmental impacts to mitigate. In addition to developing energy policies to reduce carbon emissions, policymakers must act to prepare their economies to deal with climate change. Can energy policies be designed to simultaneously provide social and environmental benefits? The answer is an emphatic yes. This chapter shares examples of a select subset of countries, reviewing their climate change agendas and targets, in conjunction with supporting environmental policies. Specifically, the economic impacts of such measures, and the long-term impacts of such policies on energy supply and demand, are assessed.

Keywords Sustainable development · Green growth · Decoupling · Climate change · Energy productivity · Energy efficiency · Decarbonisation

K. Dubey (✉)
Centre for Energy Policy, University of Strathclyde, Glasgow, Scotland, UK
e-mail: Kankana.dubey@strath.ac.uk

H. Qudrat-Ullah, M. Asif (eds.), *Dynamics of Energy, Environment and Economy*, Lecture Notes in Energy 77, https://doi.org/10.1007/978-3-030-43578-3_10

1 Introduction

We've all heard that hindsight is 20/20, and it truly is an amazing lens through which to view life. It allows all of us to say that someone should have expected some result, or should have seen it coming. Unfortunately, awareness of our present situation and the conundrums we are facing is much more difficult, especially when some of the future consequences are difficult to fathom.

There can be no reasonable doubt that anthropogenic global warming is real; this has been sufficiently documented and analyzed using both empirical data and computer simulations. The earth is warmer now than in the past millennium, and the record levels of atmospheric greenhouse gases (GHG) are evidence that the warming trend will continue.

Climate change is especially challenging vis-à-vis hindsight. Most of us did not personally observe the early warning signs of global warming, and many that did could not understand the magnitude of the consequences. For example, the Mauna Loa observatory in Hawaii began monitoring increasing levels of atmospheric carbon dioxide in the 1950s, but it was not considered generally newsworthy. Major global efforts to collect and analyze climatic data only began in earnest during the 1990s, and our understanding of the impacts of global warming are still developing. What we do know, from the analysis of various data sources (such as ice cores, ancient sediments, and even temperature records from seafaring voyagers), is that the concentrations of major greenhouse gases carbon dioxide and methane are now at the highest levels in the past 650,000 years (Dow and Downing 2011). The presence of these gases in the atmosphere are turning the earth into a planet-sized greenhouse. Further, our current understanding of the physical climate system, with its many variables, has been captured in increasingly complete computer simulation models. Their efficacy can be tested against the record of past climates, and many have managed to reconstruct past climates with surprising accuracy.

There can be no reasonable doubt that human activities have changed the chemical composition of the atmosphere, and models that account for the increased levels of greenhouse gases caused by our activities generate output consistent with real observations. Oft-cited climate change impacts, such as more extreme storm systems and more frequent droughts and heat waves are consistent with these models' predictions. While the computer simulation models capture the climate change systems well enough to replicate earth's climate history, they cannot account for current trends of global warming *without incorporating the added burden of greenhouse gas emissions caused by human actions*. The shift in the observed climate patterns has been at the center of current scientific debate, though we can no longer hope that climate change skeptics were right. Indeed, there is a growing confidence in the international scientific community that we are witnessing the first serious impacts of climate change (Dow and Downing 2011).

Global warming is affecting ecosystems in ways that are unprecedented; the consequences of climate change occur everywhere and are globally interconnected. Unlike some other forms of pollution, emissions of greenhouse gases have a global impact. Whether they are emitted in Asia, Africa, Europe, or the Americas, they rapidly disperse evenly across the globe. In addition, even local effects of climate change can become global. Impacts upon agriculture, for example, may begin with changes in crop yields at a local scale, but are ultimately felt around the world in higher prices and transportation costs. With the verity of global warming no longer a reasonable matter for debate, the proper debates are now regarding the economically- and socially-optimal ways of mitigating and adapting to global warming, as well as the best strategies for intergovernmental cooperation. The challenge of climate change is advocated through international organizations, striving to reach international collaboration, because it is a global phenomenon, and not constrained to one country. It can't be stressed enough that initiatives to tackle climate change and develop effective energy policies will require international cooperation to make a substantial impact on global warming.

2 International Cooperation

The principal forum for international climate change action has been the United Nations Framework Convention on Climate Change (UNFCCC) (Kuyper et al. 2018). Since 2005, the annual conferences, known as the Conferences of the Parties (COP), have been held to assess the progress made by countries which have ratified the Kyoto Protocol; signatory countries who are legally bound to reduce their greenhouse gases. More recently, additional international forums have become active: examples include the Asia Pacific Partnership, and agreements under the G8 (group of eight highly industrialized countries), starting with their 2005 meeting in Gleneagles, Scotland (World Nuclear Association 2017) which included the five outreach countries, known as Group of Five. In December 2015, the Paris agreement consolidated years of negotiations with agreement among 196 countries to govern climate change reduction measures from 2020, targeted at limiting global warming to between 1.5 °C and 2.0 °C above pre-industrial levels (UNFCCC 2019). Figure 10.1 shows a compressed timeline of relevant conventions.

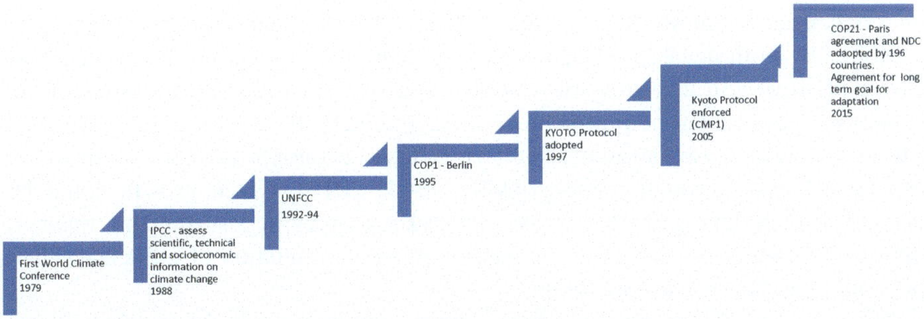

Fig. 10.1 History of major climate-change-focused conventions and forums

Although climate change agreements set carbon emission reduction targets through international approaches, policy measures to meet the obligations and objectives set by such agreements must obviously be implemented at the national or regional level. One example is the Carbon Neutrality Coalition network (CNC), founded by 16 member countries in 2017 (Carbon Neutrality Coalition 2018). EU member states dominate the coalition in a hope to meet the Paris Agreement goal of achieving carbon neutrality "in the second half of the century". CNC members are developing long-term low-greenhouse gas emission strategies to become carbon neutral by transitioning to become a net-zero emission economy. Their primary focus is restructuring the transport, energy, and agriculture sectors through technological innovation and resource-efficiency policies. The target is shared by 32 cities worldwide. While the coalition was formed to utilize collective decision-making processes to develop optimal strategies, the mandate of reducing emissions would finally lay with the member countries. Each country would go down their own situation-specific path to achieve the common objective. International cooperative initiatives between countries such as this are expected to not only reduce carbon emissions, but also boost socioeconomic welfare, create highly-skilled jobs, and enhance the interdependency of supply chains across countries.

There is much happening on regional cooperation. For example, several Latin American and Caribbean countries are coming together, as they are facing similar challenges of rising temperature leading to the destruction of coastlines by hurricanes, loss of biodiversity, and other significant effects of increased carbon emissions. These topics were all part of the agenda in the 20th Annual Caribbean Energy Conference, organized by S&P Global Platts, and held in Puerto Rico during January 2020. It is well understood by these countries that developing a strong long-term climate policy utilizing sustainable mechanisms will not only protect them from natural catastrophes, but also generate socioeconomic welfare. This is discussed in the following sections.

3 Climate Laws and Legislation

Several initiatives related to combating climate change have been – and are being – undertaken across the world in many countries. This focus on environmental policy, exhibited by both developed and developing economies, suggests the existence of strong political will driven by two factors. The first is increased awareness of the calamitous consequences of climate change, to which all countries are vulnerable. The second is increased awareness of the benefits of economic diversification which can be achieved through green growth (OECD 2019). Green growth could be defined as economic growth that uses natural resources in a sustainable manner, optimizing energy productivity; the paradigm is a combination of affluence (GDP per capita) and energy productivity.

Since the mid 1990s, there has been a significant increase in the number of laws and legislation framed for climate change across countries. A database developed by the Grantham Research Institute on Climate Change and the Environment to track governmental climate change initiatives shows a steady increase in the number of laws enacted globally. The 160+ monitored countries account for 95% of global GHG emissions.

Of course, enacted legislation does not always necessarily translate into effectively-set carbon emissions reduction targets. Strict implementation across different economic sectors, with achievable outcomes, is also required. For example, the United Kingdom (UK) was one of the first countries committed to reducing greenhouse gas emissions by setting long term, legally-binding national legislation and carbon reduction targets (Norton 2018). This has not been without controversy, as critics argue that few other countries have passed similarly stringent climate laws. The UK was also the first country to establish an independent national committee – Climate Change Committee (CCC) - to track progress on climate change targets.

Another interesting example is that of China, which is leading the international race in reducing national carbon emission by developing both short- & long-term plans (DRC and OECD 2017). China has achieved an impressive balance of industrial and environmental policies, by introducing its 11th, 12th, 13th five-year plans. These plans focused on reducing energy intensity within the nation's industrial sector by simultaneously implementing energy efficiency and allocating resources more effectively. China's efforts to restructure their industrial sector resulted in a structural shift in the economy, driving green growth. The Low Carbon Economy Index, produced by the consulting firm Price Waterhouse Cooper (Herweijer et al. 2019), shows that China is on target to cut its carbon intensity by 3.5% annually through 2030. The same report suggests that the UK's carbon budget implies annual reductions of 3.1%.

Another interesting fact highlighted by the Grantham Research Institute database is the manner in which some of the least-developed nations, which are particularly vulnerable to the consequences of climate change, are becoming increasingly active on climate change policy. While developed countries have focused on reducing greenhouse gas emissions, their counterparts have focused more on adaptation. According to the Grantham Research Institute, "Reflecting the low carbon footprint of LDC [least developed countries] and their high vulnerability to climate change, the focus of most laws has been on adaptation, but also on building frameworks for promoting and enabling green growth" (Nachmany et al. 2017).

Despite significant progress around the world, current climate policies fall far short of what is needed to prevent global warming from breaching the critical 2 °C level.

4 Climate Adaptation and Mitigation

The challenges of mitigating and adapting to climate changes are unprecedented, but not insurmountable. Key elements and institutions already exist and can be used as a foundation for change. However, there remains much more to be done, with delays resulting in higher future costs. As should be clear by now, the scale of efforts needed to return GHG emissions to safer levels require truly international cooperation. A twin goal must be to facilitate adaption among countries least able to protect themselves from climate impacts.

Adaptation means building resilience against the impacts of climate change. Adaptation and mitigation strategies are coupled together to reduce carbon emissions while developing readiness to counter unforeseen natural catastrophes. Strategies for adaptation call for increased investments in two main areas: infrastructure development for developing economies, and structural economic shifts in developed economies, such as reduced dependency on fossil fuels, and increased uptake of renewable energy. The perceived importance of adaptation by climate change policymakers has grown over recent years. In fact, it was a central pillar in the Paris Agreement. However, it remains underfinanced, especially relative to the financing available for mitigation efforts.

Financing is an important tool in designing the policies for adaptation and mitigation. According to a report by the UN Environment Programme, developing countries face enormous financial challenges to meet the requirements of advanced technological mitigation projects. Contrary to the claims of vocal citizen groups, many countries are making substantial progress towards reducing carbon emissions. However, the appropriate response must not be the instinctive reaction they demand, but a well-considered approach. We list here a few examples of countries, organizations, and institutions which are attempting to reduce the effects of global warming to protect their citizens and the environment.

The United Nations Food and Agricultural Organization has undertaken a study to assess afforestation policies across countries (Food and Agriculture Organization of the United Nations 2018). Using Global Forests Assessment data, they identified three primary types of afforestation strategies utilized by countries: natural regeneration (to restore natural forest), commercial plantations, and agroforestry (planting trees around agricultural land). In 2015, the top three tree-planting countries in absolute terms were China, the United States of America, and Russia. Relative to land area, the countries leading the race are the Czech Republic, Sweden, and Poland (Food and Agriculture Organization of the United Nations 2018). Several relevant initiatives are being promoted by NGOs, private sector organizations, government-run programs, and international institutions. One example is Treedom (www.treedom.net), an innovative online platform where an individual can plant trees in a country of their choice; common reasons are as a gift, to commemorate special events, or in remembrance of loved ones.

The World Bank's carbon finance initiatives have supported activities in nearly 80 countries and have made $2 billion in emission reduction payments since the first carbon fund was launched in 1999. In addition, the World Bank has further extended its support by starting a Climate Change Fund Management Unit (CCFMU). The CCFMU finances innovative and scalable climate and environmental action projects (Ostman 2019). With more than $6 billion in capital, these initiatives:

- create partnerships to develop new financial instruments encouraging low-carbon climate-resilient development
- build supportive policy and regulatory environments to help lower the cost of capital for relevant projects
- catalyze private-sector capital to finance and scale-up climate action

5 A Complex Network of Policies; Striking a Delicate Balance

The ever-growing world population is leading to unprecedented increases in consumerism, which is leading to increased energy consumption across every sphere of activity in everyday life. The negative impacts on the environment and climate of this situation have been sufficiently documented and analyzed through both empirical data and computer simulations. These impacts, which include melting glaciers, rising sea levels, and more extreme storm systems, cross all national boundaries and present an existential threat to all life on earth. There is a need for international cooperation, and most countries are actively attempting to develop policies that can mitigate the climate change challenge: developing a low carbon economy has become a high priority for many national governments. Policymaking is being influenced by the general public and consumer groups. The reaction favored by many of these groups, however, has been more akin to an involuntary reaction to demand a complete halt to carbon emissions.

Is it reasonable, or even possible, for policymakers to implement stringent climate policies that eliminate fossil fuel consumption for industrial use, electricity generation, and transportation entirely? Sudden cessation of fossil fuel consumption would result in a complete economic shutdown, with day-to-day life coming to a standstill. Of course, this is not what anybody desires. Further, there is substantial political risk for a nation to take radical actions to move away from fossil fuel consumption. The required additional taxation and imposition of higher fuel prices would create outrage within different economic sectors and amongst the general public – likely leading to loss of political power. What is needed is long-term structured planning to develop sustainable efforts that will benefit both the current population and future generations.

In 2019, the UK experienced a result of such long-term planning and sustainable efforts. During 1 week in May – for the first time in the post-industrial revolution history of the United Kingdom – no coal was burnt to generate electricity (Ogden 2019). This was the longest period since the inception of the industrial revolution in 1882 that electricity was produced without the use of coal. In the absence of sufficient sustainability planning, this could not have been achieved economically, short of shedding load. During 2018 and 2019, there have been several such milestones achieved globally, clearly demonstrating that progress towards reducing GHG emissions is being made (UNFCCC 2019).

Despite many claims to the contrary, policymakers have been concerned about the climate change agenda, energy security, and diversification away from fossil fuels since at least the 1970s (Fankhauser et al. 2015). Developing and planning a robust policy which meets the often-conflicting goals to simultaneously: decrease carbon dioxide emissions, reduce dependency on fossil fuels and adopt renewable energy, and reduce reliance upon energy-intensive industries that largely employ semi- and un-skilled labour; and yet: ensure continued access to energy, maintain economic development, and provide economic well-being is a complex undertaking. It requires much strategizing and compromise between the government and civic society, whilst maintaining international cooperation.

Perhaps this is not an excuse for the fact that implementation of GHG emission reduction policies around the world have taken time, but such policies are also reliant upon technological innovation. For example, renewable energy has been heavily researched since the 1970s, and strong efforts been made to adopt the technologies globally. Yet in 2018, renewable electricity generation only rose only 7%, with wind and solar PV technologies together accounting for 60% of this increase, according to the IEA (the following statistics are all due to the IEA). Although the share of renewables in global electricity generation reached 26% in 2018, renewable power as a whole still needs to expand significantly to meet the UN's Sustainable Development Goal (SDG7) of half of electricity generation by 2030. Achieving 50% share of electricity generation from renewables requires the rate of annual capacity additions to accelerate; however, renewable capacity growth stalled in 2018 for the first time since 2001. In part, this is due to several barriers that exist within the political and regulatory systems. The diverse barriers, which vary between countries and regions, include limited technical capacity, social adaptability, financial and economic mechanisms, and infrastructure limitations, to name a few (Tosun et al. 2017). Policymakers must pay substantial attention to removing these barriers prior to, and during, developing policy development. Such efforts take a substantial amount of time; commitment and target setting for climate policies were initially undertaken in the late 1970s (Kuyper et al. 2018) and the progress has been slow.

Energy is one of the most important economic resources; however, energy consumption is one of the most significant contributors to carbon emissions, whether it is for transportation or in residential buildings. It is imperative that every country balances the supply and demand of energy, and efficient use of energy can continue to provide economic and social benefits whilst securing the sustainability of the planet. The objective of climate policies should not solely be to reduce carbon emissions but also to strengthen economic performance and improve social welfare in a country. One such type of carbon emissions reduction policy is *decarbonisation*.

Decarbonisation is a process of reducing GHG emissions by deploying advanced technologies, policy legislation and regulations, and substituting for higher efficiency systems across sectors. Effective and efficient decarbonisation of an economy requires following a thoroughly-considered approach. Detailed analysis must be performed for each economic sector, so as to reach a decision of whether to implement measures to reduce carbon emissions or to completely decarbonize an individual sector. Such analysis must consider the strong inter-sectoral interdependencies which exist, and the degree to which a change in one sector can precipitate a shift in the entire economic supply chain, triggering an impact on local and international industries. To abide by international agreements, most countries will need to make major economic changes – this will need to begin with a reassessment of policymaking for the energy sector. The role of policy instruments such as legislation (primary and delegated), permits, auctions, labelling, self-regulation agreements, fees and charges, taxation / tax exemptions, loans and grants, information, etc. in developing decarbonisation policy varies across countries. Following which, the economic viability of changing policies will need to be analyzed.

Historically, energy policies were centered on the economic demands of the society, and security of supply; today's energy policies are integrated with other concerns such as: public health, industry, food production, transport, housing and infrastructure, environmental, etc. The use of energy in our daily lives has made it a vital means to create and sustain socioeconomic welfare. This has become even more true as globalization has increased, connecting and interlinking economies around the world. This is the age of international collaboration and cooperation. Action taken to shift energy generation/consumption in one country can trigger a change across the globe.

For example, a decision by Saudi Arabia to cease producing oil would cause crises in all fuel importing countries and dramatically impact global trade (Kern 2019). Oil prices would skyrocket, and many existing stocks would be redirected from their intended destination. In the absence of sufficient alternative fuel sources, a substantial increase in fossil fuel prices could lead to economic catastrophe across the world. To tackle climate change, global cooperation is necessary. Few countries could alone cut their emissions quickly and deeply enough to prevent the concentration of carbon emissions in the atmosphere from rising to further dangerous levels; due to the global interlinking of economies, none of them should attempt to do so without international cooperation.

Many developing countries are the largest carbon emitters, but expecting countries to cut carbon emissions at the expense of their economic welfare might not be practical. However, international cooperation and technological advancement/transfer can help reduce carbon emissions in developing countries such as China and India. In fact, some developed countries (the UK, USA) have successfully *decoupled* energy consumption from economic growth and transitioned towards a low carbon economy. The OECD definition of decoupling is the condition in which the growth rate of a negative environmental impact is less than that of its economic driving force (e.g. GDP) over a given period. Such economies can continue to grow without commensurate growth in energy consumption. Combating rampant climate change will require these developing countries to also decouple their economies. Strategic partnerships and joint initiatives could transfer relevant knowledge from developed to developing countries, to help them implement similar solutions for their economies, while adhering to local conditions. It is only a matter of time until all countries will be able to diversify and adapt ways to decouple energy consumption from economic growth; it is in the best interests of humanity and the planet that this happens quickly.

6 Decoupling Energy Consumption and Economic Growth

It is imperative to understand that without energy (in addition to other factors of production such as land, capital, and labor) there could be no economy. Further, we can take as a given that every government acts to promote growth and development of its own economy. These two premises lead to the conclusion that energy consumption has continually increased, and will continue to do so, leading to the current climate change crisis. With rising populations in developing and emerging economies pushing towards higher incomes and better living standards, the law of incremental returns will eventually come into play. When this happens, the drive to achieve faster growth will decelerate, reducing the growth in carbon emissions, until countries decide to start following an agenda of sustainable growth. Economic growth is commonly understood to entail increased living standards, and a social welfare system to reflect the country's economic position. However, this does not always have to come at the expense of the environment.

Wilferd Beckerman (1992), along with other economists, suggested that economies can "grow their way" out of environmental problems if they become rich. This idea is based on the theory of Simon Kuznets (1955) and was empirically tested by Grossman and Kruger (1991), showing a causal relationship between income levels and economic growth; this is known as the Environmental Kuznets Curve (EKC). The EKC hypothesis explains a two-part inverted-U shape curve. The first portion of the curve depicts the early development stage, showing a positive correlation between increased demand for per capita income and environmental degradation leading to increased emissions (as seen during the industrial revolution) and increased demand for land use (agriculture, housing, etc.). After an inflection point, as income increases, the demand for environmental quality also rises. This leads to public actions to protect the environment, technological improvements, economic restructuring, natural resource reallocation, etc.

According to this hypothesis, economic growth is both the cause and the cure of human-caused environmental deterioration. However, it's important to note that reversal of environmental degradation does not necessarily follow as a result of economic growth. As evidenced by the literature since the 1990s, this reversal only follows if a society chooses to invest their higher income to develop policies and technologies that can result in improving the environment. This phenomenon is what we see occurring today in the developed world; it has been given the moniker "green growth"!

7 Concepts of Sustainability

Defining green growth as distinct from sustainability, energy efficiency, or energy productivity is an important first step in understanding the concept. Energy-related concepts have evolved over time, as shown in the economic literature, reflecting different periods and contexts. Energy conservation emerged as a result of the oil crises of the 1970s. Energy efficiency followed, with attention on ways to reduce the amount of energy needed to run machines (e.g., cars, planes, industrial machinery). Broadening the scope to the whole economy, the concept of energy intensity emerged, shifting the focus to how much energy is consumed per unit of output (gross domestic product). More recently, the concept of energy productivity has emerged (KAPSARC 2017) as a new way of framing energy planning and policymaking. The underlying aim—i.e., to optimize the socioeconomic value of each unit of energy consumed—is of interest to multiple stakeholders, which has given rise to different ideas on how to operationalize the concept. Conceptually, energy productivity reflects what activities energy is used for in the economy (degree of structural diversification) as well as how well energy is used in specific activities (energy efficiency).

Those proposing energy productivity (KAPSARC 2017) assert the distinction that it places its emphasis more neutrally on how energy can be used to drive economic growth and increase government revenues, while also lowering overall energy consumption, reducing carbon emissions, and resulting in decreased environmental impact from energy consumption. Under energy productivity, the latter three aspects are viewed as co-benefits that can be realized through the primary drivers of economic diversification and increased energy efficiency. It can also be argued that concepts evolve to meet current anxieties and policy imperatives, and the emergence of new concepts does not need to incite competition with existing ones. Energy productivity is gaining traction around the world (KAPSARC 2017), because governments are seeking to boost growth and create jobs as first priorities, while remaining committed to reducing emissions and environmental impacts as co-benefits. Energy productivity can thus be viewed as a coherent framework that can be applied at the economy level to direct a wide range of policy domains—financial, industrial, urban, transport, innovation, etc.—toward the overarching goal of increased growth, productivity, and competitiveness.

Energy productivity is both a policy agenda focusing on how energy can best be used to create value in the economy, as well as an indicator that integrates economic development with energy consumption. The framework of energy productivity can be useful to guide energy-related policymaking in all nations, even though economic dependencies and priorities vary widely. Relevant policymaking concerns include economic development, job creation, regional development, infrastructure growth, energy access, energy security, electricity market reform, energy efficiency, pollution control, and greenhouse gas avoidance. The priority level of these, and other considerations, can be taken as indicators of the economic importance of energy productivity.

8 Decoupling Energy Consumption and Economic Development

It is imperative to understand that without energy (in addition to other factors of production such as land, capital, and labor) there could be no economy. Further, we can take as a given that every government should act to promote growth and development of its own economy. These two premises lead to the conclusion that energy consumption has continually increased, and will continue to do so, leading to the current climate change crisis, according to an article published by the Economic Policy Institute. With rising populations in developing and emerging economies pushing towards higher incomes and better living standards, the law of incremental returns will eventually come into play. When this happens, the drive to achieve faster economic development will decelerate, reducing the increase in carbon emissions, until countries decide to start following a sustainable development agenda. This is according to the theory underlying the Environmental Kuznets Curve referenced earlier (Grossman and Krueger 1991; Kuznets 1955).

The United States is the largest country to experience multiple consecutive years in which economic growth has been decoupled from growth in carbon emissions. Between 2010 and 2012, GDP grew from $14.8 to $15.4 trillion; during the same period, however, energy-related carbon emissions declined from 5.58 to 5.23 billion metric tons. While GDP grew by 4%, emissions shrunk by 6% (Aden 2016). According to the U.S. Energy Information Administration, the shift to a cleaner electricity generation and transition system, planned for the 2020s, should bring about a sustained period of economic growth and energy demand decoupling (Aden 2016). As illustrated in Fig. 10.2, total U.S. energy-related carbon emissions are forecast to decline by a further 6% between 2020 and 2025, while GDP is expected to increase by approximately 13% (in real terms).

In fact, the trend towards economic decoupling is becoming a global phenomenon. Figure 10.3 lists more than 20 countries that have successfully decoupled economic growth and greenhouse gas emissions between 2000 and 2014. With the exceptions of Bulgaria and Uzbekistan, all have done so by diversifying their economies away from the industrial sector. It is interesting to note that these two exceptions were able to reduce their reliance on energy growth to fuel economic growth, even while making only marginal reductions in the industrial sector (Aden 2016).

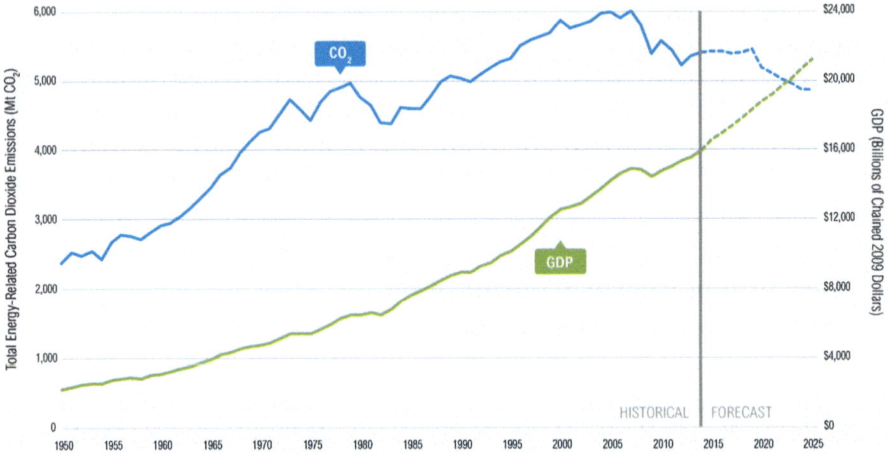

Fig. 10.2 US carbon emissions and GDP from 1950 to 2025. (Source: World Resources Institute)

COUNTRY	CHANGE IN CO_2 (2000–2014)	CHANGE IN GDP (2000–2014)
Austria	-3%	21%
Belgium	-12%	21%
Bulgaria	-5%	62%
Czech Republic	-14%	40%
Denmark	-30%	8%
Finland	-18%	18%
France	-19%	16%
Germany	-12%	16%
Hungary	-24%	29%
Ireland	-16%	47%
Netherlands	-8%	15%
Portugal	-23%	1%
Romania	-22%	65%
Slovakia	-22%	75%
Spain	-14%	20%
Sweden	-8%	31%
Switzerland	-10%	28%
Ukraine	-29%	49%
United Kingdom	-20%	27%
United States	-6%	28%
Uzbekistan	-2%	28%

Fig. 10.3 Since 2000, more than 20 countries have reduced annual greenhouse gas emissions while growing their economies. (Source: World Resources Institute)

9 Conclusion

There can be no reasonable doubt that anthropogenic global warming is real; this has been sufficiently documented and analyzed using both empirical data and computer simulations. The earth is warmer now than in the past millennium, and the record levels of atmospheric greenhouse gases are evidence that the warming trend will continue. This is largely due to the relentless pursuit of economic growth, and the resultant increase in consumption of fossil fuels.

One of the most important conundrums facing humanity today is whether there is a way for both the economy and the environment to flourish, or if we must always make trade-offs between what's best for each. Is economic growth and development a zero-sum game vis-à-vis the environment; i.e. must economic welfare always come at the cost of environmental damage? Can energy policies be designed to simultaneously provide socioeconomic and environmental benefits? The answer is an emphatic yes.

Human history is replete with energy transitions: man power to animal power, animal power to wood combustion, wood combustion to coal combustion, and the combustion of coal to oil. In each transition, relatively simple technological innovations released vast resources at low cost. Essentially, change was inevitable. Global society is currently in the midst of yet another energy transition – that of fossil fuel combustion to various renewable energy generation technologies. It is reasonable to believe that it too will be successful, provided that the transition is impelled forward by overwhelming economic benefits, instead of the fear of the future.

Energy is one of the most important resources for economic growth. In a typical economy, economic growth goes hand-in-hand with increased energy consumption. We would say that energy and the economy are coupled. Decarbonisation through energy decoupling is becoming an increasingly used concept in the context of economic production. It refers to the ability of an economy to grow without corresponding increases in energy consumption (and environmental pressure). Decoupling is an efficient way to combat climate challenges whilst sustaining balanced socioeconomic growth.

Key Definitions

Energy Productivity – reflects what activities energy is used for in the economy (degree of structural diversification) as well as how well energy is used in specific activities (energy efficiency).

Green Growth – is defined as economic growth that uses natural resources in a sustainable manner, optimizing energy productivity; the paradigm is a combination of affluence (GDP per capita) and energy productivity.

Decoupling – is the condition in a country in which the growth rate of a negative environmental impact is less than that of its economic driving force (e.g. GDP) over a given period.

Decarbonisaton – is a process of reducing GHG emissions by ways of deploying advanced technologies, policy legislation and regulations, and substituting to higher efficiency systems across sectors.

Structural Shift – occurs when countries diversify from the traditional economic sectors to new economic sectors; this change can occur due to several factors, including resource availability, shifts in supply and demand, political will, and the impact of globalization.

References

Aden N (2016) The roads to decoupling: 21 countries are reducing carbon emissions while growing GDP. Retrieved from World Resources Institute website: https://www.wri.org/blog/2016/04/roads-decoupling-21-countries-are-reducing-carbon-emissions-while-growing-gdp

Beckerman W (1992) The environment as a commodity. Nature 357(6377):371–372

Carbon Neutrality Coalition (2018, August 21) The declaration of the carbon neutrality coalition. Retrieved from https://www.carbon-neutrality.global/the-declaration/

Dow K, Downing T (2011) The atlas of climate change. Earthscan, Brighton

DRC, & OECD (2017) Industrial upgrading for green growth in China. Retrieved from OECD website: https://www.oecd.org/greengrowth/Industrial_Upgrading_China_June_2017.pdf

Fankhauser S, Gennaioli C, Collins M (2015) The political economy of passing climate change legislation: evidence from a survey. Glob Environ Chang 35:52–61. https://doi.org/10.1016/j.gloenvcha.2015.08.008

Food and Agriculture Organization of the United Nations (2018) The State of the World's Forests (SOFO). Retrieved from Food and Agriculture Organization of the United Nations website: http://www.fao.org/documents/card/en/c/I9535EN. isbn 978-92-5-130561-4

Grossman G. M, & Krueger A. B (1991) Environmental Impacts of a North American Free Trade Agreement (3914). Cambridge, MA: National Bureau of Economic Research. https://doi.org/10.3386/w3914

Herweijer C, Sura K, Davies M, Gilbert M (2019) The low carbon economy index 2019. Retrieved from Price Waterhouse Coopers website: https://www.pwc.co.uk/services/sustainability-climate-change/insights/low-carbon-economy-index.html

KAPSARC (2017) Energy dialogue 2016. Riyadh, Saudi Arabia

Kern M (2019) $100 oil? Drone strikes halt half of Saudi crude production. Retrieved from https://oilprice.com/Geopolitics/International/100-Oil-Drone-Strikes-Halt-Half-Of-Saudi-Crude-Production.htm

Kuyper J, Schroeder H, Linnér B (2018) The evolution of the UNFCCC. Annu Rev Environ Resour 43(1):343–368. https://doi.org/10.1146/annurev-environ-102017-030119

Kuznets S (1955) Economic growth and income inequality. Am Econ Rev 45(1):1–28

Nachmany M, Fankhauser S, Setzer J (2017) Global trends in climate change legislation and litigation, 2017 update. Retrieved from Grantham Research Institute on Climate Change and the Environment website: http://www.lse.ac.uk/GranthamInstitute/wp-content/uploads/2017/04/Global-trends-in-climate-change-legislation-and-litigation-WEB.pdf

Norton A (2018) Trailblazing UK climate change act is a model for the world. Climate Home News. Retrieved from https://www.climatechangenews.com/2018/11/26/uk-climate-change-act-model-world/

OECD (2019) Green growth and sustainable development. Retrieved from https://www.oecd.org/greengrowth/

Ogden C (2019) *UK's Longest Ever Coal-free Run Comes to an End*. Environment Journal. Retrieved from https://environmentjournal.online/articles/uks-longest-ever-coal-free-run-comes-to-an-end/

Ostman A (2019) Climate finance and initiatives (Brief). Retrieved from World Bank website: https://www.worldbank.org/en/topic/climatechange/brief/world-bank-carbon-funds-facilities

Tosun J, Jordan A, Maor M (2017) Governing climate change: the (dis)proportionality of policy responses. J Environ Policy Plan 19(6):596–598. https://doi.org/10.1080/1523908x.2017.1354451

UNFCCC (2019) Cooperation with the IPCC. Retrieved from https://unfccc.int/topics/science/workstreams/cooperation-with-the-ipcc

World Nuclear Association (2017) Policy responses to climate change. Retrieved from https://www.world-nuclear.org/information-library/energy-and-the-environment/policy-responses-to-climate-change.aspx

Chapter 11
The Role of Institutions in Energy Policy and Environmental Protection

Muhammad Asif and Abdul Majid

Abstract The sustainable development of a country largely depends on its energy production and an efficient utilization of available natural resources. In this regard, the role of institutions is very much imperative to set the rules of the game. In this chapter, authors firstly discussed the nature of institutions and types of social institutions. Secondly, they discussed the theories related to the institutions. Thirdly, the concept of energy and environmental constraints are discussed. Lastly, environmental protection act and energy policy of Pakistan was discussed as a case study in this chapter.

Keywords Institutions · Energy · Environment · Environmental protection · Sustainability · Social institutions · Formal and informal constraints · Pakistan

Objectives of the Chapter

The primary objective of the chapter is to discuss the role and importance of institutions in devising rules and regulation in energy production, sustainability and its efficient utilization. Some secondary objectives of the chapter include:

1. To provide understanding of institutions, institutional theories and the functions of institutions
2. To describe the role of formal and informal institutions in sustainable production of energy
3. To describe the role of formal and informal institutions in protecting the environment and controlling environmental pollution

M. Asif
Management Sciences Department, COMSATS University Islamabad, Abbottabad Campus, Abbottabad, Pakistan
e-mail: m_asif@cuiatd.edu.pk

A. Majid (✉)
Management Sciences Department, Hazara University, Mansehra, Pakistan

H. Qudrat-Ullah, M. Asif (eds.), *Dynamics of Energy, Environment and Economy*, Lecture Notes in Energy 77, https://doi.org/10.1007/978-3-030-43578-3_11

Significance of the Chapter in the Book
The development of any economy is dependent on its sustainable energy resources. The world is seriously suffering from energy crisis; therefore, it is vital not only to search for new sources of energy but also to find sustainable solution to overcome the energy crisis. In this connection, the role of institutions is imperative to set the rules and regulation for investment and in search of new means of energy sources, their secure utilization and sustainability.

This chapter "The role of institutions in energy policy and environmental protection" is well according to the theme of the book tilted as "Dynamics of energy, environment and economy: A sustainability perspective.

1 Nature of Institutions

There are a variety of definitions of institutions that are available in the literature. The contributions of Acemoglu et al. (2008), Hodgson (2007), North and Thomas (1973) and North (1990) are significant in this regard. Institutions are organizations, system of a government or well established laws. According to North (1990, p. 3), "the institutions are humanly devised constraints that shape human interactions". Institutions include both formal and informal constraints. The formal constraints are like organizations, rule and laws; whereas, informal constraints are like customs, traditions and way of life. Both formal and informal institutions determine the incentive structure for the betterment of the society. Since, institutions are key determinants of incentive to invest; therefore, they have a vital impact on economic performance of any country (Kaufmann et al. 2009, 2010, 2011; Keefer and Knack 1997).

Acemoglu et al. (2008) identified three features of institutions which include: (1) institutions are humanly devised constraints, (2) they are rules of the game with certain enforcement mechanism and (3) institutions also determine the incentive structure. Societies improve the living standard through the prosperity of the country with sound rules and regulations. Institutions are very important element for economic development of any country. According to Engerman and Sokoloff (2002), countries with better institutions and more property rights are fairly better in economic performance than countries with worse institutions. Better institutions increase productivity and attract more foreign investment, whereas, weak institutional structure discourages foreign investment in one form or other (Busse and Hefeker 2005).

1.1 Formal Institutions

Formal institutions are organizations, rules, regulations, policies, constitution, charter or set of laws (Matthews 1998; North 1990; North 1994). These are well organized system with clear directions and procedures, which applicable for each and every member of the society, regardless of the age, gender, race and religion. These formal institutions keep individuals socially, morally and psychological bound to do the things which are legitimate according to the law and for the betterment of the society. The common examples of formal institutions are government, public offices, judiciary, school, colleges, universities, constitution of the country, traffic rules, industrial and environmental policies etc.

1.2 Informal Institutions

Informal institutions are the customs, traditions, norms, mores, society, civilizations and the way of life (Matthews 1998; North 1990). These institutions are not well organized or documented but uniformly acceptable for each member of the society. The family is the main institution that known as one of the main form of informal institute. According to Anyakoha and Eluwa (1991), family is a group of persons united by the ties of marriage, blood or adoption; constituting a single household, interacting and inter-communicating with each other in their respective social roles of husband and wife, mother and father, son and daughter, brother and sister by creating a common culture. The families formed a society; which has its own norms and traditions. Traditions and cultures vary from society to society, however, some ways of life are equally shared by most of the societies, like 'respect the elders', 'switch off the lights when they are not in use', 'keep your environment clean', 'use water carefully', 'always speak the truth' etc.

1.3 Social Institutions

In sociology, there are five major institutions which include: family, education, religion, economic and political (Fig. 11.1).

1.3.1 The Family

Family is the fundamental social institution in the society. The family is the biological social unit comprised of family members including father, mother and their children. The father is the head of the family, responsible for economic and social needs of the family (Hornby 2006; Lotempio 2002). The mother, the counterpart of the head of the family is responsible for moral and psychological needs of the family. If the family is economically, socially and morally strong, then each member of the family is consider as an asset of the society.

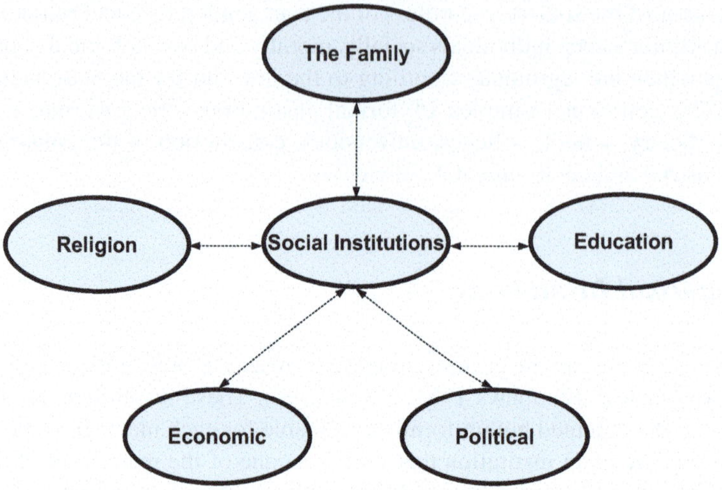

Fig. 11.1 Major five social institutions

1.3.2 Religion

The second most important element of the social institutions is the religion. Religion is concerned with the faith and belief in supernatural (Little and McGivern 2012). Religion includes beliefs and practices that serve the needs of society. Religion is also an example of a cultural universe because it is found in all societies of the world in one form or another. This provides the set of beliefs regarding Oneness of the God, the ultimate power in the Universe. The religion bound the member of the society to right things to do and do the things right, because the God rewards for right things and punish for evil things. Therefore, this belief keeps the society confined for ideal and proper ways or behaviors to performed duties for the betterment of the society. When society has a strong belief and properly practicing religion, then it has clear rules of the game and member of the society abide these rules by the core of their heart.

1.3.3 Education

Education as well as educational institutions is the most important social institutions in a society (Hornby 2006; Little and McGivern 2012). Education is considered as a formal constraint as it is structured on the basis of proper psychological, philosophical, social and cultural norms. The process of formal education starts from schools and ends at university level. In educational institutions we can learn about the laws, rules, regulations, policies and other ways of life.

1.3.4 Economy

The economy is the social institution responsible for the production and distribution of goods (Little and McGivern 2012; North 1990). The two dominant economic systems in the world are capitalism, under which resources and means of production are privately owned, and socialism, a system under which resources are owned by the society as a whole. Well-established arrangements and structures that are the part of the culture or society are: competitive markets, banking system, kids' allowances, customary tipping, and a system of property rights. Institutions strongly affect the economic development of countries and act in society at all levels by determining the frameworks in which economic exchange occurs. They determine the volume of interactions available, the benefits from economic exchange and the form which they can take.

1.3.5 Political (Government)

Political institutions are the organizations in a government which create, enforce, and apply laws (Hodgson 2006; Little and McGivern 2012). They often mediate conflict, make (governmental) policies on economic and social systems, and otherwise provide representation for the population. Political institutions and systems have a direct impact on the business environment and activities of a country. For example, a political system that is straightforward and evolving when it comes to political participation of the people and focused on the well-being of its citizens contributes to positive economic growth in its region.

Every society must have a type of political system so it may allocate resources and ongoing procedures appropriately. Along with the same concept, a political institution sets the rules in which an orderly society obeys and ultimately decides and administers the laws for those that do not obey appropriately.

2 Institutional Theories

2.1 Early Institutional Theories in Economy

The word *institutions*, is the one of the oldest ideas and thought that has continued to take diverse meanings over time. According to Scott (2008), "One of the early institutionalists, the Carl Menger, the Viennese economists, argued that institution is a social phenomenon which demands a comprehensive theoretical explanation. Menger insisted on the utility of simplifying assumptions and the value of developing economic principles". He further revealed that, in mid-nineteenth century, three institutional economists had become quite influential: Thorstein Veblen, John Commons and Westley Mitchell. Veblen defined institutions as "settled habits of thought common to the generality of man". He insisted that much behavior was governed by habits and convention. According to Commons the "rules of conduct" are the basics to define the limits within which individuals and firms could pursue their objectives. He further stated that "the institutions are the solutions that consist of set of rights and duties, an authority for enforcing them and some degree of adherence to collective norms of prudent reasonable behavior". Mitchell has similar views about the institutions and he is pioneered in collecting empirical data (Scott 2008, p. 12).

2.2 Institutional Theories in Social and Political Sciences

Early institutionalists in political economy emphasized in constitutional law and moral philosophy (Scott 2008). According to them, institutions are formal structures and legal system therefore, they focused their attention on formulating formal political institutions, legal codes and administrative rules.

The sociologists paid much attention on defining and explaining institutions than the economists and political theorists. "The work of social theorists like Spencer and Sumner; Cooley and Hughes; Marx, Durkheim, Weber and Parsons; and Mead and Luckmann were widely considered in literature (Scott 2008, p. 7)". According to Scott (2008), "an institution consists of a concept and a structure". The "concept" defines the purposes or functions of the institutions, whereas the "structure" symbolizes the idea of the institution and furnishes the mechanism by which the idea put into action. He further stated that the theorists like Cooley and Hughes also believed interdependence of individual and institutions. The great institutions like language, government, the worship places, laws and customs of property and of the family are all appeared to be independent and they are developed through the interactions among the individuals. They further stated that "the individual is always a cause and effect of the institution". "Durkheim and his colleagues considered "religious and cultural beliefs" as institutions. For Durkheim, systems of knowledge, belief and oral authority are the social institutions. The social theorist like Mead and Luckmann focused on the process of institutionalization (Scott 2008, pp. 8–18)".

Whatever the definition and explanation of the institution, according to the early institutionalists is, but Douglas North defined and explained the institutions in quiet professional manners. According to North (1990), "the institutions are humanly devised constraints that shape human interactions". Institutions are formed of both formal and informal constraint; whereas, the work of the North and other modern institutionalists is described at the beginning of the chapter.

3 Understanding Energy Policy and Environmental Constraints

It is essential to know about the energy policy, energy production, energy consumption and environmental concerns related to energy. Energy being the backbone of the economy is the primary element to devise the rules of the game for regional as well as global economy. Economies which are self sufficient in production of energy either renewable or non-renewable energy are now setting the rules and regulations, laws and policies for rest of the economies.

3.1 Energy and Economic Growth Nexus

Energy sources are the assets and are essential for any economy as they stimulate the economic production (Shahbaz et al. 2015; Soytas and Sari 2003). Being a necessary input, it is utilized not only in production but in many consumption activities as well. For development process, its importance cannot be negated. It is not only an input used in production process but also supports other factors like labor and capital. Earlier conventional growth theories were symbol of the idea that human capital and machinery are two main and primary inputs for industrial production while later theories like new growth theory highlighted the role of technology. The history of energy crisis proved that whenever there is a downfall in energy supply or increase in its prices it hampered the growth of many countries. Whether it is the oil crisis of 1970s and 1980s or Asian financial crisis of 1997–1999 the world had experienced sizeable drop in economic production (Dorgan 2015; Hondroyiannis et al. 2002; Lee and Chang 2007).

After the oil crisis of 1970 it gained huge importance and discussed widely in the literature followed by the formative work on energy growth nexus by Kraft and Kraft in 1978. Thus, until these disasters were not happened, energy was not considered as a separate input for production due to the reason that the cost incurred on this input represents quite small percentage in total cost as compared to other factors like labor and machinery (Dorgan and Seker 2016). However, it is another additional important factor of production like labor and capital. It does not have passive role to play in production rather co-equal with other factors or more important than these being accessory to them.

3.2 Energy and Sustainable Development Goals

In order to attain a better, sound and more sustainable future, UN has established Sustainable Development Goals (SDGs). It is also known as 2030 agenda since it is decided to achieve these goals by 2030. These are 17 goals interconnected to each other and 7th of these goals is affordable and clean energy (WB 2019). Since energy is necessary for human, social and economic life as it is utilized in investments, industrial and agricultural productions which enhance employment and growth of country. Thus, not only this goal itself is important to achieve to secure energy system but also important to achieve other goals. The main objective of this goal is not only to save existing global energy sources but to achieve a clean and inexhaustible energy system (Fig. 11.2).

For many years traditional or limited energy sources like coal, fossil fuels, gas and diesel are used extensively for both production purposes and consumers' welfare (Dorgan, 2015). Humans have used these resources continuously as they are a necessary element of operating their machines, devices, automobiles or other means of transportation and lighting up their places. Similarly, farmers have used them for different activities like fertilizing, irrigating lands and harvesting and industrialist for assisting their labor and capital machinery (Huang et al. 2008). Thus this prolonged use of coal, oil and other energy kinds is eventually resulting in their diminution. Over utilization of these and other energy sources is due to increasing demand for them by specially manufacturing and transportation sectors as both are more energy dependent than other sectors so that's why each country's effort to flourish further, increase living standard and incomes of their people is what provoking this demand to rise (Chandran et al. 2010). This situation of increased demand and use of energy can result in their complete extinction and brought serious impacts not only for development but for human continuity also. Since the history witnessed

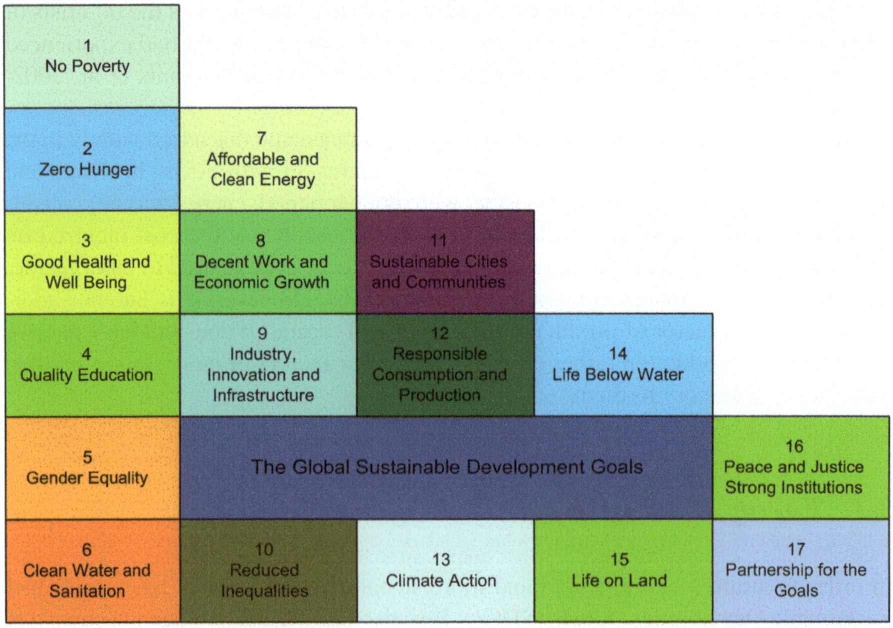

Fig. 11.2 Global sustainable development goals

that all energy crises had disturbed the growth of countries going through it. Thus it is needed to come up with those sources that are not fixed in their quantity.

3.3 Need for Renewable Energy

Renewable energy sources (e.g. hydro, wind, solar, biomass etc.) will not only save the limited available natural energy sources that are in danger of complete dying out but also save world from serious environmental threats due to its negative externality (RGSR 2017). IEA (2017) statistics showed that share of world energy use in GHG emission is 60%. That's why uninterrupted supply of clean energy is not only necessary for current and future progress but also guarantee the dirt less surroundings as well. On the one hand energy is necessary for economic activities while on the other environmental pollution also come along with its use. Burning of fossil fuels and wood is major reason for ever increasing CO_2 emission level (Wang et al. 2011; Warr and Ayres 2010). Due to easy availability and low cost these are used massively by people irrespective of its consequences on environment (Lin and Moubarak 2014; Sadorsky 2009; Squalli 2007). According to report of IEA (2017) fuel energy use has increased CO_2 emissions up to 33 $GtCO_2$ in 2015 due to industrial revolution. Now, emission emitter economies particularly and others generally are pressurized by environmental agency to review and revise their energy policies and portfolio after energy scarcity, global warming and other changes in world climate are becoming the focus of their attention (Fig. 11.3).

Without reducing current global warming, the world can lead to severe environmental disaster along with sharp decline in GDP. Since, the economic cost of global warming is greater than the amount of GDP needed to improve environmental condition. It would incur just 1% of world's gross domestic product in order to mitigate the greenhouse gases (GHG) while its 25% could be economic cost of deteriorating the climate. Production and installation of renewable energy plants can be one of the corrective actions that will lead to the sustainable development (Jaforullah and King

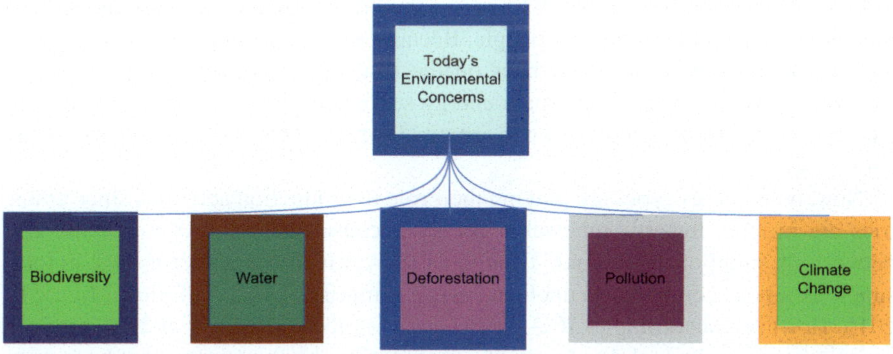

Fig. 11.3 Five main environmental concerns

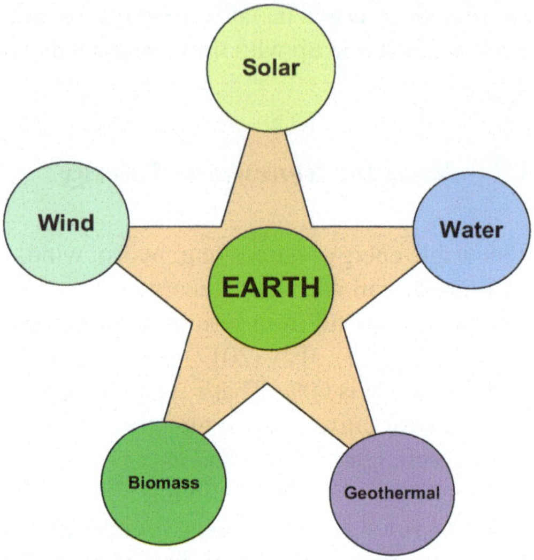

Fig. 11.4 Renewable energy sources

2015). So, major reason for introducing alternate clean energy in country's energy portfolio can be twofold. One is that its utilization will not deteriorate the climate and second is its non-exhaustible characteristic which highlights its importance not only in improving environment but also sustaining growth (Menyah and Wolde-Rufael 2010).

Since, it is renewable energy that can strike a balance between economic, social and ecological development, therefore, in order to cut down increasing energy use, increasing carbon level and to maintain green sustained economic growth, renewable energy is a possible solution. Because this energy has endless and beneficial sufficiency along with possible means of achieving environmentally sound growth. There are investment incentives as well to install renewable energy production plants since world renewable energy capacity has increased from 2,012,430 MW in 2016 to 2,179,099 MW in 2017 with Asia being first having capacity 918,655 MW and china being second with 618,803 MW capacities (IRENA 2018). Moreover, investment in renewable energy will add to the growth by creating jobs as in 2016 this sector employed 9.8 million people. Because of this, the need aroused to check its impact on growth and thus, later on the researchers focused their studies to uncover the significance of alternate energy sources in growth process and debate has converted from energy-growth nexus to renewable energy-growth nexus (Fig. 11.4).

Now, both energy types are playing their role in production activities since countries are moving towards renewable energy because of its ability to fill the gap between its supply and demand. Though its share which accounts for 34% in total supply is less as compared to unclean energy comprised of 82% of global figure in 2015. However, according to IEO (2017) statistics they will grow at 2.8% rate per year as the technological development and incentives by the government body support their use over the years of 2015–2040. Moreover, being fastest growing sources their collective share including hydro power will rise to 57% in 2040.Thus the effect of both fossil (non-renewable) and non-fossil (renewable) energies on the growth of world's economies is needed to capture in order to establish that which contributes more.

4 Role of Formal Institutions and Environmental Protection

The concept of environmental protection is to protect the existing natural resources and natural environment. For this purpose, the individual as well as organization based activities to conserve the natural resources and repair the damage by any means are considered as environmental protection. The environmental protection is also concerned with the protection of ecosystem from unwanted changes in the environment. The protection of the environment and the ecosystem is the dire need of the day all over the world due to rise in industrial production and deforestation. To cope up with the rest of the world, the industrial revaluation was also felt in Pakistan and the gradual rise of industrial production lead to air pollution and water pollution, deforestation, natural resources degradation, urbanization and damage of pretty natural environment. To safeguard the natural resources and environment, the government of Pakistan has passed the legislation to protect the natural resources and environment and also devised a policy regarding energy production and its consumption.

4.1 Pakistan Environmental Protection Act 1997

The Pakistan Environmental Protection Act 1997 was passed by the National Assembly of Pakistan on September 3, 1997, and by the Senate of Pakistan on November 7, 1997. The Act received the assent of the President of Pakistan on December 3, 1997. The text of the Environmental Protection Act 1997 is as follows:

Act No. XXXIV of 1997

> An Act to provide for the protection, conservation, rehabilitation and improvement of the environment, for the prevention and control of pollution, and promotion of sustainable development.

The Punjab Provincial Assembly and Khyber Pakhtunkhwa Provincial Assembly also passed the environmental protection act in 1997 and 2014 respectively. The purpose of these Acts is the practice of protecting the natural environment by individuals, organizations and governments. Its objectives are to conserve natural resources and the existing natural environment and, where possible, to repair damage and reverse trends (Fig. 11.5).

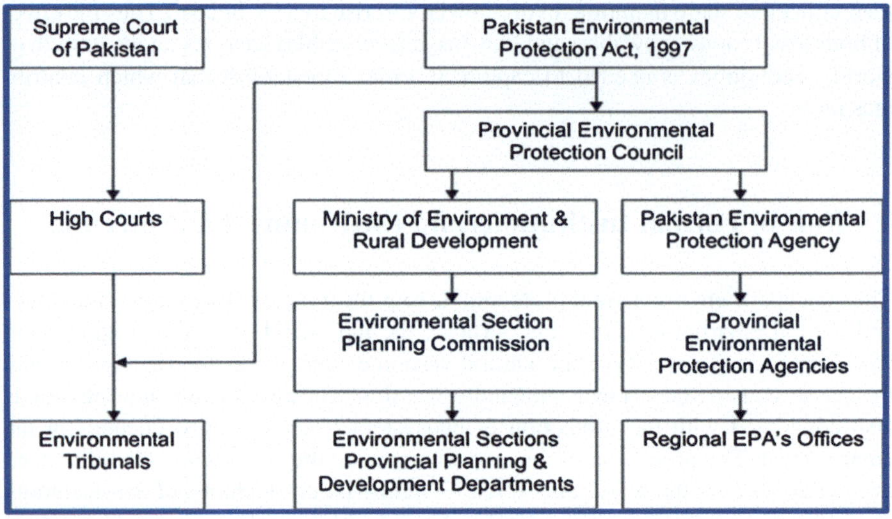

Fig. 11.5 Working hierarchy of Pakistan Environmental Protection Agency

4.2 Pakistan Energy Policy

The energy policy of Pakistan is formulated and determined by the federal, provincial, and local institutional entities in Pakistan, which address the issues of energy production, distribution, and consumption of energy, such as gas mileage and petroleum standards. Energy policy requires the proper legislation, international treaties, subsidies and incentives to investment, guidelines for energy conservation, taxation and other public policy techniques.

Several mandates and proposals have been called over the years to overlook the energy conservation, and reducing the electricity load used by industrial units by 25% during peak hours but no comprehensive long-term energy strategies were implemented. Since 1999, many legislative provisions were adopted for energy conservation including the seeking energy from various renewable energy sources. There is also an intense criticism about the unequal distribution of energy, the irresponsible usage of energy sources, and the country's new plan which is aimed to raise country's dependence on imported oil for power generation to 50% by 2030. After much public criticism, the long-term energy security policy was announced in 2013 through the introduction of equal cutting edge energy transmission network, minimizing financial losses across the energy system and aligning the ministries involved in the energy sector as well as improving the governance of energy sources (Wikipedia 2019).

Studies and policy implementation recommended by Alternative Energy Development Board (AEDB), Water ministry (as policy enforcer), the National Electric Power Regulatory Authority(NEPRA) regulates the energy sources network as well as determining the financial prices of the usage of energy.

Table 11.1 Institutions and their functions in promoting clean and green energy, and environmental protection

Institutions		Major Functions
Formal	Political	Policy making, legislations, forming rules and regulations, power delegations, deciding penalties
	Education	Educating formally in Schools, Colleges and Universities about the rules and regulations, laws, policies of clean energy and sustainable environmental development
	Economic	Providing cost benefit analysis of using renewable energy resources, scarcity of non-renewable resources
Informal	Family	Educating children informally at home about efficient utilization of resources like energy, water, paper, educating about protecting environment, how to save water, how to keep the environment clean, how to save energy
	Religion	Providing religious information regarding energy efficiency, cleanliness, optimal utilization of resources

Government-specific energy-efficiency incentive programs also play a significant role in the overall energy policy of Pakistan. As of 2013 Prime Minister has announced a determined and aggressive energy policy to meet the energy challenges and energy management.

Now the Federal Government has approved renewable energy policy 2019 with the aim to produce cheap and clean energy by using indigenous resource. Giving the targets set in the policy, the Ministry for Power aimed to produce 8000 megawatts of power from renewable sources including solar and wind by 2025 and 20000 megawatts by 2030. By the inclusion of hydel generation the share of clean and green energy will be raised to 60–65%. Under this policy, manufacturing of solar panels and wind turbines will also be locally manufactured and this will give a major boost to the industrial base in the country in coming future (Table 11.1).

5 Conclusion

The role of institutions are very much imperative in energy efficiency, environmental protection and sustainable development. It is clear from the discussion above that the institutions are the humanly devised constraints that helps the member of the society without discrimination. These constraints are equally acceptable for each member of the community. These constraints are formal as well as informal. Both kinds of institutions are important in shaping the society. Formal institutions provides the rules, regulations, policies and legislations, while informal institutions provides ethical and moral norms, habits and cultural values of the society.

Energy being the primary source of industrial production is oxygen for the economic development of the economy. Over utilization of energy- a scarce resource is a big challenge for the economies but also the industrial waste is creating problem for the environment. In this regard the regulations about the environmental policy of Pakistan were discussed in detail. The government of Pakistan has passed a regulation (i) for the protection, conservation, rehabilitation and improvement of the environment, (ii) for the prevention and control of pollution, and (iii) for promotion of sustainable development. Similarly, Pakistan has formulated the national energy policy by the federal, provincial, and local government, which addresses the issues of energy production, distribution, and consumption of energy. So, there is a strong need for sustainable and clean energy resources not only in Pakistan but all over the world. In this regard, the role of institutions is very much critical to provide proper rules and regulations, policies and laws to protect the earth and environment from pollutions, global warming, and deforestation. Also educate the society about efficient and optimal utilization of resources, keeping the environment clean and green. In this way we may be able to become self sufficient in energy production and have a green and clean future.

References

Acemoglu D, Johnson S, Robinson J, Yared P (2008) Income and democracy. Am Econ Rev 98(3):808–842

Anyakoha RU, Eluwa MA (1991) Home management for schools and colleges. Africana Publishers Ltd., Onitsha

Busse M, Hefeker C (2005) Political risk, institutions and foreign direct investment. HWWA discussion paper no. 315

Chandran VGR, Sharma S, Madhavan K (2010) Electricity consumption–growth nexus: the case of Malaysia. Energy Policy 38(1):606–612

Dorgan E (2015) The relationship between economic growth and electricity consumption from renewable and non-renewable sources: a study of Turkey. Renew Sust Energ Rev 52:534–546

Dorgan E, Seker F (2016) Determinants of CO_2 emissions in the European Union: the role of renewable and non-renewable energy. Renew Energy 94:429–439

Engerman SK, Sokolof KL (2002) Factor endowments, inequality, and paths of development among the new world economies (NBER Working Paper, No. 9259). Cambridge, MA: National Bureau of Economic Research

Hodgson GM (2006) What are institutions? J Econ Issues XL(1) 1–25

Hodgson GM (2007) The Revival of Veblenian Institutional Economics, Journal of Economic Issues, 41(2): 324–340, https://doi.org/10.1080/00213624.2007.11507019

Hondroyiannis G, Lolos S, Papapetrou E (2002) Energy consumption and economic growth: assessing the evidence from Greece. Energy Econ 24(4):319–336

Hornby AS (2006) Oxford advanced learner's dictionary of current English. Oxford University Press, Oxford

Huang BN, Hwang MJ, Yang CW (2008) Does more energy consumption bolster economic growth? An application of the nonlinear threshold regression model. Energy Policy 36(2):755–767

IEA (2017) CO_2 emissions from fuel combustion highlights 2017 edition. Online available at: https://www.iea.org/publications/freepublications/publication/CO2EmissionsfromFuelCombustionHighlights2017.pdf

IEO (2017) International energy outlook 2017. Online available at: https://www.eia.gov/outlooks/ieo/pdf/0484(2017).pdf

IRENA (2018) International Renewable Energy Agency. Renewable capacity statistics 2018. Online available at: https://www.irena.org/-/media/Files/IRENA/Agency/Publication/2018/Mar/IRENA_RE_Capacity_Statistics_2018.pdf

Jaforullah M, King A (2015) Does the use of renewable energy sources mitigate CO_2 emissions? A reassessment of the US evidence. Energy Econ 49:711–717

Kaufmann D, Kraay A, Mastruzzi M (2009) Governance matters VIII: aggregate and individual governance indicators for 1996–2008, World Bank policy research working paper no. 4978. World Bank, Washington, DC

Kaufmann D, Kraay A, Mastruzzi M (2010) The worldwide governance indicators: methodology and analytical issues, World Bank policy research working paper no. 5430. World Bank, Washington, DC

Kaufmann D, Kraay A, Zoido P (2011) Governance Matters II: updated indicators for 2000-01, World Bank policy research working papers no. 2772

Keefer P, Knack S (1997) Why don't poor countries catch up? A cross-national test of an institutional explanation. Econ Inq XXXV:590–602

Lee CC, Chang CP (2007) Energy consumption and GDP revisited: a panel analysis of developed and developing countries. Energy Econ 29(6):1206–1223

Lin B, Moubarak M (2014) Renewable energy consumption–economic growth nexus for China. Renew Sust Energ Rev 40:111–117

Little W, McGivern R (2012) Introduction to sociology, 2nd Canadian edition. OpenStax College, Rice University, Houston

Lotempio DJ (2002) The concept of family in family concept and issues. In James Robbinson's Starman. Clan Walker Wet

Matthews RCO (1998) The economics of institutions and sources of growth. Econ J 96(384):903–918

Menyah K, Wolde-Rufael Y (2010) CO_2 emissions, nuclear energy, renewable energy and economic growth in the US. Energy Policy 38(6):2911–2915

North DC (1990) Institutions, institutional change, and economic performance. Cambridge University Press, New York

North DC (1994) Economic performance through time. Am Econ Rev 84(3):359–368

North DC, Thomas RP (1973) The rise of the Western world. Cambridge University Press, Cambridge

RGSR (2017) Renewable global status report 2017. Online available at: http://www.ren21.net/wp-content/uploads/2017/06/17-8399_GSR_2017_Full_Report_0621_Opt.pdf

Sadorsky P (2009) Renewable energy consumption and income in emerging economies. Energy Policy 37(10):4021–4028

Scott WR (2008) Institutions and organizations: ideas and interests, 3rd edn. Sage, Los Angeles

Shahbaz M, Loganathan N, Zeshan M, Zaman K (2015) Does renewable energy consumption add in economic growth? An application of auto-regressive distributed lag model in Pakistan. Renew Sust Energ Rev 44:576–585

Soytas U, Sari R (2003) Energy consumption and GDP: causality relationship in G-7 countries and emerging markets. Energy Econ 25(1):33–37

Squalli J (2007) Electricity consumption and economic growth: bounds and causality analyses of OPEC members. Energy Econ 29(6):1192–1205

Wang Y, Wang Y, Zhou J, Zhu X, Lu G (2011) Energy consumption and economic growth in China: a multivariate causality test. Energy Policy 39(7):4399–4406

Warr BS, Ayres RU (2010) Evidence of causality between the quantity and quality of energy consumption and economic growth. Energy 35(4):1688–1693

WB (2019) World Bank energy progress reports 2019 highlights. Online available at: https://trackingsdg7.esmap.org/downloads

Wikipedia (2019) Energy policy of Pakistan. https://en.wikipedia.org/wiki/Energy_policy_of_Pakistan. Accessed on 1 Sept 2019

Chapter 12
Access and Limitations to Clean Energy Use in Nigeria

Ayobami Abayomi Popoola and Bamiji Michael Adeleye

Abstract The energy situation in Nigeria has always been a paradox. Despite having abundant energy resources in the country, widespread energy poverty is faced by the citizenry. About 60% (74 million) are not served with electricity, while another 94% (171 million) do not have access to clean energy. In a bid to cushion the effect of energy poverty, households and business enterprises in Nigerians relied on the constant use of generators, which is not eco-friendly, is costly and harmful to human health. The study adopted a think-through thematic methodological analysis, which involves the mapping of the country's potential clean energy sources. Thematic literature reviews were integrated to investigate the clean energy experience in the country. Taking into consideration the geopolitical classification of the country, interviews were conducted to examine the energy conditions in the country and the limitation to the maximization of clean energy within their locality, as well as the perception of its acceptability within the country. Study findings show that the main factors limiting the use of clean energy in Nigeria are exorbitant costs of installation and maintenance, inadequate investment in the energy sector; non- involvement of the private sector, and the subsidies granted to generators of energy from fossils.

Keywords Access · Clean energy · Rural · Urban · Local renewable resources · Electricity

A. A. Popoola (✉)
Department of Town and Regional Planning, School of Built Environment and Development Studies, Shepstone Building, Howard Campus, University of KwaZulu-Natal, Durban, South Africa

B. M. Adeleye
Department of Urban and Regional Planning, Faculty of Environmental Technology, Federal University of Technology, Minna, Niger State, Nigeria

H. Qudrat-Ullah, M. Asif (eds.), *Dynamics of Energy, Environment and Economy*, Lecture Notes in Energy 77, https://doi.org/10.1007/978-3-030-43578-3_12

1 Background to the Study

Cities globally are experiencing rapid urbanization. It is envisaged that by the year 2050, the world population would have risen by 70%, with cities attaining sizes not experienced before (United Nations (UN) 2017; Food Agriculture Organisation (FAO) 2009). This implies that the global demand for energy will continue to increase since man's daily activity depends on energy. Energy is seen as an essential component of development that affects the peace, security, well-being, physical environment, and the socio-economic growth of a nation (Jatau et al. 2006; Urban 2009; Ajah 2013). The energy situation in Nigeria has always been a paradox (Eleri et al. 2012) because despite having abundant energy resources in the country, widespread energy poverty is faced by the citizenry. Approximately 60% (74 million) are not served with electricity, while another 94% (171 million) do not have access to clean energy (REN21 2018).

In a quest to achieve economic, technological, and industrial advancement, cities in both developing and developed countries depend primarily on energy from fossil fuel and other sources. Nnaji et al. (2010) reported that in a bid to cushion the effect of energy poverty, households and business enterprises in Nigeria relied much on the constant use of generators. This alternative (generator or fossil fuel) energy use is not eco-friendly, costly, and harmful to human health. REN21's (2016) report stated that 78.3% of global energy consumption was produced from fossil fuel, while 19.2% and 2.5%, accounts for renewable and nuclear energy, respectively. Studies (Uduma and Arciszewski 2010; Oyedepo 2014) have identified that the continuous use of generators and fossil fuel poses a massive obstacle to sustainable economic and social development, both in urban areas and at the grassroots level.

The increasing rate of global warming has been attributed to cities' generation of greenhouse gases owing to their enormous dependence on fossil fuel (Satterthwaite 2008). This shows that the colossal dependence on fossil fuel by cities over the world is unsustainable in terms of its global impact and depleting resources. Amidst this rising global energy use, Merem et al. (2017) opined that access to renewable energy would catalyze global and regional economic advancement. Renewable energy or clean energy is a form of energy generated from the natural resources that are constantly replenished. This form of energy includes Solar, hydroelectricity, wind, tides, and geothermal energy. According to Newman et al. (2011), renewable energy enables cities to create a healthy and liveable environment while minimizing the use and impact of fossil fuels.

Despite having high potentials of renewable energy in Africa, the continent is yet to fully harness these potentials, thus making access to energy a challenge (Kerrigan 2001). For instance, 58% of the energy supply of Africans comes from fuelwood and charcoal, and these sources (fuelwood and Charcoal) rival other sources of industrial energy such as electricity (Specht et al. 2015). This scenario is not far-fetched in terms of the situation of access to clean energy in Nigeria. With the abundance of fossil fuel and renewable energy resources in Nigeria, Nigerians still experience acute energy poverty. This connotes that Nigerians either lack access to clean energy sources or have to cope with insufficient choice in accessing adequate, affordable, reliable, high quality, and environmentally energy services to support economic and human development (Nnaji et al. 2010).

Reports are that 55% of Nigerians are wholly reliant on charcoal, fuelwood, biomass, and animal waste for heating and cooking (Maduka 2011). This signifies that Nigerians are climbing down the energy rungs. REN21 (2018) affirms this claim by asserting that 171 million people in Nigeria do not have access to clean cooking energy, while another 74 million do not have access to electricity in the country. The "sporadic" access to clean energy experienced in Nigeria threatens the realization of goal number seven of the sustainable development goals.

Different studies and approaches have been taken by various scholars to address the issues of sustainable energy in Nigeria. Most of the studies conducted have dwelt more on exploring the potentials and conditions for renewable energy adoption in Nigeria. Additionally, ample studies carried out on sustainable energy in Nigeria were conducted in the field of engineering, with a sole focus on access to clean energy in the urban areas, thus neglecting the social aspect of the rural (grassroots) dwellers. Slightly aligning with the multi-tier energy matrix, this study using a bird's eye view, the country-wide experience shall investigate conventional energy access in Nigeria, access to clean energy, and also identify the factors that limit the use of clean energy in the country.

The driving keyword and question of this study is if local resources can be maximised to alienate a communal energy crisis. The study justification lies in and implies that the study on access to clean energy services in Nigeria with a focus on rural areas will help ameliorate the issues that accompany energy poverty at the grassroots level. This study will guide policy-makers to put in place necessary mechanisms that will aid the uptake of clean energy service development in Nigeria. This will, in turn, help in the swift realization of the SDGs across communities and the entire country at large.

2 The Energy Situation in Africa and Nigeria

The existence of a global energy crisis has been reported by Vahid-Pakdel et al. (2017). The heavy reliance of developing countries on fossil fuels as a significant source of energy presents a negative consequence of increased CO_2 emission (Jebli and Youssef 2017). In North Africa, it was reported that 92% of the energy used is from fossil fuel (coal, gas, and oil), with only 8% from renewable energy sources (United Nations Economic Commission for Africa 2012). The importance of energy in the socio-economic, political, and industrial development of any nation cannot be over-emphasized (Oyedepo 2014) because energy provides an essential ingredient for virtually all human activities ranging from transportation to communication, schools, industrial activities, agricultural activities, and other domestic activities.

According to the UN (n.d), energy is believed to be central to every major challenge and opportunity the world faces today, be it challenges and opportunities for jobs, security, climate change, food production increasing incomes. Steurer et al. (2016) reported that electricity project challenges in Africa could be traced to political instability and limited private sector involvement, which can, in turn, be traced to a highly burdened regulatory framework. Although Africa is characterized by a difference in the level of connection and access to electricity, the common denominator is that many households remain unconnected to electricity (Mas'ud et al. 2015; Alnaser and Alnaser 2011). According to UNDESA (2004), one limitation to Africa's imbalance and the poor state of electricity connectivity is the imbalance between electricity generation and consumption.

Limited access to electricity services and power infrastructure is experienced in Africa (Taliotis et al. 2016). Across the rural and urban spaces, the dichotomy in electricity and energy access remains a common reality. Aliyu et al. (2018) wrote that Africa (especially Nigeria, Libya, Mozambique, Algeria, and Egypt) is blessed with a potential for energy from non-renewable sources, many of which remain under-utilized and optimized. Taliotis et al. (2016) reported that variability as much as low access is reported across Africa (85% national electrification in South Africa) and more evidently experienced in sub-Saharan Africa (3% in the Central African Republic and 4% in Chad). Hence the spatial disparity across urban and rural communities cannot be ignored. For example, in Cameroon, 88% of urban areas and 17% for rural communities are connected. Trotter's (2016) study on rural electricity experience from 46 sub-Saharan African countries, which included Ghana, Swaziland, Uganda, Senegal, and Rwanda, shows that political, economic, and demographic variables were a significant determinant of rural electrification. The study identifies the roles played by country democracy in increasing rural electrification and directly reducing electrification inequality between rural and urban areas. Furthermore, and as narrated by Pachauri et al. (2012), Fig. 12.1 presents the level of rural connectivity to electricity, which differs across global and African spaces.

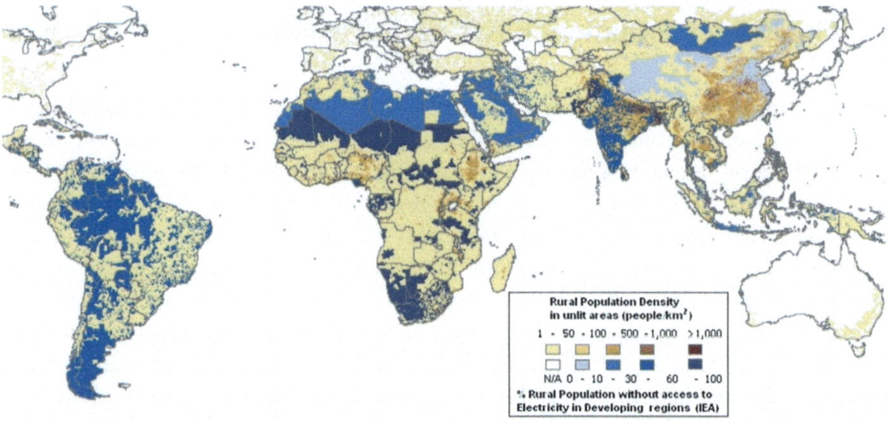

Fig. 12.1 Global Rural Electricity Access. (Source: Pachauri et al. 2012)

In Africa, sub-Saharan, Eastern, and Central African countries are characterized by the least rural population electricity accessibility ratio (Fig. 12.1).

According to Popoola and Magidimisha (2019), the solution to the increasing energy demand for Africa is for African counties, with a particular focus on Nigeria, to fully maximise and explore the integration of renewable clean energy sources into the national grid and for sparsely arranged rural households and urban poor.

3 Clean Energy in Africa and Nigeria

The global energy crisis (Vahid-Pakdel et al. 2017) led to the maximization of and diversification into the use of renewable and clean energy sources. Various scholars have carried out studies on access to clean energy. These tend to share the same perception that clean energy is a crucial component to achieving sustainable development. Oyedepo (2014) further opined that working with the goal number seven of the SDGs (affordable and clean energy) is essential as it interlinks with other sustainable development goals, which include eradicating extreme poverty and hunger; achieving universal primary education; promoting gender equality and empowering women; reducing child mortality; improving maternal health; combating diseases; and ensuring environmental sustainability.

Increased access to clean energy services in Nigeria has been identified to help ameliorate the issues that accompany energy poverty in the country. In a study carried out by Osunmuyiwa and Kalfagianni (2017) on the adoption and variation in renewable energy in Nigeria's 36 States using three analytical lenses (niches, regimes and landscape), it was revealed that a combination of regime and landscape characteristics enables states to overcome dependence on fossil fuel while triggering the adoption of renewable energy. The study reveals that States with high income and a regime featuring institutions and coalitions supporting transitions establish themselves as pioneers, while States with medium/low income and a regime characterised by a weak pro-renewable energy political coalition support emerge as laggards.

Explaining the wealth and income effect, Edomah (2016) examined the barriers of cost, pricing, legal, regulatory, and market performance to explain sustainable development in Nigeria. The study concluded that policy recommendations that will bring about infrastructure upgrades, curb pipeline vandalism, increased fossil fuel consumption, increasing demand, and increasingly scarce resource constraints are imperative to addressing the economic, social, and environmental dimensions of sustainable energy. Places forwarding a policy of bilateral arrangements in enhancing access through clean energy sources, Jebli and Youssef (2017) identified that an example of the clean energy initiative between the European Union (EU) and some North African countries which is expected to improve clean and renewable energy within the regional markets is the Mediterranean solar plan (MSP).

In the same vein, Oyedepo's (2014) opinion on energy and sustainable development in Nigeria is that energy policy and strategy for delivering access to modern elemental energy need to be put in place by the government. Akorede et al. (2017) reported that overcoming the energy poverty in Nigeria can only be achieved through an increased exploration of the abundant renewable energy resources in the country across energy demanding sectors. It is therefore imperative for the Federal Government of Nigeria to develop a "Sustainable Clean Energy Future Framework" aimed at increasing the deployment and innovation of renewable energy in the country (Oyedepo 2014).

In Nigeria and Cameroon, despite solar radiation and good wind speed, poor leadership and weak governance have limited the maximization of the available clean energy sources and connection to the national grid (Mas'ud et al. 2015) for more commercial usage. Buttressing this, the International Energy Outlook report by the US Energy Information Agency (2013) reported that Africa's reliance on conventional energy sources of biomass and hydroelectricity generation is a limitation to her potential improved access to increasing energy demands. Using sources from SolarGIS (2011) analytics, Mas'ud et al. (2015) presented the solar energy potential of Nigeria using the country's irradiation map (see Fig. 12.2).

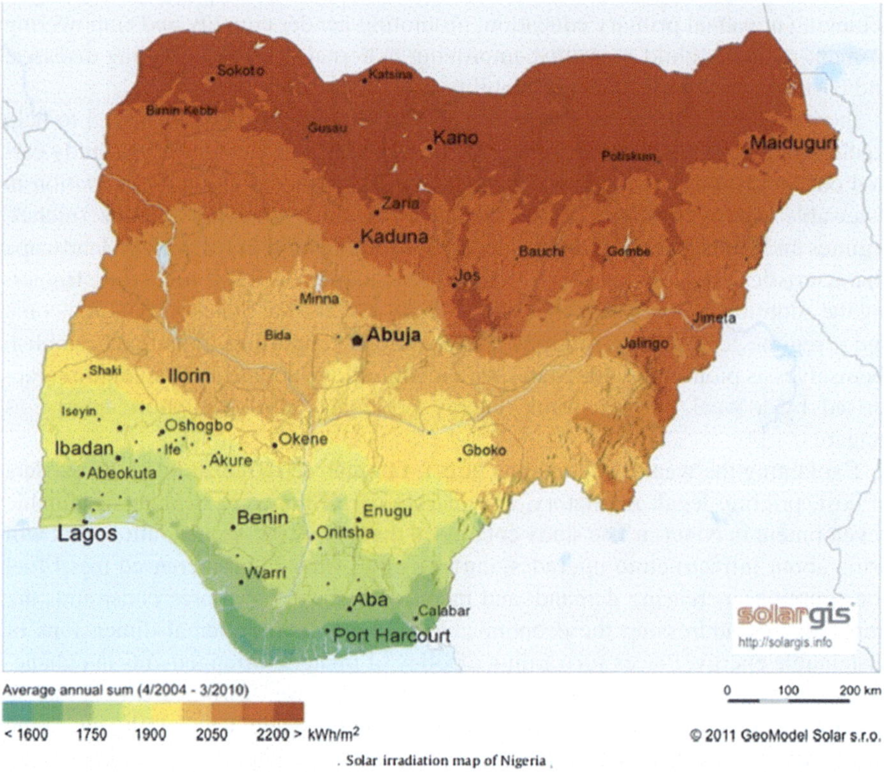

Fig. 12.2 Solar Energy Potential of Nigeria showing Sun Irradiation. (Source: Mas'ud et al. 2015 data from SolarGIS 2011)

Popoola and Magidimisha (2019) argued that with the abundance of solar radiation in Nigeria, the country limits its use to some politically motivated "streetlight show-off projects." They argue that few households, many of which are within urban areas, use solar energy sources. Aliyu et al. (2018) suggested that with South Africa, Egypt, and Nigeria being the major African energy users, renewable sources such as the sun (photovoltaic and solar thermal), hydro and wind fuel can be optimized for African consumer demand. In Northern Africa, Jebli and Youssef (2017) and the United Nations Economic Commission for Africa (2012) reported that with an over 8% annual increase in the demand for energy and, fossil fuel domination in energy sources, there is a need for an energy mix to explore the use of renewable and clean energy fully.

4 Methodology and Materials

This study involves a mixed-method approach. Primary data was obtained from the interviews conducted, and secondary data was gathered from monographs and other existing literature. The methodology involves a Think-Through Thematic Analysis, which involves the mapping of the country's potential clean energy sources. In analyzing, the geopolitical classification of Nigeria was taken into consideration in the analysis of the potential of maximizing clean energy for the easing of energy deficiency in the country. A Thematic literature review shall be integrated to investigate the clean energy experience in the country. Interviews were conducted for twenty (20) conveniently sampled stakeholders across Nigeria on the conditions of electricity, limitation to the maximization of clean energy within their locality, and perception of its access across the country (Table 12.1). The interviewee responses were transcribed and analysed using thematic analysis. Electricity access across the country was investigated and mapped using secondary data from the Nigerian Bureau of Statistics.

Table 12.1 Sampled respondents for the interview

S/n	Location/geopolitical zone	Occupation
1.	Northern Nigeria	1 Academic/lecturer
2.	South-Western Nigeria	2 Private consultant and 1 IT personnel
3.	South Nigeria	1 Spatial scientist and health consultant
4.	South-Western Nigeria	1 Rural farmer and solar panel user
5.	South-Western Nigeria	2 Researchers
6.	Eastern Nigeria	2 Academics/lecturer
7.	South-Western Nigeria	1 Planner
8.	Northern Nigeria	1 Planner
9.	Middle belt of Nigeria	1 Private consultant
10.	South-Western Nigeria	4 Peri-urban/Urban dweller
11.	South-Eastern Nigeria	3 Peri-urban dweller

Source: Authors' Compilation (2019)

5 Access to Energy (Electricity) in Nigeria

The interviewed respondents report a mix-grill experience of the electricity situation in Nigeria. Responding to the situation within the country, the energy situation in Nigeria as being deficient in energy resource generation investments and legislation was identified. The Nigerian energy sector generates far less than is required for daily use in the country, thus tremendously affecting other sectors such as the industrial, power and transportation sector that run predominantly on electricity defined energy. Nigeria's extreme electricity deficiency has been argued to be multi-facet as it represents a vehemently grossly below the national average demand. Participant 11 opined that financial, structural, and political factors cause this multifaceted deficiency, none of which are mutually exclusive. Analysing the deficiency and the geographical place of deficiency, it was identified that the urban poor and isolated rural areas (in the words of Participant 10 "the masses") are the most affected by the electricity deficiency.

Participant 4 stated that there is a "*…general knowledge that the situation of energy in Nigeria is appalling. Though there are improvements on multiple fronts in the chain of production and supply of energy, the sustainability of such improvement is virtually impossible…*" This statement presents a 'seem positive' energy experience and attempts to improve the condition.

As presented by the Nigeria Bureau of Statistics (2015) and represented in Fig. 12.3, the level of access to electricity in the country continues to decline and is at the lowest in the Northern part of Nigeria.

The Western part of the country is characterised by and recorded the highest access to electricity. Reinstating the idea that poor electricity energy access across the Nigerian space cannot be unexpected based on the "welfarist approach" with which the government has handled the generation and consumption mechanism. This approach is continually characterised by massive subsidization of electricity and energy changes, pricing, and costing systems.

Awoyinfa et al. (2019) reported the thoughts of stakeholders (such as the former Governor of the Central Bank of Nigeria; the Nigerian Employers' Consultative Association; the Chartered Institute of Bankers of Nigeria; the Corporate Affairs Director; the Manufacturers Association of Nigeria; the Centre for Social Justice; and other academia) that the subsidy implementation on electricity and fuel remains unsustainable and unrealistic taking into consideration the cost implication of the subsidy on the country's debt profile. The Daily Trust (26 June 2019) reported that the complete removal of the subsidy, but in a gradual process, remains the route to solving the electricity distribution, generation, and consumption variance. This argument aligns with the views of participants who attributed the below-par energy condition of the country to its subsidization.

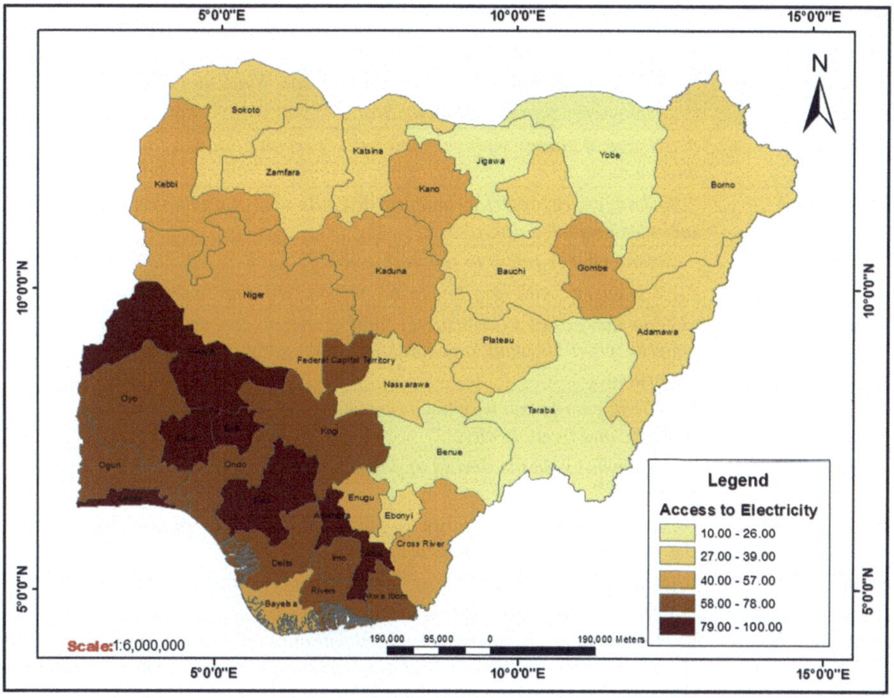

Fig. 12.3 Access to electricity in the Nigerian States (in percentage %). (Source: Authors' Mapping. Adopted from the Nigerian Bureau of Statistics, 2015)

6 Clean Energy Access and Potential Distribution in Nigeria

6.1 Understanding clean energy and access

The study attempted to investigate the understanding of the conveniently sampled respondents of what clean energy entails. The responses present four main themes, which are based on typology, limited or no waste, the capacity to not emit CO2, and the source of generation from natural sources (Table 12.2). From the responses received, clean energy is an eco-friendly form of energy that is naturally derived from sources such as sunlight, wind, and hydro and is pollution-free with no greenhouse gas emissions such as CO2.

Having presented a general perception of the understanding of interviewees of clean energy, the study attempted to evaluate the country's access to clean energy. The evidence reveals that accessibility, which is rated to be poor, to clean energy remains limited owing to the significant financial cost of installation (Participant 1); and is exacerbated by low/under-efficient facilities (Participant 3); few providers (Participant 4); is capital-intensive; lack of affordability and poor awareness (Participant 5); and weak purchasing power (Participant 13). Narrating the reasons

Table 12.2 Understanding of what clean energy entails

Participant	Response to what clean energy means
Participant 1	"…*Energy from renewable sources like Solar, Wind, and Hydro*…"
Participant 2	"…*The energy that is carbon neutral*…"
Participant 3	"…*Usage of energy with very low or no carbon pollution to the environment*…"
Participant 4	"…*Clean energy refers to an earth-friendly way of producing energy. This includes biofuel, solar*…"
Participant 5	"…*clean energy refers to renewable energy, such as wind and solar*…"
Participant 6	"…*The energy with very low or no carbon pollution to the environment*…"
Participant 7	"*Clean energy has to do with the kind of energy generated from natural sources, vis-à-vis wind, and sunlight. Moreover, little to no waste is generated*…"
Participant 8	"…*Clean energy is a way towards achieving a sustainable, healthy, friendly, and liveable city*…"
Participant 9	"…*The energy that is devoid of pollution or at low level / environmentally friendly*…"
Participant 10	"…*clean energy is the energy supplied without any detrimental effects on the people and the environment at large*…"
Participant 11	"…*The clean energy produced through means that do not pollute the atmosphere, e.g., sun, water*…"
Participant 12	"…*Clean energy is energy generated without polluting/causing harm to the environment. It can also be said to be an energy void of carbon emissions. Examples of clean energy include solar energy and wind energy*…"
Participant 13	"...*Energy from the sun and others*..."

Source: Authors' Analysis (2019)

why access can be limited, Participant 4 reported that with the availability of affordable solar energy solutions from the leading providers of Mobile Telecommunication Network (MTN), Solar (Yellow Box) and Novel Ltd., it is safe to assume that clean energy is relatively available for access by Nigerians.

Although it was identified that the long-term cost implications of the use of clean energy are beneficial. Based on his experience, Participant 5 reported that, "…*true-green energy is expensive to set up*…" The context of "true-green energy," as reported by the interviewee, was based on experiences of installation over the years from less technically capable companies, and sometimes that clean energy facility becomes faulty before the expected warranty period. Participant 7 stated that the installation is not subsidized, so most households and individual bears the full cost of supplementary gadgets such as batteries, Solar panels and cables and in most instances, installation technicians' end up buying sub-standard materials which make the clean energy installed not sustainable.

Participant 11 and 12 states that why there is an increase in the exposure of Nigerian society to clean energy usage; only a few have easy access to it. Participant 10 said that clean energy access to the Nigerian masses is miserable. She observed that "…99.9% of the energy used in the country is detrimental to our environment because there is emission of carbon monoxide. Be it cooking with gas, cooking with stoves, automobile emissions, and even our crude oil extraction points generate much carbon monoxide, which is detrimental to the environment…".

Bringing to fore the place of energy diversification amongst Nigerians, an interviewee observed that while the reliance on solar energy has increased over the past 3 years, solar energy is still not a product for the common man, as fuel and diesel generators locally tagged "…*I pass my neighbor*…", with all their carbon emissions and noise are still primarily used as alternatives to sporadic power from the national grid. Questioning the sustainability of some of these alternative sources, such as generators, Oyedepo et al. (2019) was of the view that increasing demand for energy; inadequate generation capacity from conventional sources; and dilapidated and limited electricity extension/distribution infrastructure mainly to rural areas are limitations to clean energy use in Nigeria. The study states that the promotion of renewable energy remains the route to promoting industrialization and sustaining the nation's economy.

6.2 *Potential Clean Energy Sources*

Despite the massive potential of renewable energy resources that Nigeria is endowed with, energy issues still abound in the country. Authors (Nnaji et al. 2010; Bamisile et al. 2017) believe that harnessing the full potential of renewable energy would go a long way to reducing the current energy "poverty" experienced in the country. Nigeria's renewable energy resources can be classified into four main types, namely are solar energy, hydropower, wind energy, and biomass/biogas. While the potential towards the optimization of these energy sources is undoubted, the capacity of the energy generated from these resources and sources differs in time. Table 12.3 shows the renewable energy generated in the country between the years 2008 and 2017. Table 12.3 reveals that between 2008 and 2017, a total of 20,271 Mw of energy was generated from renewable energy sources in Nigeria. The renewable energy generated was classified into off-grid and on-grid; total capacity of 131.24 Mw was generated off-grid while 20,140 Mw was generated on-grid.

6.3 *Hydropower*

The only renewal energy used for commercial power generation in Nigeria is hydropower (Bamisile et al. 2017). The hydropower state in Nigeria is classified into small and large hydropower stations. According to the Renewable Electricity Action

Table 12.3 Renewable energy generated in Nigeria between 2008 and 2017

	Renewable energy sources					Total off-grid generation
Year	Hydropower	Solar	Wind	Biomass/biogas	Total	Renewable energy
2008	1941	na	2	na	1943	2.200
2009	1941	na	2	na	1943	2.400
2010	1941	na	2	na	1943	2.400
2011	1941	na	2	na	1943	2.400
2012	2042	15	2	na	2060	18.540
2013	2042	15	2	na	2060	18.740
2014	2042	16	3	na	2061	19.940
2015	2042	17	3	na	2062	21.340
2016	2042	18	3	na	2063	21.540
2017	2042	19	3	na	2064	21.740
Total (MW)	**20,016**	**100**	**24**	**na**	**20,140**	**131.24**

Source: International Renewable Energy Agency (2018)

Plan (2006), hydropower stations that generate less than 30 Mw are considered to be small, while hydropower stations that generate above 30 Mw are regarded as large hydropower stations. Nigeria has six major small hydropower stations with an aggregated capacity of 30 Mw and three major large hydropower stations with an aggregated capacity of 1930 Mw (Table 12.4).

The six major small hydropower stations were all constructed by the National Electricity Supply Cooperation Limited (NESCO), a private company (Osunmuyiwa and Kalfagianmi 2017), while the Federal Government of Nigeria constructed the three major large hydropower stations. Aderoju et al. (2017) opined that the small hydropower stations are more eco-friendly than the large hydropower stations because they do not require severe deforestation, rehabilitation, and submergence.

In harnessing the vast potential of hydropower in Nigeria (Fig. 12.4), five additional large hydropower stations are under construction in the country. These include the Zungeru Hydropower plant (700 Mw), Mambilla Hydropower plant (3050 Mw), Kashimbilla Hydropower plant (40 Mw), Dadin Kowa Hydropower plant (40 Mw) and the Gurara Hydropower plant (30 Mw). Of all the large Hydropower stations under construction, the Mambilla Hydropower plant would be Nigeria's biggest power plant. This study revealed that construction work is yet to commence at the proposed site of the Mambilla hydro-power plant since the project was approved in 2017. Adetayo (2017) envisaged that the Mambilla Hydropower plant, when completed, will reduce the reliance on fossil fuel and cut the issues of climate change in Nigeria.

6.4 Solar Power

Solar energy can be described as energy generated from the sun in the form of electric or thermal energy (Energysage 2019). Solar energy can be captured either through the use of a photovoltaic solar panel or solar thermal conversion (Aderoju et al. 2017). Studies have shown that Nigeria is situated within a high sunshine belt

Table 12.4 Small and large hydropower stations in Nigeria

No.	Hydro power station	State	Installed capacity (MW)
Small hydropower station			
1	Bagel I	Plateau	1
	Bagel II	Plateau	2
2	Ouree	Plateau	2
3	Kurra	Plateau	8
4	Lere I	Plateau	4
	Lere II	Plateau	4
5	Bakalori	Sokoto	3
6	Tiga	Kano	6
Total			**30**
Large hydropower station			
1	Kainji	Niger	760
2	Shiroro	Niger	600
3	Jebba	Niger	570
Total			**1930**

Source: Renewable Electricity Action Plan (REAP) (2006)

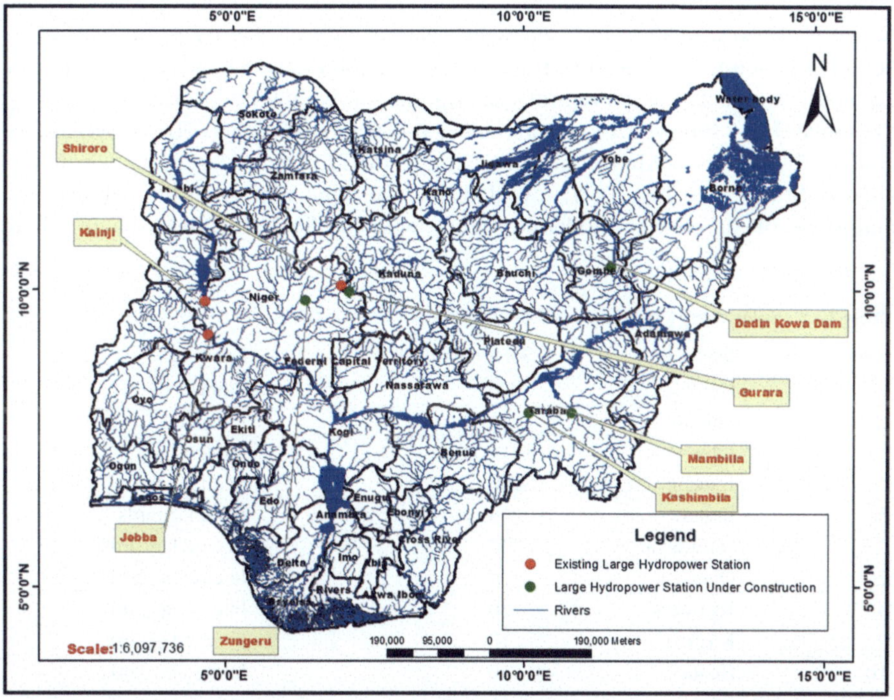

Fig. 12.4 Hydropower Sites and Potentials in Nigeria. (Source: Authors' Mapping 2019)

and the solar radiation is equitably distributed across the States of the federation (Oyedepo 2014; Adebayo 2014; Bamisile et al. 2017). According to Oyedepo (2014), annual average solar radiation ranges from 12.6 MJ/m^2-day (3.5 kWh/m^2-day) at the coast to about 25.2 MJ/m^2-day (7.0 kWh/m^2-day) in the far north. This implies that Nigeria's solar potentials, if adequately harnessed has the possibility of generating 1850 × 10^{3}GWh of solar electricity per year, which is a hundred times the current electricity grid consumption in Nigeria (Uzoma et al. 2011 in Adebayo 2014).

Amidst the vast potential of solar radiation that abounds in Nigeria, only 12 States out of the 36 states in the country have been able to generate electricity from Solar energy. These states include; Delta, Sokoto, Lagos, Bauchi, Edo, Enugu, Nasarawa, Benue, Jigawa, Bayelsa, Katsina, and Ogun (see Table 12.5). Of all these States, Osunmuyiwa and Kalfagianni (2017) referred to Delta, Sokoto, and the Lagos States as the pioneer states because of their comprehensive policy on renewable energy. The laggard states were referred to as states with no visible form of renewable energy adoption or programs (Osunmuyiwa and Kalfagianni 2017).

6.5 *Wind Energy*

Nigeria falls within the moderate wind regime with an energy reserve at 10 m height, which implies that some regions in the country have a wind regime between 1.0 and 5.1 m/s. The wind regimes in Nigeria can be classified into four, namely 44.0 m/s; 3.1–4.0 m/s; 2.1–3.0 m/s; and 1.0–2.0 m/s (Oyedepo 2014). Wind data collated by Nigerian Metrological Agency (NIMET) from 44 stations across the Nigeria States (See Fig. 12.5) revealed that the wind regime in the country lies majorly between poor to moderate regime with the Southern States having their

Table 12.5 Solar energy generated in Nigerian states

S/no	States	Solar energy (KW)	Ranking
1	Delta	8000	Pioneer
2	Sokoto	2045	Pioneer
3	Lagos	1814	Pioneer
4	Bauchi	600	Semi laggards
5	Enugu	64	Semi laggards
6	Nasarawa	30	In case laggards
7	Ogun	5	In case laggards
8	Jigawa	10	In case laggards
9	Kastina	10	In case laggards
10	Benue	10	In case laggards
11	Edo	36	In case laggards
12	Bayelsa	10	In case laggards

Source: Osunmuyiwa and Kalfagianni (2017)

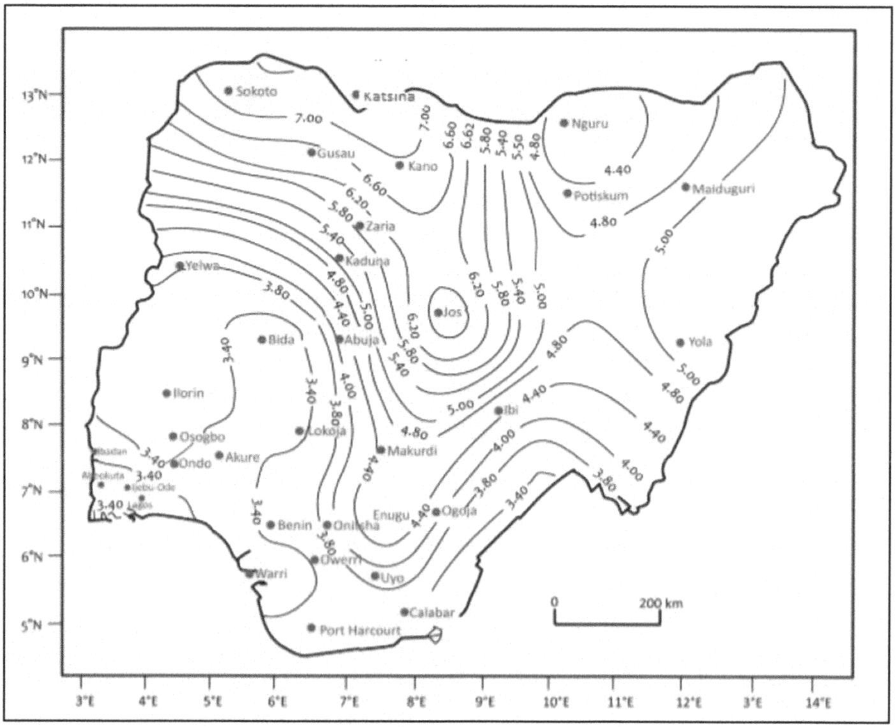

Fig. 12.5 Isovents (M/s) Across Nigeria. (Source: NIMET 2009)

mean wind profile at 10 m height between 3.0 and 3.5 m/s and the Northern States having a mean wind speeds of between 4.0 and 7.5 m/s (Okoro et al. 2007 in Adejoru et al. 2017). Oyedepo (2014) also asserted that the coastal areas from Lagos State through Ondo, Delta, Rivers, and Bayelsa States to Akwa-Ibom State have strong potential for wind energy throughout the year. Although the coastal and Northern areas of Nigeria are perceived to have great potentials of wind energy. Despite this, little effort has been made by these States to harness the potentials of wind energy. In a study carried out by Osunmuyiwa and Kalfagianni (2017), only Enugu State was able to generate 26 kw in 2010, amongst the States with high wind speeds.

6.6 *Biomass/Biogas*

Actively or passively, Biomass is the largest renewable energy source used in Nigeria (Bamisile et al. 2017). The biomass resources identified in Nigeria include wood, forage grasses and shrubs, animal waste, agricultural and forest residue, municipal and industrial activities as well as aquatic biomass. Biogases are fuel gotten from plant biomass which is often fermented by anaerobic bacteria to produce a very versatile and cheap fuel (Nnaji et al. 2010). To improve biomass utilization in Nigeria, the government has encouraged the use of biomass in the transportation sector, which prompted the 2007 bio-ethanol policy that calls for a 10% inclusion of ethanol into petroleum products in the country (Oyedepo 2014). It is estimated that around 61 million tonnes/year of animal waste can be obtained and which translates into about 83 million tonnes/year of crop residue (Oyedepo 2014). The vast potential of biomass/biogas in Nigeria does not have a significant effect on power generation. It was revealed that of the 36 States of the federation, only Lagos, Oyo, Sokoto, Niger, and Enugu had explored the biomass potentials (Osunmuyiwa and Kalfagianni 2017). The study further revealed that 17.9 Kw, 10 Kws, 35 Kws, and 63 Kws of electricity were generated in Lagos, Enugu, Sokoto, and Niger States, respectively. Oyo state was the only State to have generated 500 Kw within the period under study (2010–2014).

6.7 *Household Electrification Rate in the Six Geo-Political Zones of Nigeria*

The household electrification rate in the six geopolitical zones of Nigeria is explained in Table 12.6. The table revealed that 55.6% of the total households in Nigeria have access to electricity, while 44.2% do not have access to electricity. This implies that there is tremendous energy poverty in Nigeria. This is evident in the sporadic power supply experienced in the Country. Oyedepo (2014) also affirmed this claim. According to Oyedepo (2014), household access to electricity services in Nigeria is low, and about 60% of the population (over 80 million people) are not served with electricity. Table 12.6 explains the household electrification rate across the six geo-political zones of the federation.

Table 12.6 Household electrification rate in the geo-political zones of Nigeria

Zones	Have electricity (%)	No electricity (%)	Missing	Household surveyed
North Central	48.7	51.2	0.1	5942
North East	29.3	70.1	0.3	5115
North West	42.2	57.7	0.1	9992
South East	66.4	33.6	0.0	4687
South South	68.3	31.3	0.4	5239
South West	81.1	18.8	0.1	7546
Total	55.6	44.2	0.2	38,522

Source: NBC (2014) in GIZ (2015)

7 Limitations to Clean Energy Access and Utilization in Nigeria

There exists a dichotomy in rural and urban access to clean energy. Based on transcribed data, evidence shows that access to information and finance between urban and rural areas has been identified to be the reason for this scenario. Participant 16 revealed that one limitation to Nigerians accessing clean energy is poor access to information, lack of awareness, and poverty, which has limited the capacity of the people to access this energy source. He argued that where there might be an abundance of renewable sources to be used, the limited capacity of households (most notably in rural areas) limits their access. This was also mentioned in the SWOT analysis of clean energy systems by Dincer and Acar (2015). In their analysis, public perception, lack of information and training, and infrastructure changes were identified as weaknesses while market enhancement, climate change effects, and energy security are the opportunities mentioned. Financial investments, complexity, over-burdened regulatory guidelines, and the low price of conventional energy systems were seen as threats.

In this vein, an interviewee stated that the limited capacity is a reflection of weak government policies and energy process corruption. An interviewee reported that energy corruption is mainly a reflection of the lack of political will, which is often regime based rather than a system of government that is incremental and continual in policy and funding processes. The role of the government as an advocator and major marketer of clean energy was reiterated by Participant 20. She responded that with the government not encouraging the use of clean energy amongst households through sensitization and subsidies, the use would continually remain limited amongst the households. In Nigeria, Corfee-Morlot et al. (2019) and the Nigeria Gas Policy (2017) reported that a quick approach to the implementation of the LPG Availability Intervention Fund of about $160 million and, a reduction in the capital cost of using the LPG stove, gas, and cylinder for household cooking must be enhanced to bring about an easy route for clean energy use.

Further interviews with Participant 5, found that Nigerian society is stuck in non-renewable energy source/use. The majority of rural dwellers still rely on firewood for cooking, while the majority of urban dwellers rely on fossil fuels. She further stated that while there is an abundance of energy from renewable sources, a lack of awareness regarding the affordability of clean energy especially amongst the rural population along with poverty levels depriving them access to credit and ultimately the weak perception of the benefit of the energy source continues to erode the citizens and country in terms of maximizing its use. Participant 6 mentioned that dwellers might be limited by inadequate funding; the cost of maintenance and weak purchasing power for prior implementation that urban dwellers like him might be limited by. In his words, *"...for instance solar energy - without a solar panel, there is no way to access this energy. The question is, how many people can afford the solar panel. I use the inverter in my house, and I spent almost 2million (5,714 USD at 350/USD. How many people can afford it?..."* This same issue was seen as a challenge to using renewable energy in Nigeria (Newsom 2012).

Participant 9 mentioned that the limitation to rural people making use of clean energy might be based on their cultural beliefs. He used an example of the use of firewood in a modern kitchen without a chimney, as in the case of resettled fishermen in New Bussa, Niger State, Nigeria. Bisu et al. (2016) identified that religious beliefs are a significant determinant of the use of fuelwood in the rural areas of Bauchi, Northern Nigeria. The study shows that large Muslim households and cultural alignment during festive periods are the factors that account for increased fuelwood use and the preference for conventional energy usage by these households. In their study, household size, dwelling ownership status, change of season, income, level of education, dwelling location, availability, and affordability are factors found to influence household cooking energy choice. It was further narrated that instances where cultural attachments are not the limiting factor; rural spatial alienation limits exposure to clean energy. This was evident in the views of Participant 11, 18 and 20 who were of the views that clean energy is needed in urban areas than rural areas as demand for electricity and energy is more in the urban space owing to the concentration of industries in these areas and the location of administrative offices where policies are introduced.

In summary, one of the interviewees (a private consultant) states that the main factors that limit the use of clean energy in Nigeria are the high cost of installation and maintenance; poor investment in the energy sector; weak and limited involvement of the private sector; and subsidies granted to generators of energy from fossils. These factors have shaped his perception that access to clean energy is more in urban than in rural areas. Explaining further, he observed that urban areas are associated with better education, higher standards of living, and individuals who could facilitate the installation and maintenance of these forms of energy without the government's assistance or subsidy. However, for these forms of energy to be accessible in the rural areas, it would involve significant investments from the government. Which is presently limited and slow. Brew-Hammond (2010) also pointed this out. He argued that for Africa to muddle through the energy crisis, there is a need for increased involvement of local actors (entrepreneurs, companies, capacity developing institutions) in energy production, generation and consumption markets. He also advocated for improved utilization of a wide range of modern technologies and resources available within the space.

8 Conclusion and Way Forward

Electricity generation in Africa still leaves Sub-Saharan Africa with the lowest generation capacity. This is evident in the electricity condition of Nigeria. The general perception of the people shows that many households remain under-served when energy access is taken into consideration. While the country's energy condition is considered to be below average, the study identified that Nigeria's energy situation remains sporadic, and variance exists in the access of states in the country to electricity infrastructure. This variance in electricity access amongst states can be traced to available energy resources and the location (urban, the capital city, historical administrative capacity, and proximity to energy generation sites). Concerning location, the Northern region of the country remains largely under serviced by the electricity infrastructure.

Expressing the views of the research evidence on the state of access to clean energy remains restricted to neighbourhood and circulation lighting systems and water infrastructure (mainly SDG water projects in rural communities and health centers). The pointers, as identified in the study, show a need for increased investments in the sensitization of access to clean energy in the country. The relevance of energy governance to explore the decentralization of Nigerian settlements away from the national grid towards the maximization of the location of specific clean energy resources needs to be explored and implemented.

It is argued that with the maximization of "place-specific clean energy resources," the drive at eliminating the urban-urban or South-Western and Northern energy dichotomy can be achieved. Based on the interview results, it is identified that achieving this will depend on the introduction of energy policies that reduce clean energy costs of purchase, maintenance and also improve on public sensitization to the opportunities available for households, settlements and sectoral activities in the use of clean energy at a subsidized local level in their various households or small-scale businesses. This assertion of better electricity access if a clean energy source was invested upon is shown in Fig. 12.6.

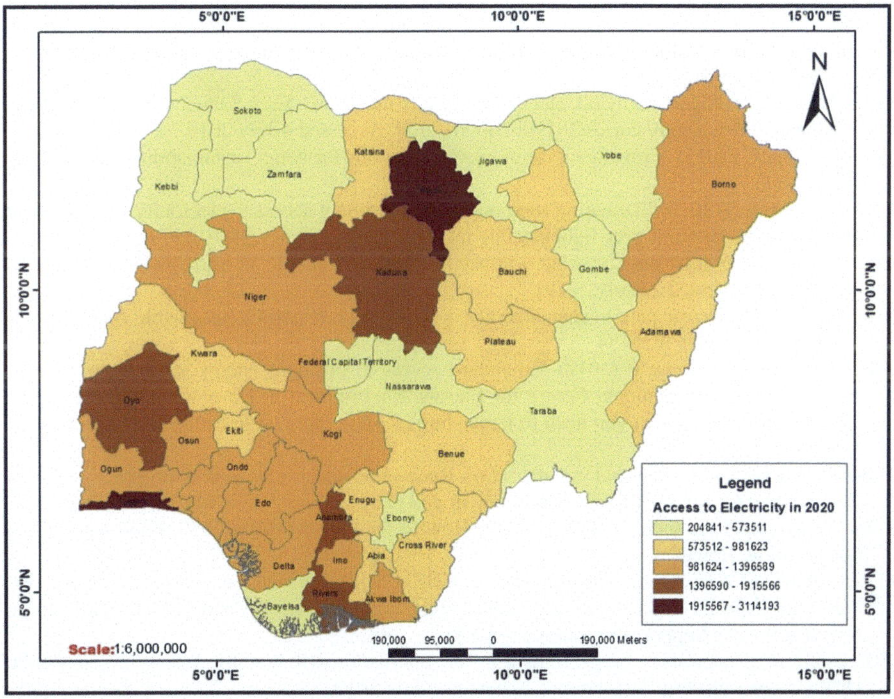

Fig. 12.6 Projected access to electricity in Nigerian States by 2020 (in %). (Source: Authors' Mapping 2019. Adopted from The Nigeria Bureau of Statistics, 2015)

References

Aderoju OM, Dias GA, Echakraoui Z (2017) Assessment of renewable energy sources & municipal solid waste for sustainable power generation in Nigeria. IOP Conf Ser Earth Environ Sci 95(4):042043

Adetayo O (2017) FG awards $5.8bn Mambilla power plant contract. https://punchng.com/fg-awards-5-8bn-mambilla-power-plant-contract/. Accessed 23 July 2019

Adebayo C (2014) How is 100% renewable energy possible for Nigeria? Global Energy Network Institute (GENI). Available at: http://geni.org/globalenergy/research/renewable-energy-potential-of-nigeria/100-percent-renewable-energy-Nigeria.pdf [Accessed 23 July 2019]

Ajah J (2013) Households' access and preference to cooking fuels in Abuja, Nigeria. J Environ Sci Technol 6(2):91–98

Akorede M, Ibrahim O, Amuda S, Otuoze A, Olufeagba B (2017) Current status and outlook of renewable energy development in Nigeria. Niger J Technol 36(1):196–212

Aliyu A, Modu B, Tan C (2018) A review of renewable energy development in Africa: a focus in South Africa, Egypt and Nigeria. Renew Sustain Energy Rev 81:2502–2518

Alnaser W, Alnaser N (2011) The status of renewable energy in the GCC countries. Renew Sustain Energy Rev 15(6):3074–3098

Awoyinfa S, Popoola N, Onuba I, Okon A (2019) NECA, CIBN, MAN, others back Sanusi's call for fuel subsidy removal. Available at: https://punchng.com/neca-cibn-man-others-back-sanusis-call-for-fuel-subsidy-removal/. Accessed 4 July 2019

Bamisile O, Dagbasi M, Babatunde A, Ayodele O (2017) A review of renewable energy potential in Nigeria; solar power development over the years. Eng Appl Sci Res 4(4):242–248

Bisu D, Kuhe A, Iortyer H (2016) Urban household cooking energy choice: an example of Bauchi metropolis, Nigeria. Energy Sustain Soc 6:15–27. https://doi.org/10.1186/s13705-016-0080-1

Brew-Hammond A (2010) Energy access in Africa: challenges ahead. Energy Policy 38:2291–2301

Corfee-Morlot J, Park P, Ogunleye J, Ayeni F (2019) Achieving clean energy access in sub-Saharan Africa, Report in financing climate futures: rethinking infrastructure. OECD, UN Environment and the World Bank Group

Daily Trust (2019) Nigeria: why oil, electricity subsidy must end – Emir Sanusi, June 26. Available at: https://allafrica.com/stories/201906260061.html. Accessed 4 July 2019

Dincer I, Acar C (2015) A review on clean energy solutions for better sustainability. Int J Energy 39:585–606

Doll C, Pachauri S (2010) Estimating rural populations without access to electricity in developing countries through night-time light satellite imagery. Energy Policy 38(10):5661–5670

Energysage (2019) Solar energy: what you need to know. Available at: https://www.energysage.com/solar/ [Accessed 23 July 2019]

Edomah N (2016) On the path to sustainability: key issues on Nigeria's sustainable energy development. Energy Rep 2:28–34

Eleri E, Ugwu O, Onuvae P (2012) Expanding access to pro-poor energy services in Nigeria. Abuja, Nigeria: International Centre for Energy, Environment & Development Energy sage. 2019. Solar energy: what you need to know. https://www.energysage.com/solar/. Accessed 23 July 2019

Federal Ministry of Power and Steel (2006) Renewable electricity action program (REAP). International Centre for Energy, Environment and Development. www.iceednigeria.org

Food Agriculture Organisation (FAO) (2009) How to feed the world in 2050. Food Agriculture Organisation, Rome

Gesellschaft für Internationale Zusammenarbeit (GIZ) (2015) The Nigerian energy sector: an overview with a special emphasis on renewable energy, Energy Efficiency and Rural Electrification. Nigerian Energy Support Programme (NESP)

IRENA (2018) Renewable energy statistics. The International Renewable Energy Agency, Abu Dhabi. www.irena.org/Publications. Accessed 5 July 2019

Jatau D, Dishan E, Bonkary H (2006) Domestic energy consumption pattern in Zinder State, Niger Republic. In: Proceedings of the 31st annual conference of the Forestry Association of Nigeria, Markudi, November 20–23, 2006, Nigeria, pp 351–358

Jebli M, Youssef S (2017) The role of renewable energy and agriculture in reducing CO2 emissions: evidence for North Africa countries. Ecol Indic 74:295–301

Kerrigan K (2001) Benefits of and challenges to energy access in the 21st century. Small Bus Entrep Counc 16(703):140–147

Maduka JO (2011) Popularizing the use of Liquefied Petroleum Gas (LPG) as a substitute for fuel wood among women in Nigeria.In: Proceedings of the 3rd international conference of the African renewable energy alliance on renewable energy and gender, June 29–July 1, 2011, Abuja, Nigeria

Mas'ud A, Wirba A, Muhammad-Sukki F, Mas'ud I, Munir A, Yunus N (2015) An assessment of renewable energy readiness in Africa: case study of Nigeria and Cameroon. Renew Sustain Energy Rev 51:775–784

Merem EC, Twumasi Y, Wesley J, Isokpehi P, Shenge M, Fageir S, Crisler M, Romorno C, Hines A, Hirse G, Ochai S, Leggett S, Nwagboso E (2017) Regional assessment of energy trends in West Africa using GIS. Int J Energy Eng 7(1):1–27

Newman P, Rauland V, Holden D (2011) Creating resilient cities: how a new generation of tools can assist local governments in achieving carbon their abatement goals. In state of Australian cities national conference state of Australian cities national conference.

Newsom C (2012) Renewable energy potential in Nigeria: low-carbon approaches to tackling Nigeria's energy poverty. International Institute for Environment and Development, London

Nigeria Gas Policy (2017) Nigeria gas policy, government policy and actions. Available at: http://www.petroleumindustrybill.com/wp-content/uploads/2017/06/National-Gas-Policy-Approved-By-FEC-in-June-2017.pdf. Accessed 4 July 2019

Nnaji CE, Uzoma CC, Chukwu JO (2010) The role of renewable energy resources in poverty alleviation and sustainable development in Nigeria. Cont J Soc Sci 3:31–37

NIMET (Nigeria Meteorological Agency) (2009) Meteorological data. Nigeria Meteorological Agency, Oshodi, Nigeria

Osunmuyiwa O, Kalfagianni A (2017) Transitions in unlikely places: exploring the conditions for renewable energy adoption in Nigeria. Environ Innov Soc Trans 22:26–40

Oyedepo S (2014) Towards achieving energy for sustainable development in Nigeria. Renew Sustain Energy Rev 34:255–272

Oyedepo S, Dunmade I, Adekeye T, Attabo A, Olawole O, Babalola P, Oyebanji J, Udo M, Kilanko O, Leramo RO (2019) Bioenergy technology development in Nigeria – pathway to sustainable energy development. Int J Environ Sustain Dev 18(2):175–205

Okoro O I, Chikuni E, Govender P (2007, April) Prospect of wind energy in Nigeria. In Proceedings of the International Conference on Domestic Use of Energy, Cape Town, South Africa (pp 10–13)

Pachauri S, Brew-Hammond A, Barnes D, Bouille D, Gitonga S, Modi V, Prasad G, Rath A, Zerriffi H, Dafrallah T, Heruela C, Kemausuor F, Kowsari R, Nagai Y, Rijal K, Takada M, Wamukonya N, Sathaye J (2012) Energy access for development. Global energy assessment – toward a sustainable future. Cambridge University Press/The International Institute for Applied Systems Analysis, Cambridge/New York/Laxenburg, pp 1401–1458

Popoola AA, Magidimisha HH (2019) Rural energy condition in Oyo state: present and future perspectives to the untapped resources. Int J Energy Econ Policy 9(5):1–14

REN21 (2018) Renewable 2018 global status report. REN21, France

REN21 (2016) Renewable energy global status report. REN21, France

Satterthwaite D (2008) Cities' contribution to global warming: notes on the allocation of greenhouse gas emissions. Environ Urban 20(2):539–549

Specht M, Pinto S, Albuquerque U, Tabarelli M, Melo F (2015) Burning biodiversity: fuel-wood harvesting causes forest degradation in human-dominated tropical landscapes. Glob Ecol Conserv 3:200–209

SOLARGIS 2011 Solar irradiation map of Nigeria. Available at: http://solargis.info/doc/free-solar-radiationmaps-DNI. Accessed 16 June 2019

Steurer E, Manatsgruber D, Jouégo E (2016) Risk clustering as a finance concept for rural electrification in Sub-Saharan Africa to attract international private investors. Energy Procedia 93:183–190

Taliotis C, Shivakumar A, Ramos E, Howells M, Mentis D, Sridharan V, Broad O, Mofor L (2016) An indicative analysis of investment opportunities in the African electricity supply sector-using TEMBA (The Electricity Model Base for Africa). Energy Sustain Dev 31:50–66

Trotter P (2016) Rural electrification, electrification inequality and democratic institutions in sub-Saharan Africa. Energy Sustain Dev 34:111–129

Uduma K, Arciszewski T (2010) Sustainable energy development: the key to a stable Nigeria. Sustainability 2:1558–1570

United Nations (UN) (n.d.) Ensure access to affordable, reliable, sustainable and modern energy. Available at: https://www.un.org/sustainabledevelopment/energy/. Accessed 11 Nov 2019

United Nation (UN) (2017) World population projected to reach 9.8 billion in 2050, and 11.2 billion in 2100. United Nations, New York

United Nations Economic Commission for Africa (2012) The renewable energy sector in North Africa: current situation and prospects. United Nations economic commission for Africa, Rabat

Urban F (2009) Sustainable energy for developing countries: modelling transitions to renewal clean energy in rapidly developing countries. Available at: http://www.rug.nl/research/portal. Accessed 10 Nov 2019

UNDESA (2004) Sustainable energy consumption in Africa. New York, United Nations Department of Economic and Social Affairs (UNDESA)

US Energy Information Agency (2013) International energy outlook. U.S. Energy Information Administration (EIA), New York

Uzoma C, Nnaji C, Ibeto C, Okpara C, Nwoke O, Obi I, Unachukwu G, Oparaku O (2011) Renewable Energy Penetration in Nigeria: A Study of the South-East Zone. Cont J Environ Sci 5(1): 1–5

Vahid-Pakdel M, Sayyad-Nojavan B, Mohammadi-ivatloo K (2017) Stochastic optimization of energy hub operation with consideration of thermal energy market and demand response. Energy Convers Manag 145:117–128

Chapter 13
The Nexus of Climate Change and Increasing Demand for Energy: A Policy Deliberation from the Canadian Context

Hirushie Karunathilake, Kh Md Nahiduzzaman, Tharindu Prabatha, Kasun Hewage, Rehan Sadiq, Shahria Alam, and Pamela Shaw

Abstract Canadian energy demand has been increasing due to population, industrial, and economic growth, and the effects of climate change have gained more visibility. Energy use is a major contributor for anthropogenic climate change. Therefore, global scale energy management strategies are paramount in climate change mitigation. However, the complicated 'marriage' between the climate change, energy demand and consumption, and the policy instruments are not sufficiently investigated. Therefore, this paper attempts to study the effect of policy instruments on energy demand and to identify other causes behind the demand trends. A comprehensive review of governmental policies assesses the consistency and effectiveness of existing policy instruments. Communication models for participatory involvement of stakeholders in mitigation initiatives as well as the financial benefits and offsets are critically evaluated. The findings indicate that often, the views of some stakeholder groups, including the individual households and citizens, are not successfully reflected in policies. There is an apparent gap between the regulatory instruments and policies of the territorial, provincial, and local governments.

H. Karunathilake
School of Engineering, The University of British Columbia (UBC), Kelowna, BC, Canada

Department of Mechanical Engineering, University of Moratuwa, Moratuwa, Sri Lanka
e-mail: hirushiek@uom.lk

K. Md Nahiduzzaman (✉) · T. Prabatha · K. Hewage · R. Sadiq · S. Alam
School of Engineering, The University of British Columbia (UBC) Okanagan, Kelowna, BC, Canada
e-mail: Kh.Nahiduzzaman@ubc.ca; Tharindu.Prabatha@alumni.ubc.ca; Kasun.Hewage@ubc.ca; Rehan.Sadiq@ubc.ca; Shahria.Alam@ubc.ca

P. Shaw
Community Planning, Vancouver Island University (VIU), Nanaimo, BC, Canada
e-mail: Pam.Shaw@viu.ca

H. Qudrat-Ullah, M. Asif (eds.), *Dynamics of Energy, Environment and Economy*, Lecture Notes in Energy 77, https://doi.org/10.1007/978-3-030-43578-3_13

Most stakeholders possess limited knowledge due to missing or partial information about energy demand and the outcomes of various policies. This paper aims to trigger a scholarly discussion focusing on the dynamics of energy demand and reguatory instruments and policies for climate change mitigation.

Keywords Energy use trends · Energy demand · Climate action · Energy policy · Emissions mitigation instruments · Carbon pricing · Stakeholder involvement

Glossary

BC	British Columbia
GDP	Gross domestic product
GHG	Greenhouse gas
IEA	International Energy Agency
MJ	Megajoule
OCP	Official community plan
PIG	Price increment gradient
PJ	Petajoules
RE	Renewable energy

1 Climate Change and Energy Use – A Global Issue

In March 2019, the International Energy Agency (IEA) reported that the global energy demand increased by 2.3% in the year 2018, a rate nearly twice the average growth rate since 2010, thus reaching the fastest pace of growth in the current decade (International Energy Agency 2019a, b). It has been predicted that by 2040, the global energy demand will increase by 28% over the 2017 levels, mainly due to the growth in global population and the emerging economies in India, China, and Africa (U.S. Energy Information Administration 2017). With the rising need for energy, other issues have come to play, creating a complicated economic, social, and political dynamic all over the world.

Around 80% of the world's primary energy demand is supplied via conventional fossil fuel resources, mainly coal, petroleum, and natural gas (Höök and Tang 2013). The large-scale demand for energy initiated with the beginning and spread of industrialisation. With the rise of rail travel, mass manufacturing of goods, motorisation, and access to modern energy supplies for large segments of population, the energy use grew exponentially at global level (Vanek and Albright 2008). During the century from 1850 to 1950, the per capita energy intensity doubled, and in the twentieth century, both energy production and consumption experienced rapid growth with the post-World War II economic boom and the population growth in the latter part of the century (Vanek and Albright 2008).

As economies grow, so does the associated energy consumption with the boost to people's life style and industrial activity. It has been indicated in several studies that in general, the per capita gross domestic product (GDP) of a country has a positive correlation with the per capita energy consumption (Vanek and Albright 2008; Brown et al. 2011). It is argued that that not only does energy use increase with increasing GDP and a growing economic, but access to readily available and low cost energy is actually a major contributor to economic growth (Ayres et al. 2013; Sorrell 2015). However, an interesting phenomenon has been observed in the more recent times regarding energy use and economic growth. While overall primary energy consumption has grown, the global energy intensities have declined, with the steady rise in energy productivity (Sorrell 2015). It is estimated that the annual decline in global energy intensity was 1.3% between 1990 and 2000, which is generally attributed to an awareness about the global energy crisis and the interest in efficient energy use. On the other hand, this declining trend has started to change with the advent of the emerging economies, particularly China and India into major roles in the global energy markets (Sorrell 2015).

As the key global economic players compete for the finite energy resources, communities across the world are facing challenges due to high costs of energy, decline in energy resources, energy poverty, lack of supply reliability and availability, as well as climate change and other detrimental environmental impacts of energy use (Hernández and Bird 2010; Höök and Tang 2013; Green et al. 2016). It is predicted that the production of world's oil and gas fields decline approximately 4–6% annually (Höök et al. 2009). More than 1.3 billion people in the world do not have access to electricity, and many more cannot afford the energy prices even in developed countries (Worldwatch Institute 2019). The communities without access to their own fossil fuel supplies are affected by the lack of energy security, which is defined "as the uninterrupted availability of energy sources at an affordable price" by the International Energy Agency (IEA) (International Energy Agency 2018). Being at the mercy of the global energy market forces outside of their control also creates energy dependence for communities. In the environmental front, around 70% of the world's anthropogenic greenhouse gas (GHG) emissions are caused by the use of fossil fuels (International Energy Agency 2015). GHG emissions are a main contributor to climate change, resulting in problems, including global warming, melting ice caps and sea level increase, and changes in weather patterns (United States Environmental Protection Agency 2016; Environment and Climate Change Canada 2015). In addition to GHG emissions and the associated climate change concerns, energy use has been linked to other negative environmental impacts such as damage to eco-systems and habitat alteration, in addition to human health risks (Karunathilake et al. 2019). With all of the above concerns, governments and other decision makers have turned towards the development of sustainable energy and emissions management policies. At the United Nations climate change conference goals in Paris (COP21) held in 2015, it was agreed upon to "avoid dangerous climate change" by capping the global average temperature well below 2 °C of pre-industrial levels, and to further attempt to limit the increase to 1.5 °C (European Commission 2016).

Managing energy more efficiently, effectively, and sustainably has now become a priority to address all of the above concerns. However, this is not such a simple aspiration to attain in reality, due to the complex nature of energy supply, demand, and the interlinked socio-economic and political phenomena associated with them (deLlano-Paz et al. 2017; Burke and Stephens 2018; National Energy Board 2016). At the heart of this issue is the need for a cheap, reliable, and accessible supply of energy, to support communities' livelihood and economic development (Sorrell 2015). This goal is assumed to be in constant conflict with the drive towards cleaner energy alternatives and reduced consumption to mitigate the environmental concerns surrounding energy use. The concept of energy planning, both supply and demand sides, has gained attention in the past few decades due to these very reasons. The latest shift towards energy efficient technologies, passive constructions, energy conservation, renewable energy (RE), and energy sharing are all attempts at solving the multitude of issues associated with energy use. Policymakers and researchers in all quarters of the world are engaged in the attempt to identify the best energy solutions and strategies. These policies and initiatives range from simple rebates and incentive schemes for energy efficient and clean energy technologies, to high-level policies that aim to deliver energy equity and energy security to communities, develop novel economic models, and reshape the landscape of community and regional development. To do this effectively, further studies are necessary to understand all the dynamics associated with energy use and related policy impacts.

The overall goal of this chapter is to study nature of energy supply and demand, especially in the Canadian context, and to discuss about the policy requirements for tackling the economic and environmental issues associated with rising energy demand. The dynamics of energy demand and supply, climate change related issues, and the mitigation policies aimed at curbing the undesirable effects of energy use are investigated in the course of this chapter. The current policies for energy management and promotion of clean technologies will be analysed in details, with a discussion on the existing gaps and potential solutions. The knowledge compiled here is expected to benefit policymakers in Canada and in other parts of the world in developing the much-needed policy prescriptions to tackle energy issues, leading to solutions that are simultaneously economically viable, environmentally responsible, and socially acceptable.

2 Energy Demand and Supply – A Canadian Perspective

Canada is a key player in the global energy market, and is the sixth largest energy producer in the world accounting for 3% of the global energy supply (Natural Resources Canada 2019c). Canada's main energy products are Uranium, natural gas, and crude oil. However, hydropower is also a major energy source in the country. The oil and gas sector provides 0.3% of the Canadian jobs, and energy sector as a whole accounts for 4.4–4.9% of the total employment in the country (Natural Resources Canada 2019a; Natural Resources Canada 2018a). It is a relatively high energy intensive country, ranking 39th out of 237 countries in the world with 7.3 MJ/$2011 purchasing power parity GDP (National Energy Board 2018).

However, Canadian energy intensity has shown a declining trend with increasing GDP in the past decades, although the overall primary energy consumption has increased (The Conference Board of Canada 2019). Several factors are assumed to be contributing to the above observations, which will be discussed in detail within this section.

2.1 Energy Landscape in Canada

Canada's energy supply and consumption scenarios have evolved considerably over the last century. Today, in comparison with other international peers in the same range of economic and social development, Canada ranks as a high energy consumer (The Conference Board of Canada 2019). Currently, Canada is in an attempt to shift towards a lower carbon future within the next few decades. This has resulted in an ongoing diversification of the Canadian energy mix, with the addition of renewable energy. At present, the energy demand growth is slowing down along with the reducing carbon intensity of the energy supplies (National Energy Board 2018). Even though Canada is country with natural energy resources, these supplies are concentrated in some regions, and the other regions, especially the territories in the North, have to rely on external supply sources for their needs. The variations in Canadian energy across time and space use can be attributed to several factors, including weather conditions and climate change, changes in lifestyle, as well as the emergence of new industries and technologies. In order to explore the energy policy needs of Canada, it is necessary to gain an understanding of the energy supply and demand dynamics of the country and the current challenges faced in managing the above.

Canada currently has the highest per capita energy supply among the member countries in the IEA (International Energy Agency 2020). Its primary energy production mix is currently made up of Uranium (32%), crude oil (31%), natural gas (24%), coal (5%), hydro (5%), other renewable energy (3%), and natural gas liquids (2%) (Natural Resources Canada 2018b). Secondary energy use accounts for approximately 70% of the primary energy use (Natural Resources Canada 2016a). At present, Alberta and Saskatchewan are the biggest energy producers among the Canadian provinces. However, Saskatchewan's energy trade is mostly based on Uranium, and when that source is excluded, British Columbia (BC) is the next biggest energy producer after Alberta. Both these provinces have a large industry in oil, gas, and coal extraction and trade. In 2017, 22% of the total Canadian goods exports was made up of energy products, and energy's nominal GDP contribution to economy was 10.6% (Natural Resources Canada 2018b). While renewables have been growing rapidly in the Canadian energy supply mix, with a RE capacity growth amounting to 8.3% by 2016 and a supply share of ~30%, fossil fuels still control the Canadian energy outlook by far (National Energy Board 2017; Hughes 2018). With the above scenario, Canada has a socio-political partiality towards the fossil fuel industry. The abundant availability of secure and reliable energy supplies coupled with high standards of living has contributed to the generally elevated energy consumption of the country (Hughes 2018). At 9.1 tonnes of oil equivalent (toe) per annum, the per capita energy consumption of Canada is five times the average global per capita consumption (Hughes 2018).

However, the above vision of Canadian energy use is an overly simplified generalisation, and there are various energy use patterns to be observed across various geographic regions, communities, and economic sectors in Canada, which are coupled with geological, economic, political, and social factors. Energy efficiency and emissions reduction policy prescriptions cannot be adequately tackled without looking through the lens of the above factors and trends.

2.1.1 Canada's Energy Use Patterns

The Canadian energy consumption increased by nearly three times (183%) during the 50-year period from 1965 to 2015. During the most recent times, energy consumption has been increasing at a relatively slow rate compared to the rest of the world, and during 2010–2015 energy consumption has increased at 0.13% per annum (Hughes 2018). However, the fossil fuel consumption has declined in comparison, with the advent of renewables (Hughes 2018). With a growth rate of 1.4%, the mounting population also creates a higher demand for energy, with new communities being developed across the country (Statistics Canada 2018). It is further predicted that Canada's energy demand will annually grow at 0.7%, ultimately reaching 13,868 PJ by 2040 (Robins 2017).

Figure 13.1 depicts the energy use of the main demand sectors in Canada during the 25-year period from 1990 to 2015 (Natural Resources Canada 2019d). The industrial and transportation sectors show the greatest growth in terms of total energy use, with a combined increase of nearly 1500 Petajoules (PJ) during this period. However, when the sector consumption growth is considered, the agricultural sector shows the greatest growth at 48.72%. This is followed by the transportation sector with a growth of 40.39%. The industrial and commercial/institutional sectors have growths of 26.98% and 35.12% respectively, and the residential sector shows the least growth at 8.88%. The growth in total energy demand during this period was approximately 28.4%. It can be seen that the Canadian energy consumption is heavily dominated by the industrial and transportation sectors, while the agricultural energy consumption is quite low. This is a common trend in many countries, where the agricultural energy intensity reduces or stays roughly the same while industrial activities grows along with the industrial/transportation energy demand.

In 2016, Canada's total energy use was 8786.4 PJ, split across the above end use sectors, with the sectorial and sub-sectorial energy use mix illustrated in Fig. 13.2 (Natural Resources Canada 2019d). The industrial and transportation sectors dominate the energy demand at nearly 70% of the country's total consumption. However, the residential and commercial/institutional sectors, which mainly require energy for maintaining building occupancy and essential services, together contribute to 38% of the total energy use.

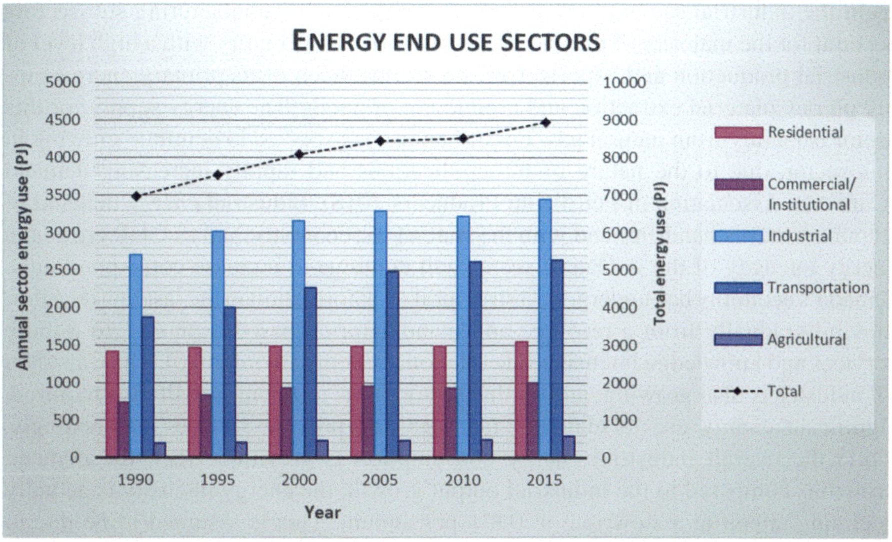

Fig. 13.1 Energy use variation in main end use sectors across the years

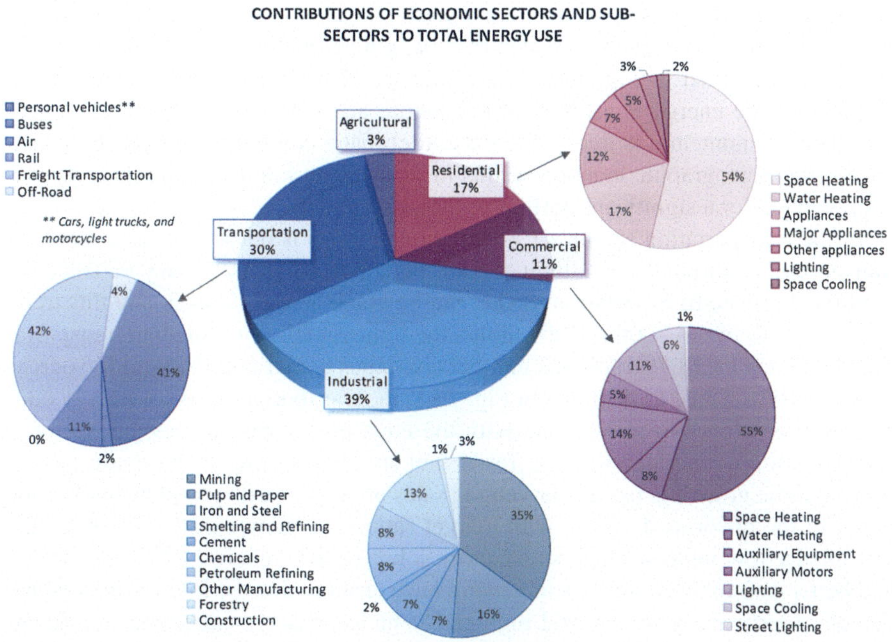

Fig. 13.2 Sector contributions to overall energy use in Canada

In the industrial sector, mining, pulp and paper, and manufacturing sub-sectors account for the majority of the energy use. In short, as a country with a high level of industrial production and exports, Canada spends much of its primary energy supply on raw material extraction and production of goods. The energy supply for this sector is mainly from natural gas. Fossil fuel use is expected to continue growing in this sector, due to the nature of the applications and rapidly increasing demand (Canadian Association of Petroleum Producers 2018). Industrial energy demand of a country moves hand-in-hand with the state of the economy, and as GDP grows the energy intensity of the industrial sector will continue to increase correspondingly. Canada's economy has undergone a structural transformation in the last few decades, moving gradually from a resource and manufacturing based economy, to a more services and knowledge intensive side (National Energy Board 2010). The Canadian IT industry is also growing, and as this sector starts to boom, it will also require a significant energy use to maintain facilities and provide services. Interestingly, while the overall industrial energy consumption is growing, when the demand growth is compared to the industrial output growth, the energy intensity is actually declining, albeit at a slow rate at 0.8% per annum. This is assumed to be due to economic and structural changes, energy efficiency, and variations in activity levels (National Energy Board 2010). Moreover, e-commerce, cannabis, and storage and warehouse leasing sectors are poised to grow in the Canadian industrial sectors, and some of these applications have their own inherent and unique energy needs. The energy needs of the emerging economic sectors is an area that needs much critical attention in the next decade, to match the changing global economic landscape.

As can be noted in Fig. 13.2, space and water heating is responsible for the majority of the energy consumed in residential and commercial/institutional sectors. This phenomenon is due to the general climatic conditions in Canada, where the Northern geographic location results in seasonal variations, with relatively low temperatures for a significant portion of the year.

In the transportation sector, 41% of the energy use is due to the use of private transportation, through cars, light trucks, and motor cycles. Heavy reliance on motorised private transportation modes can be noted across Canada, and this trend is growing. Between 2000 to 2009 alone, the light vehicles in Canadian transportation fleet grew by 18.7%, and this number continues to increase (Natural Resources Canada 2011). Thus, as depicted in Fig. 13.1, the transportation sector energy consumption continues to grow along with the emissions, especially as this sector is overwhelmingly operated through fossil fuel supplies. However, there are indications that the growth is passenger vehicle sales are now starting to plateau after the boom in the previous decades, with market saturation (Lewis 2017). Even so, the historically low interest rates, satisfactory job creation, relatively low gas prices, and better economic conditions in general have ensured that the interest in personal vehicles, particularly luxury and high-end vehicles, has not experienced a downward trend in the Canadian market (Lewis 2017).

Energy use also shows regional variations within Canada, due to factors, such as varying climate conditions between the provinces, different levels of industrial

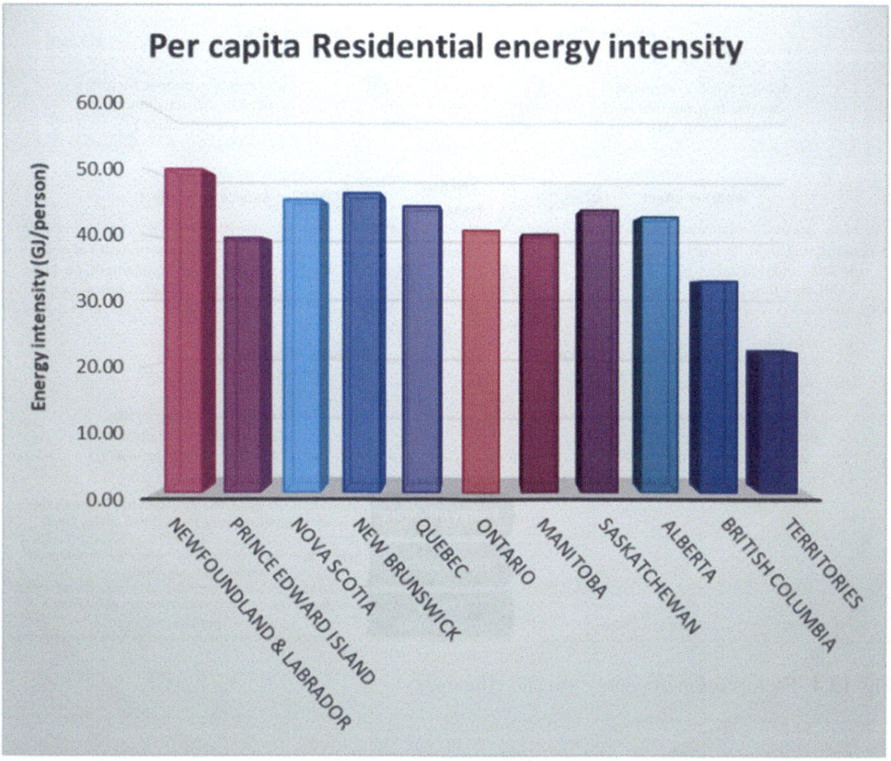

Fig. 13.3 Per capita residential energy use intensity across the provinces and territories

activity and economic development, and inherent transportation and lifestyle patterns among the populace. In building applications, such as residences and commercial buildings, this variation is notable across different climate zones due to variations in heating loads, as depicted in Fig. 13.3. Here, it can be seem that BC has a significantly lower per capita residential energy intensity, due to the temperate climate conditions and milder winters. Interestingly, while the Northern territories are located in the coldest regions of Canada, the per capita energy intensity is much lower, possibly due to the relatively lower socioeconomic development in those regions, the higher costs of energy, as well as the lack of access to technological advancements. Most of the provinces with per capita higher energy intensities such as Alberta, Saskatchewan, Nova Scotia, and New Brunswick all have high fossil fuel fractions in their energy capacity mix. It is definite that energy patterns are subject to a definite socioeconomic component as well as the effect of climate.

In order to investigate the possibility of reducing energy demand, it is necessary to look at the challenges faced by Canada with regards to energy use.

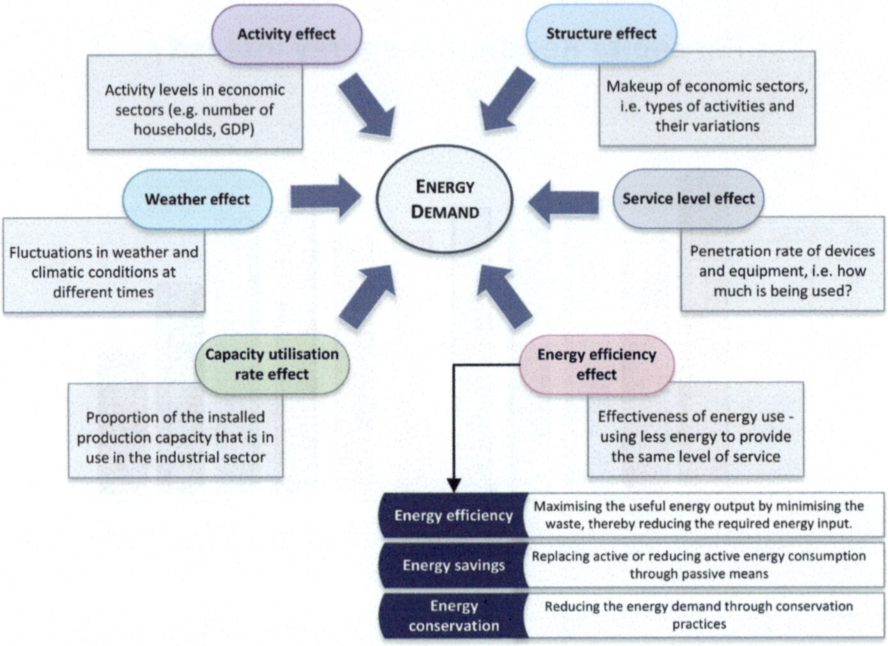

Fig. 13.4 Factorisation of energy use and efficiency

2.1.2 Canada's Energy Use Challenges

Figure 13.4 summarises the nature of energy use and efficiency based on the various factors contributing to demand, and the modes of demand reduction (Natural Resources Canada 2016a; Karunathilake et al. 2018). The effect of all of the six contributing factors to energy use denoted in the figure is interlinked. For example, even if the appliance technologies reach very high efficiency levels, if the service levels also increase with higher device use, the gain of efficiency increase will be offset by the increased device demand. Moreover, many of these factors represent conditions that cannot be changed to reduce energy use. The economic activity and service levels required by growing population due to improvements in quality of life cannot be cut back, simply due to a desire to reduce energy use by numbers. Energy efficiency acts as an impeding factor to the growing energy use, with various technological, behavioural, and regulatory means, resulting in active and passive interventions.

In Canada, the climate and weather conditions pose the greatest challenge to energy demand reduction targets. As a "Northern" country associated with extreme drops in temperature and high heating requirements, Canada is highly reliant on its energy supplies to counter the weather-related challenges and maintain acceptable living conditions. While the impact of climate conditions is direct and obvious building sector, both residential and commercial/institutional, what is less apparent is the effect of climate/weather on other energy end use sectors, such as transportation. The fuel efficiency of Canadian vehicle fleet reduces in the colder winter months. Higher vehicle usage, increased heating needs due to extreme winter temperatures, and more users letting their vehicles idle in the "on" position in winter also lead to higher vehicle energy consumption in winter (Natural Resources Canada 2007). In addition to the extreme temperatures, the vast landscape and the dispersed population also contribute to the high energy use (Natural Resources Canada 2019b). The dispersed nature of Canadian communities correlates to higher transportation requirements for both passengers and goods, leading to increased energy consumption.

While Canadian population growth remains on the lower end compared to most countries, the residential and commercial building stock has grown significantly in the recent times, especially in the urban centres. The building energy intensities vary by building type, building location, and application. This makes planning and policy making for building energy efficiency challenging, as there are no "one size fits all" solutions. A policy or regulation that is developed for the temperate climatic regions of the West Coast, where the above 80% of the electricity grid mix consists of hydro, does not necessarily fit the remote and frigid Arctic Canada, where external fossil fuel supplies are the lifeline for communities.

The energy prices, both electricity and fuel, has shown an increasing trend in the past years (Green et al. 2016). It has been identified that by 2013, 7.9% of the Canadian households were suffering from energy poverty,[1] an increase from 7.2% in 2010. When gasoline expenses are also factored in this level of energy poverty increases further, with 19.4% of the households spending above 10% of their expenditures on energy (Green et al. 2016). This compares with the Canadian average household energy expenditure of 5.3%.

Energy related emissions are a major challenge for Canada, and at present, over 81% of the Canadian emissions are caused by energy (Natural Resources Canada 2019b). In spite of the low-carbon energy supply mix with the significant contributions by hydro and nuclear sources, the oil and gas sector emissions increased by 14% between 2005 to 2013 (International Energy Agency 2020). As previously discussed, many Canadian provinces rely on high emission energy sources for their supply, and this fact is not likely to change soon due to various factors.

The remote Northern communities of Canada suffer additional burdens compared to the rest of the country in terms of supply reliability, energy security as well as energy prices. These communities are not connected to the North American electricity grid, resulting in a lack of guaranteed, reliable, and affordable energy (Natural Resources Canada 2019e). The unavailability energy supplies in these regions make it necessary for them to rely on high-emissions fossil fuels, mainly externally sourced diesel, for their energy needs (Arriaga et al. 2013). This also creates economic problems in these regions, as the high energy costs of the diesel-generated power hinders the community growth. Thus, the remote communities lack many of the benefits and the quality of life taken for granted as basic necessities in other

[1] A benchmark of 10% or more of the household expenditure going towards the purchase of energy goods was used as the measure of energy poverty (Green et al. 2016).

parts of Canada (Natural Resources Canada 2019e). The phenomenon of disadvantaged communities lagging behind more and more with the fast-paced growth of the world's economy and technology is becoming increasingly familiar in all regions of the globe. The governmental policies need to play an integral role in transitioning these communities towards a low-emission, secure, and affordable energy future.

One key challenge facing Canada as well as other parts of the world in energy demand reduction is climate change phenomenon itself. The increase in global temperatures and other environmental variations will likely result in elevated energy demand. It has been suggested that possible impacts of this climate change-induced temperature increase can be higher cooling loads in the hotter seasons, as well as reduced power production efficiencies in existing fossil fuel and nuclear power plants. Increasing water scarcity can also contribute to elevated energy demand (United States Environmental Protection Agency (USEPA) 2016).

Energy use in Canada, similar to many other parts of the world, has undergone a certain level of reduction due to the energy efficiency initiatives in the recent times. Many initiatives are being adopted across Canada to reduce the energy intensity of the building sector as well as the municipal services, in addition to the industrial and transportation sectors. The National Energy Board of Canada has forecast that the energy consumed per a square meter of residential floor space will decrease by 0.7% per annum between 2016 and 2040. This trend is attributed to efficient technologies, energy saving building envelope construction, and new energy efficiency standards for appliances and buildings (National Energy Board 2016). Between 1990 to 2013, the energy intensity per unit activity of GDP reduced by 25%, while per capita energy intensity increased around 1% during the same period. If the economy had not grown in GDP value by 2013, it would have consumed much less energy than the 1990 levels (Natural Resources Canada 2016a). This indicates that the industrial energy efficiencies have also increased significantly in the past decades. Overall, energy efficiency has increased by 24% between 1990 and 2013, and reduced GHG emissions by 85.4 Mt. while delivering a cost saving of $37.6 billion (Natural Resources Canada 2016a). However, in 2017 it was identified that Canada could reduce energy consumption by 15% by 2035 if energy efficiency improvements were pursued more aggressively. In striving towards this end, effective strategies, regulations, and policy frameworks are critical.

3 Policy Development for Energy Efficiency, Energy Security, and Climate Change Mitigation

Policy development is a complex game, involving many stakeholders and priorities at different levels of decision making. While the end goal is ostensibly to reduce energy demand and emissions, the pathway towards this is not necessarily a straight line with simple choices. Many factors and players have to be balanced and satisficed in energy policy development. This necessitates a consideration of the triple bottom line factors, i.e. economy, environment, and society. In order to be most effective and equitable, policies need to take an inclusionary and multi-objective approach. To explore the policy environment surrounding energy demand and emissions management, a review of the current regulations, policies, and goals can be helpful.

Canada is currently undergoing an energy transition, which is expected to continue until mid-century. The Pan-Canadian Framework on Clean Growth and Climate Change released by the Canadian government in 2016 (Pan-Canadian Framework) outlines Canada's commitment to meeting the 2030 climate action target to achieve a 30% below 2005 levels of emissions, and the action plan to achieve this goal (National Energy Board 2018). The Government of Canada has now taken this a step further, by setting a highly ambitious goal to reduce 80% of the 2005 levels of GHG emissions by 2050 (Doluweera et al. 2017). To support this aim, the provincial governments of Canada have set their own emissions reduction targets, and emissions reduction has been integrated to municipal government strategies and official community plans (OCP) across Canada. Meeting these targets in reality is far from easy, and most current projections deem these targets unachievable if the present conditions are not drastically changed (Ruparathna et al. 2017). Table 13.1 lists the latest emissions targets (by 2018) of the Canadian provinces and territories. Such emission target setting is common across the world, to integrate climate change and environmental impact mitigation into official government visions and strategies.

It is interesting to note that while accounting for over one third of Canada's GHG emissions (Boyd 2019), Alberta has one of the least ambitious emissions reduction target among the provinces. This indicates the sociopolitical play that impacts climate mitigation endeavours. Alberta accounts for 80% of Canada's crude oil production as well as more than 50% of Canada's natural gas (National Energy Board 2019; Government of Alberta 2019), and Albertan economy and job market are heavily reliant on the fossil fuel industry (Boyd 2019). Therefore, the province as a whole is reluctant to adopt clean energy initiatives, and there is a strong social resistance towards replacing fossil fuels with other alternatives. These sociopolitical factors need to be taken into account in attempting emissions mitigation endeavours, to be effective and acceptable to all layers of the society. Interestingly, some Canadian provinces have already met their 2030 emissions reduction targets with efficacious mechanisms and policies, indicating that climate action targets can have successful outcomes (Natural Resources Canada 2017). A thorough investigation of what was "done right" can help in replicating such successes and avoiding potential pitfalls in energy policy making.

3.1 Energy and Emissions Reduction Mechanisms

As previously depicted in Fig. 13.4, energy demand reduction efforts can be categorised into energy efficiency, energy saving, and energy conservation interventions (Public Works and Government Services Canada 2001). Energy efficiency involves maximisation of useful energy output by reducing the waste, thus ensuring that the same activity level and service levels may be maintained with a lower

Table 13.1 Emissions mitigation targets of Canadian provinces and territories

Region	Province	Emissions mitigation targets
Atlantic region	Newfoundland & Labrador	Reduce emissions by 10% below 1990 levels by 2020 and 70–85% below 2001 levels by 2050 (Government of Newfoundland and Labrador 2015)
	Prince Edward Island	Reduce emissions by 10% below 1990 levels by 2020 (Government of Prince Edward Island 2019)
	Nova Scotia	Reduce emissions at least by 10% below 1990 levels by 2020 (Nova Scotia Environment 2013)
	New Brunswick	Reduce emissions by 10% below 1990 levels by 2020, and by 75–80% below 2011 levels by 2050 (Government of New Brunswick 2013)
Central Canada	Quebec	Reduce emissions by 20% below 1990 levels by 2020 (Government of Quebec 2012)
	Ontario	Reduce emissions by 15% below 1990 levels by 2020, followed by 37% reduction in 2030 and 80% in 2050 (Government of Ontario 2016)
Prairie provinces	Manitoba	Reduce emissions by 1/3 over 2005 levels by 2030, by half in 2050, and be carbon neutral by 2080 (Government of Manitoba 2015)
	Saskatchewan	Reduce emissions by 20% below 2006 levels by 2020 (Stastna 2015)
	Alberta	Reduce emissions by 14% by 2050 compared to 2005 levels (50% below business as usual scenario) (Government of Alberta 2008)
West coast	British Columbia	Reduce emissions by 40% below 2007 levels by 2030, 60% by 2040, and 80% by 2050 (Government of British Columbia 2018)
Territories	Yukon	Reduce the emissions intensity of existing residential, commercial and institutional buildings across Yukon by 5% and the emission intensity of on-grid diesel power generation by 20% by year 2020 (Government of Yukon 2015)
	Nunavut	No targets are set for reducing GHG emissions (Auditor General of Canada 2018)
	Northwest Territories	To limit emissions increases to 2500Kt by 2020, before stabilizing emissions at 2005 levels by 2030 (The Government of Northwest Territories 2011)

energy use. Efficiency interventions can be categorised as technological and regulatory, both of which go hand in hand. Advancements in equipment manufacturing have ensured that the energy use in appliances is more efficient. The amount of energy estimated to be saved in Canada from all shipped appliances was over 66 PJ between1992 and 2011. With this, a customary set of main household appliances use below 2800 kWh per year, which approximates to a 50% reduction from the energy used in 1992) (Natural Resources Canada 2014). However, technological efficiency advancements are now reaching the saturation point, and further efficiency improvements beyond a certain level may not be economically viable due to prohibitive development costs. Regulatory interventions in energy efficiency mainly involve setting minimum efficiency standards and labelling programs.

Energy saving interventions are generally applied to replace or reduce active energy consumption through passive technological means, using alternative technologies, products, or designs. Building retrofitting for better insulation or increased natural lighting and ventilation are examples of this. Integrating energy saving features in building design has been done through regulatory means, resulting in building energy codes and standards (Du et al. 2014; Berardi 2013; International Energy Agency 2013). Green building rating tools with passive energy saving techniques incorporated in them are now in use to evaluate the environmental impacts of the building sector (Chen et al. 2015; Nejat et al. 2015). The main barrier to energy saving technology adoption at present is the additional cost during construction and retrofitting. A challenge that affects technological interventions for demand reduction across the board is the *principal agent* issue, especially in a context such as Canada (Kelly 2012; Davis 2011). The building construction is primarily undertaken by property developers and then sold to occupants. Apartment-condominium constructions account for a significant fraction of the housing stock at 32% (Natural Resources Canada 2019f). In 2016, 32.2% of the households were rentals (Statistics Canada 2017). In such a setting, the direct beneficiary of energy demand reduction may not be the party who is incurring the costs of energy efficiency and savings enhancements. Thus, property developers and landlords may be unwilling to invest higher costs on energy saving designs and efficient appliances (Karunathilake et al. 2018).

In contrast, energy conservation is a user-centered and low cost approach to reduce energy use, by changing behavioural patterns of consumption (Karunathilake et al. 2018). Energy demand can be reduced significantly by eliminating non-essential instances of energy use with more energy-conscious behavioural choices at zero cost. It has been indicated that conservation-oriented behaviours can reduce heat, electricity, and water consumption by 51%, 37%, and 11% on average respectively (Huebner et al. 2013). Technological and regulatory interventions can play a role in promoting energy-conscious behaviours. Regulations and information focused programs can incentivise people and motivate them to be more conscious and conservation-focused energy users (Lindén et al. 2006; Campillo et al. 2016). Smart control technologies and building automation can remove the burden on users on eliminating unnecessary energy use by simply mechanising the process based on the existing conditions (Karunathilake et al. 2018).

In battling climate change, more and more regions are now trying to regulate emissions from all economic sectors. In addition to energy demand reduction, RE integration is promoted globally as an instrument of emissions reduction as well as increased energy security. While the transformation from fossil fuels to renewables is slow, RE installed capacity has been steadily increasing. Another fascinating development in the recent times for emissions mitigation has been the advent of carbon capture, storage, and utilisation technologies into mainstream markets. Due to the lack of maturity of the above technologies, they have not yet been integrated into large scale emissions reduction planning strategies and policy development.

Currently, Canada is a world leader in both RE and carbon capturing (International Energy Agency 2020). As fossil fuels are unlikely to go out of use in the near future, further investigations on clean energy carbon capturing technologies are extremely important for emissions reduction efforts.

3.2 *Policies and Regulations to Battle Adverse Scenarios of Energy Use and Climate Change Levels of Policymaking*

Various policies and instruments have been adopted to manage energy demand effectively and mitigate climate change in Canada. These policies focus on demand-side management through technical and procedural interventions, as well as behavioural transformations. Figure 13.5 summarises some of the policy instruments used in energy and emissions management and promotion of clean energy technologies (Boza-Kiss et al. 2013; The Climate Policy Info Hub 2019). Policies and instruments are developed at all levels of governance. Certain policy instruments can only be defined and applied at specific levels of government, depending on the level of authority and enforcement power required.

The positive fiscal instruments and chiefly targeted towards reducing the cost barrier towards increased penetration of demand reduction and emissions mitigation measures. Tax rebates, deductions, and reliefs are expected to incentivise both commercial and industrial sectors for demand reduction and emissions mitigation (Price et al. 2005). Subsidies, grants, loans, and funds are aimed at making energy efficiency more affordable. In Canada, many such financial incentives and affordability enhancement schemes are available at federal, provincial, and municipal levels (Natural Resources Canada 2016b). Price signals are another mechanism used to indicate energy conservation goals to the general public (Campillo et al. 2016).

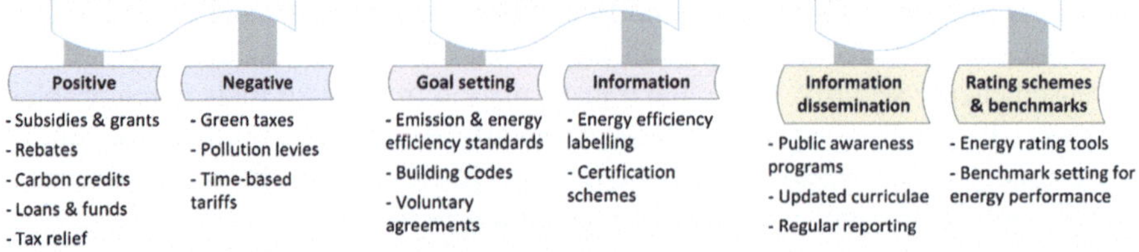

Fig. 13.5 Policy instruments for effective energy and emissions management

The negative fiscal instruments are more controversial. The economic benefits of emissions inducing activities have to be carefully balanced against their negative environmental impacts in imposing these (Price et al. 2005). There are concerns that these policies are have negative impacts, particularly on the industrial sector, hampering competitiveness and economic growth. Similarly, commercial and residential energy users complain that carbon taxes place an unfair economic burden especially on the low-income households and small businesses (Tax Policy Center 2019). The danger is that this can lead to indirect negative impacts, such as increased energy poverty, reduced industrial activity, and loss of jobs unless carefully managed. This thinking has led the Canadian Government to scale back some guidelines for taxation counter to the interests of climate change mitigation, such as reducing the carbon tax thresholds (Quinn 2018).

Imposing minimum energy efficiency requirements and building energy performance standards are also expected to reduce inefficient energy use and losses, thus leading to a reduction in energy-related emissions. This is a common approach adopted across the world (Du et al. 2014; Berardi 2013; International Energy Agency 2013). IEA has reported that the effectiveness of building energy codes is dependent on mandatory enforcement (International Energy Agency 2013). Moreover, for maximum effectiveness, these codes need to be extended to cover all building types in construction, extension, and renovation phases, with minimum performance levels set based on best technologies available in market. Building energy codes are classified as "performance-based" and "prescriptive". Prescriptive codes define minimum performance levels for individual building system components. The criticism against this type of codes is that they do not allow enough flexibility and autonomy to developers and other stakeholders to make investment decisions in the most effective manner considering economic, environmental, and social aspects together (Timmons et al. 2016). Performance based codes, on the other hand, specify overall energy and emissions performance based on the different load types instead of components (International Energy Agency 2013; Nejat et al. 2015). This addresses the "inflexibility" limitation associated with prescriptive codes, but may be too difficult to interpret in actual application due to having only high level mandates without specifying how to get there. The National Energy Code of Canada for Buildings is constantly evolving with the changes in technology and building sector, and sets requirements for heating, lighting, ventilation, envelope, system components etc. (Natural Resources Canada 2018a).

The concept of Energy STEP Codes that look beyond the minimum requirements of the national code is taking root in Canada at present. The province of British Columbia has taken a lead in this, by introducing a performance-based standard with measurable energy efficiency requirements for new construction in consecutive steps, claimed to be *North America's most innovative beyond-code energy efficiency standard* (Frappé-Sénéclauze 2018). Builders must demonstrate compliance by meeting a set of defined metrics for building systems, envelope, and airtightness. The overall goal is to move towards net-zero ready buildings by 2032. With each step, achieving a higher level of performance is required. Individual local governments (i.e. municipal level) have the choice of mandating or incentivising the building sector to meet the stipulations. The BC Energy STEP code also attempts to consider the effects of geography and climate zones, building types, and cost-effectiveness, and is currently under further research and development (Government of British Columbia 2017).

Currently, "*ENERGY STAR*" and "*EnerGuide*" labelling is used within Canada to inform the buyers on the energy efficiency performance of equipment and appliances (Natural Resources Canada 2015). There are also green building certification schemes widely used in Canada, such as BOMA EESt, BREEAM, BuiltGreen, EnerGuide, ENERGY STAR for New Homes, R-2000, LEED, and Living Building Challenge (Gamalath et al. 2018). While many of these are recognized and promoted by policymaking entities, such as the federal and provincial governments, they are not yet mandated. Instead, these are implemented as voluntary programs and informative measures. Natural Resources Canada, the federal department responsible for natural resources, energy etc., also provides access to various data analysis software and modelling tools for performance rating, such as RETScreen, EE4, HOT2000, and CAN-QEST to further facilitate energy efficiency endeavours (Natural Resources Canada 2019a).

Compliance-based energy codes and minimum efficiency mandates seem to be an effective idea at face value, particular as it is expected that energy efficiency improvements between 10.3 and 14.4% are achieved under the current standards (National Research Council of Canada 2019). However, this too can pose an unfair penalty on the low-income earners. It should be kept in mind that high efficiency appliances, energy saving technologies, and green constructions all come with an added price tag. Therefore, these efficient and energy saving options may be unaffordable to the lower-income segments of the society under current conditions. Making minimum efficiency and energy performance standards mandatory may well drive the low cost options out of the market, further contributing to economic inequality, loss of quality of life, and energy poverty. In addition, the high cost of green residential construction has the potential to drive the housing prices higher, when the developers attempt to pass on the costs to the buyers, in a market that is already facing a housing affordability crisis, thus indirectly contributing to poverty and homelessness.

When the actual outcomes of energy efficiency and emissions mitigation policies or strategies are considered, some interesting observations can be made, sometimes contrary to popularly-held beliefs. *Rebound effect* is one such example, where the actually realised energy saving due to efficient technologies can be much lower that expected, or in fact may be negative. This is simply because the perception of energy efficiency results in more carefree energy use patterns in the users, who this that they consume less energy due to energy saving measures and more efficient technologies (Berry and Davidson 2016). This indicates that policies focusing on technological interventions need to take into account the social and behavioural components as well, and they should be supplemented with education and awareness initiatives. Income levels and socio-cultural demographics too make this issue further complicated. When energy users do not have a sense of the actual energy consumption and where they stand with reference to the expected levels, it is difficult to modify behaviours and usage patterns. Information, self-evaluation mechanisms, and benchmarks are tools that can be used to help with the above issue. To support the above, the Canadian federal and provincial government bodies engage in regular statistical surveys and analyses, which are reported in the form of publicly available databases and information (Statistics Canada 2016).

Education and knowledge can play a significant role in promoting conscious behaviour (Gyberg and Palm 2009). Information and outreach focused programs have been successfully adopted in different places to improve public awareness and knowledge on energy efficiency (Lindén et al. 2006; Campillo et al. 2016). Moreover, it has also been identified that providing efficiency education at a young age to children through means such as school curricula, can be a highly effective mechanism for promoting demand reduction. Children educated early in their life grow up to be advocated of energy saving, and families can be influenced through them. Learning new practices is much harder as adults, and energy-consciousness is better transmitted to intellect at early ages. Further, it has been demonstrated that children have considerable influence on their parents purchasing decisions (Fell and Chiu 2014).

3.2.1 Carbon Economy – Fiscal Instruments for Emissions Mitigation

One of the most popular climate action tools today is carbon pricing, a mechanism that seeks to attribute a monetary price to the release of GHG emissions with the assumption that this will drive low-carbon innovation in all economic sectors (National Energy Board 2018; Environment and Climate Change Canada 2017). It is anticipated to be the most practical and cost-effective mode of reducing GHG emissions as well. These carbon pricing systems are currently in the process of being adopted by various provincial and territorial governments in Canada, in line with the Pan-Canadian Framework on Clean Growth and Climate Change. It is assumed that by 2025, the nominal carbon price will level out at \$50/tonne in all provinces and territories (National Energy Board 2018). Figure 13.6 summarises the carbon pricing mechanisms and instruments used in propagating carbon economy. The two main strategies proposed for carbon pricing by the Canadian

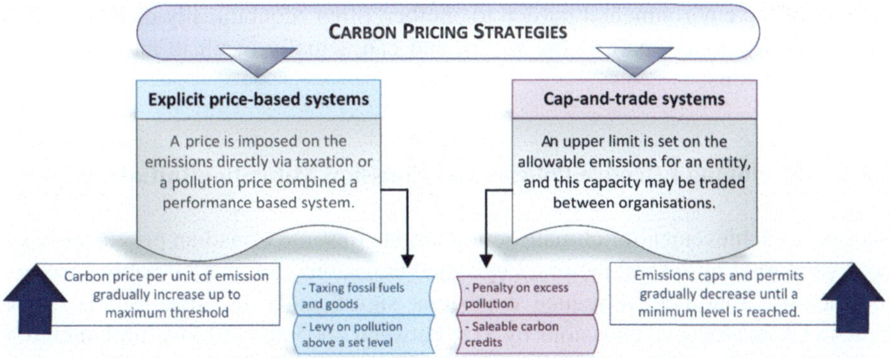

Fig. 13.6 Carbon pricing systems in Canada

government are explicit price-based systems (e.g. BC and Alberta) or cap-and-trade systems (e.g. Ontario and Quebec) (Government of Canada 2016). Carbon credits go hand in hand with such pricing mechanisms where entities that have low, zero, or negative emissions operations receive tradeable credits. Under the cap-and-trade systems where maximum allowable emissions levels are imposed, exemptions and output-based permit allocations in order to keep certain industries, such as steel and mining competitive in the world market. This is an important aspect to be considered in defining carbon pricing policies. While the end goal of carbon pricing is to achieve a positive environmental impact, the economic side-effects should be mitigated as far as possible. Carbon pricing also opens up new avenues of commerce through the new trading opportunities between high and low emitters. This is an international market, where carbon credits can be acquired from other countries as well, and the next stage of Government of Canada's carbon pricing is to integrate international credit import and export in the pricing framework after the initial five-year review in 2022 (Government of Canada 2016).

By allowing provinces and territories to decide on their own carbon pricing system and ensuring that the revenues generated in levying the prices are reallocated to the jurisdiction of origin, the Government of Canada ensures that the localities have some control over the process and benefit from it. A side effect of the carbon pricing is the higher fuel prices (heating, electricity, transportation etc.) and the negative impact they may have on already disadvantaged individuals and communities. However, the government intends that climate tax refunds to citizens will ensure that the higher energy costs are offset eventually for around 70% of population (Joseph 2018). Provinces have also adopted their own variations on the carbon pricing mechanisms. For example, BC has an incentive scheme that reduces carbon-tax costs for entities that have high performance with reference to emissions benchmarks, and invests the industrial carbon tax revenue in promoting direct emissions reduction projects.

Carbon leakage is one adverse situation that can occur as a result of climate policies and carbon pricing mechanisms. Here, as a result of stringent emissions control policies, businesses transfer emissions inducing activities to other countries with lax standards and constraints (European Commission 2019). This results in a mere transfer of the environmental burden to another, often economically disadvantaged and under-developed, part of the world, and can actually result in an increase of overall emissions.

3.2.2 Developing Effective Policies and Emission Mitigation Initiatives

Successful achievement of climate action targets in some Canadian provinces have been supported by several factors. Legislated emissions caps are a notably successful measure that helped reduce emissions significantly in Nova Scotia. New Brunswick has reduced emissions by 31% between 2005 and 2015 by implementing policies in favour of importing hydro and increasing wind energy generation to replace oil and coal (Natural Resources Canada 2017).

Policies for energy demand management and climate change mitigation focus on various goals, and as such, should consider economic impacts, environmental benefits, and social acceptance (Karunathilake et al. 2018). However, these goals are often conflicting. While a considerable effort is being made across Canada for energy and emissions management policy development, one key limitation in most of the current policies is an integrated approach that considers economic, environmental, and social aspects simultaneously and comprehensively in policy development. Moreover, in Canada, energy demand patterns vary heavily based on climatic regions, as previously discussed, and any developed policy frameworks should take this variation into account. Further, policy making that happens at one level of administration can have impacts on various strata of government and society. This disconnect between the different interests can lead to complaints about policies being too distant from practical reality. Taking an inclusive and participatory approach to policymaking can address the above challenges and limitations, which is especially important in a diverse and geographically vast country like Canada. Equity is a key aspect that needs to be kept in mind when developing policies. A prescriptive approach that focuses on *equal treatment* may not necessarily be the most *equitable* course of action.

4 The Dynamics of Carbon Pricing and Emission Policies

Given the ongoing and potential adverse effects of climate change, carbon pricing in the form of tax and cap and trade systems are understood to be ideal tools that can potentially strengthen the current mitigation battle against climate change. In order to increase the political legitimacy and social acceptance of these tools, information needs to be transmitted to all stakeholders, highlighting their benefits to different layers of the society, especially to individual families. As previously mentioned, carbon pricing will elevate the costs for fossil fuel, electricity, ands gas, yet it is the "price increment gradient (PIG)" that will control the behavior of the individual consumers (Chen 2019) (Nahiduzzaman et al. 2018). The federal tax prices carbon at the rate of $20 per tonne, which is equivalent to a PIG of 4.4 cents per one litre of gasoline in 2019, is projected to be 11 cents a litre by 2022 due to the gradual increase of the carbon tax (Chen 2019) (The Globe and Mail 2019). This essentially makes PIG a critical price factor that is expected to curb the fossil fuel consumption while changing the behavior in favor for non-fossil fuel based choices. In the existing carbon pricing program, specific elements are critically designed and directed to minimize the stresses related to finances and market competition at the individual family level. Rebates and tax credits are among the most well designed elements (The World Bank Group 2019). The tax paid in the form of carbon pricing is expected to be reimbursed to the individual families through rebate programs. While tax could be categorically perceived as a "stick" tool, rebate turns out to be a "carrot" measure that paves out the inspirational pathway to embrace as well as practise "stick" in the daily behavior of the consumers.

Table 13.2 Average expenditure for carbon tax versus net benefits for families

Province	Average expense for each family (C$)	Rebate (C$)	Net benefit (C$)	Net benefit Percentage	Average net benefit (C$)	Average net benefit Percentage
Ontario	244	300	56	22.95	100.25	34.74
Saskatchewan	403	598	195	48.38		
Manitoba	232	336	104	44.87		
New Brunswick	202	248	46	22.77		

Source: Adapted from (The Globe and Mail (2019))

Carbon price is not constant across the provinces. It is modified with changes and variations in line with the regulations of the provincial government that support the federal regulatory goals. Each jurisdiction relies on different amounts of fossil fuels while the payments in each province will be based on the number of people in a family and paid to one tax-filer. This way, the families will counter cost with a general annual rebate based on the average expenses of a province and evenly divided across the board (Chen 2019). For examples, Ontario household is expected to pay $244 in direct and indirect costs for carbon, while it will receive $300 under the "climate-action incentive" with a net benefit of 22.95%, equivalent to $56. Table 13.2 shows an average cost that each family is going to pay under carbon tax versus net potential gain through carbon rebate for Ontario, Saskatchewan, Manitoba and New Brunswick (The Globe and Mail 2019).

Regardless of differences, each family is going to gain an average amount of $100.25, which will be about 35% appreciation on the amount paid as carbon tax. Ideally, this should be an encouraging stimulus to get the families streamlined in favor of carbon taxation. However, there is still a considerable resistance amongst the families and other stakeholders across the provinces against the federally imposed regulatory tool for the territories and provinces, largely due to misinformation and widely held misconceptions surrounding this initiative. Next section attempts to unearth the reasons behind and critically discusses the current policy gaps and pertinent challenges that tend to make carbon pricing less than popular among the general public.

4.1 Lack of 'Right' Information: Addressing the Prevailing Pitfalls

There has been a tremendous financial and intellectual endeavor to effectively address the negative consequences of the surging demand for fossil-fuel based energy (National Round Table on the Environment and the Economy 2009). Fiscal and regulatory instruments along with education and awareness strategies are put forward along the line in order to achieve the maximum efficiency (see Fig. 13.5).

While the first two domains of supportive interventions are prominent and highly visible in favor of carbon pricing, "education and awareness" segment largely lags behind in disseminating clear information to the Canadians on:

- Detailed price breakdowns of carbon tax and corresponding monetary incentives in the form of rebates and subsidies;
- The operational method for carbon pricing – how does it work?
- Short and long-term benefits of carbon tax (e.g., people will drive less and choose more-efficient cars or sustainable modes of transportation, such as electric vehicle, public transit, etc.) and contributions from each family as per the annual income and other pertaining attributes;
- Manifestation on how the continued contributions through carbon tax is going to address the current and future adversities posed by rapidly changing climate;
- Key modules of energy, environment, and sense of security for the future generations where families could see direct and indirect benefits of their contributions.

Public perception about any regulatory instruments in Canada relies more on perceived fairness and equity than actual efficiency that the designed instruments are aiming to achieve (Cross 2019). Arguably, the apparent "perception" and embedding knowledge about carbon tax and the expected contributions from the families are some what vague. Carbon tax has also become mired in politically adverse landscape, as it comes to the forefront of highly controversial arguments based on political agendas and philosophies. The phenomenon has echoed in various formats and occasions across the provinces. For instance, only 47% of the British Columbian residents opined that they were in favor of carbon tax, although the BC government offered a climate action tax credit with GST/HST returns in order to offset the embedded cost. However, the provincial government recently declared the CleanBC plan that incorporates industrial incentives and a "Clean Industry Fund" to help industries stay competitive by being innovative, which in turn is expected to reduce both the price of commodities and tax burden on the families (The Canadian Press 2019).

The benefits of carbon tax and the consecutive price burdens on families have also been misinformed and slanted. Many of such attempts could be attributed to politically motivated propaganda, as indicated by recent evidence in Ontario. The Ontario Provincial government contends that carbon tax is likely to cost the average family $648 a year in 2022, in contradiction to the federal government currently held (2019) plans to set it at $244 (The Globe and Mail 2019) (The Canadian Press 2019). While the figure for 2022 is yet to be authenticated, the public information turned out to be only partial without the fact that 80% of the families are expected to receive a higher return through a rebate delivered on their income tax return than what they paid as climate-related taxes, as shown in Table 13.2. Furthermore, the Ontario government also insists that the carbon tax is going to elevate prices for both gas and groceries (The Canadian Press 2019), which does not seem to have a factual ground. Ironically, many people tend to react to such negative information, and oppose regulatory provisions i.e. carbon taxing based on the prevailing (distorted or partial) information without any attempt for triangulation (McCarthy 2019). While the dissemination of such deceiving and fabricated information might be entertained in the political dogma, the role of the federal government in addressing such practices cannot be ruled out. In other words, the federal government must play a responsible role in order to disseminate the "right" and "complete" information about carbon tax to all governmental tiers and families across the provinces and territories.

On the other end of the spectrum, there is a growing need to evaluate the current level of knowledge among the Canadians about the existing regulatory tools and extent of urgency to tackling climate change in order to reduce GHG emissions. A poll released by Abacus Data suggests that one third of the Canadians believe that climate change is not caused by human and industrial activities. While the above finding is surprising, a group of economists suggest that this survey was not broadly understood by Canadians (Zimonjic 2018). However, a recent poll conducted by Nanos Research further advocates that though majority of the Canadian families support climate change mitigation, they are not ready to pay more than $200 per family annually through carbon tax (Gordon 2019). Interestingly, the number of people who wanted government to focus less on policies to reduce carbon emissions has doubled – from 8% in 2015 to 16% in 2018. Another staggering figure suggests that only 42% Canadians are familiar with the concept of carbon pricing and its resulting benefits to the environment, with 10% being "very" familiar and 32% being "quite" familiar (Zimonjic 2018). When the level of perception about carbon pricing, its core benefits, the individual contributions, and backstop year round return (rebate) policy is poor, gaining the necessary support to implement this vital regulatory tool would be immensely challenging. Moreover, it will be cumbersome to get all the tiers of government, including territories, provinces, regional districts and municipalities on board to support the initiative. This will further complicate the fact that without having a desired consensual opinion from the individuals, any level of government would be reluctant to proceed with carbon pricing, although it might yield exceptional merit and better prospects to the current and future generations. Given the prevailing status quo along with the intellectual and political impasses, it is significant for the federal government to design a context-specific, clear and all-encompassing package for the individuals and families to receive "right" information in order for them to take informed decisions.

4.2 The Need for 'Right' Information: What Needs to Be Done?

As discussed in the previous section, information dissemination strategy has been very weak and not well thought out. As a result, a significant number of families is still either completely unaware or only partially informed about carbon tax and pricing concepts (Zimonjic 2018). The only forms of communication methods used for public education were mass media, notably TV channels, YouTube, and social media. Moreover, the content of communication has been brief enough to adequately convey the key essence about carbon pricing tool (Carattini et al. 2017). This stems a pressing need to convey the "right" information package about the backward and forward linkages associated with carbon pricing. Canada being a country with significant socio-cultural and ethnic diversities demands a well designed communication strategy that takes into account the sensitivity and significance associated with such diversity to fruitfully educate the families. Clearly, the current endeavor to disburse the right information falls short of what is adequate (Carattini et al. 2017). Below are some of the thoughtful communication strategies that would potentially bridge up the prevailing knowledge gap(s) among the Canadians. They are likely to contribute in addressing the asymmetries of (*mis-*) information. A well-designed communication strategy aims to increase the visibility of potential benefits of carbon economy, by offering "accurate" and "complete" message with evidence, which in turn would overcome the fundamental issues of distributional fairness, political trust, and policy effectiveness at the federal, provincial, and local levels (Carattini et al. 2017; Mattauch et al. 2017).

The information package must clearly entail

- Price breakdown while specifying the price increment gradient (PIG) with actual contributions of carbon tax in regular gas price and energy bills with a clear message on rebates against the family income and other pertaining attributes.
- GHG reductions achieved versus the expected when carbon tax rates are increased over time along with the co-benefits gained in reducing congestion, air pollution and health costs, improving health and quality of life, among others.
- The aggregate impacts of carbon tax on family income and economy with a highlight on the potential competitive effects and job prospects with a specific focus on the consequence of rebates, subsidies and any social cushioning measures that are used to minimize these impacts (Carattini et al. 2017). The modes for these information package would include printed stickers, leaflets, brochures, etc. that are also going to be publicized through online (social) media;
- Awareness sessions need to be organized at the level of communities, schools, and universities with a concerted effort to facilitate dialogues with the people from all age-cohorts, socio-cultural and income diversities.
- Supplementary programs on consumption behavior and pertaining directives are to be designed on the systemic acts to curtail fossil fuel based consumption (Nahiduzzaman et al. 2018) (Chelleri et al. 2015).
- Energy policies being impactful to everyone – from the federal government to individual families and citizens – it is vital that everyone voices their opinions, suggestions and advocative directives in the ensuing discussion. Therefore, a participatory approach needs to be pursued to consult with the wide arrays of stakeholders, notably federal, provincial, territorial, and municipal governments, industries, residents, and academia.

This may help revise both federal and local regulatory instruments for carbon pricing and cap to effectively meet the stakeholders' expectations and needs.

5 Conclusion: Take Away Lines and Way Forward

While climate change and its adverse consequences appear to the forefront of intellectual and political discourses, carbon pricing and cap seem to be effective regulatory tools for the emission mitigation endeavours. However, implementation of these tools is going to be immensely daunting due to lack of consensual support from a range of stakeholders, including the government and regulatory bodies at different tiers and industrial and commercial partners. Moreover, these tools may not seem very appealing to the Canadians when the immediate financial compensations hastily overshadow the long-term gains. The phenomenon of acceptance is further exacerbated by the inability of the stakeholders to comprehend the key essence of the temporary financial compromises in order to reap later gains through rebates, subsidies and other forms of compensations along with the promise for a better environment. This is primarily attributed to the apparent failure of the federal government to communicate the "right" information to the lower tier of governance and mass Canadians. The detailed architecture of transfer for "complete" and "right" information concerning carbon pricing is under discussion at the federal policy. However, the current "customary" practice of its dissemination among the stakeholders has been staggeringly weak. Due to lack of complete and holistic knowledge of carbon pricing, the discussions on this issue often turn out to be partial and incomplete. The instance of lack of participation by the Canadian families further perplexes the status quo. Because they are not provided with the needed information and functional platforms in order to contribute to the current debates towards (*re-*) shaping the carbon pricing regulations and stemming policies.

In pursuit to remove the current layers of barriers to achieve the goals for carbon pricing, the current status quo essentially calls for a "pathway" of stakeholder engagement. The trail to deliver a package of accurate and complete information on climate pricing to help the stakeholders take informed decision and policy is conceptually illustrated in Fig. 13.7. On the same note, it is imperative to provide details on the encouraging factors as "carrots" while stressing on the price increment gradient (PIG) as "stick" to depict the financial penalties that eventually pay off in the forms of reimbursements and environmental benefits. This paper argues that the success of carbon pricing tool to address the climate change induced adversities does not solely rely on mandatory penalties, but also on stakeholder's acceptance. It is apparent that the short- and long-term benefits outweigh the costs, although this fact is inaccurately rendered in the current intellectual and political deliberations.

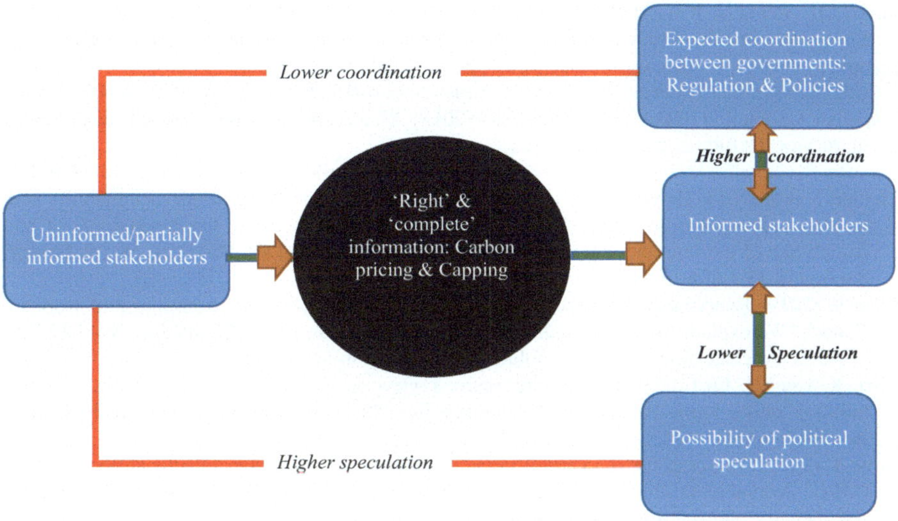

Fig. 13.7 A conceptual pathway to the implementation carbon pricing

Acknowledgments The authors gratefully acknowledge the funding and support provided by FortisBC Inc., Green Construction Research and Training Centre (GCRTC) at the University of British Columbia (UBC), and Mitacs Canada to conduct this research.

References

Arriaga M, Canizares CA, Kazerani M (2013) Renewable energy alternatives for remote communities in Northern Ontario, Canada. IEEE Trans Sustain Energy 4(3):661–670. https://doi.org/10.1109/TSTE.2012.2234154

Auditor General of Canada (2018) Climate change in Nunavut. Available at: http://www.oag-bvg.gc.ca/internet/English/nun_201803_e_42874.html. Accessed 15 July 2019

Ayres RU et al (2013) The underestimated contribution of energy to economic growth. Struct Chang Econ Dyn 27:79–88. https://doi.org/10.1016/j.strueco.2013.07.004

Berardi U (2013) 'Stakeholders' influence on the adoption of energy-saving technologies in Italian homes. Energy Policy 60:520–530. https://doi.org/10.1016/j.enpol.2013.04.074

Berry S, Davidson K (2016) Improving the economics of building energy code change: a review of the inputs and assumptions of economic models. Renew Sustain Energy Rev 58:157–166. https://doi.org/10.1016/j.rser.2015.12.220

Boyd B (2019) A province under pressure: climate change policy in Alberta. Can J Polit Sci 52(1):183–199. https://doi.org/10.1017/S0008423918000410

Boza-Kiss B, Moles-Grueso S, Urge-Vorsatz D (2013) Evaluating policy instruments to foster energy efficiency for the sustainable transformation of buildings. Curr Opin Environ Sustain 5(2):163–176. https://doi.org/10.1016/j.cosust.2013.04.002

Brown JH et al (2011) Energetic limits to economic growth. Bioscience 61(1):19–26. https://doi.org/10.1525/bio.2011.61.1.7

Burke MJ, Stephens JC (2018) Political power and renewable energy futures: a critical review. Energy Res Soc Sci 35(October 2017):78–93. https://doi.org/10.1016/j.erss.2017.10.018

Campillo J et al (2016) Is real-time electricity pricing suitable for residential users without demand-side management? Energy 109:310–325. https://doi.org/10.1016/j.energy.2016.04.105

Canadian Association of Petroleum Producers (2018) 2018 economic report series: Canada's role in the world's future energy mix. Ottawa

Carattini S, Carvalho M, Fankhauser S (2017) How to make carbon taxes more acceptable. London

Chelleri L et al (2015) Exploring the user-driven implications in building urban sustainability and resilience: lessons from OURS CITIES global network study cases. In: Proceedings of 8th conference of the International Forum on Urbanism (IFoU). MDPI, Basel, pp 58–69. https://doi.org/10.3390/ifou-A006

Chen R (2019) Everything you need to know about a carbon tax—and how it would work in Canada. We explain the policy—and what it means for Canadians, Chatelaine. Available at: https://www.chatelaine.com/living/politics/federal-carbon-tax-canada/. Accessed 15 Nov 2019

Chen X, Yang H, Lu L (2015) A comprehensive review on passive design approaches in green building rating tools. Renew Sustain Energy Rev 50:1425–1436. https://doi.org/10.1016/j.rser.2015.06.003

Cross P (2019) The case for the carbon tax – what went wrong?, Macdonald-Laurier Institute (MLI). Available at: https://www.macdonaldlaurier.ca/went-wrong-canadas-carbon-tax-mli-commentary-philip-cross/. Accessed 11 Nov 2019

Davis LW (2011) Evaluating the slow adoption of energy efficient investments: are renters less likely to have energy efficient appliances?, The Design and Implementation of US Climate Policy, pp 301–316. Presented at 'The Design & Implementation of US Climate Policy', May 13–14, 2010

deLlano-Paz F et al (2017) Energy planning and modern portfolio theory: a review. Renew Sustain Energy Rev 77(March 2016):636–651. https://doi.org/10.1016/j.rser.2017.04.045

Doluweera G, Hosseini H, Sow A (2017) Greenhouse gas emissions reductions in Canada through electrification of energy services. Calgary

Du P et al (2014) Barriers to the adoption of energy-saving technologies in the building sector: a survey study of Jing-jin-tang, China. Energy Policy 75:206–216. https://doi.org/10.1016/j.enpol.2014.09.025

Environment and Climate Change Canada (2015) The 2015 Progress report of the Federal Sustainable Development Strategy. Gatineau QC

Environment and Climate Change Canada (2017) Pricing carbon pollution in Canada: how it will work. Available at: https://www.canada.ca/en/environment-climate-change/news/2017/05/pricing_carbon_pollutionincanadahowitwillwork.html. Accessed 8 Aug 2019

European Commission (2016) Paris agreement, climate action. Available at: http://ec.europa.eu/clima/policies/international/negotiations/paris_en. Accessed 27 Dec 2016

European Commission (2019) Carbon leakage, climate action. Available at: https://ec.europa.eu/clima/policies/ets/allowances/leakage_en. Accessed 8 Aug 2019

Fell MJ, Chiu LF (2014) Children, parents and home energy use: exploring motivations and limits to energy demand reduction. Energy Policy 65:351–358. https://doi.org/10.1016/j.enpol.2013.10.003

Frappé-Sénéclauze T-P (2018) Is B.C.'s Energy Step Code a blueprint for Canada?, Sustainable Architecture & Building Magazine. Available at: https://www.pembina.org/op-ed/canada-energy-step-code. Accessed 8 Aug 2019

Gamalath I et al (2018) Energy rating system for climate conscious operation of multi-unit residential buildings. Clean Technol Environ Policy 20(4):785–802. https://doi.org/10.1007/s10098 018-1510-x

Gordon G (2019) Most Canadians unwilling to pay much for carbon tax: poll, Toronto Sun. Available at: https://torontosun.com/news/national/most-canadians-unwilling-to-pay-much-for-carbon-tax-poll. Accessed 15 Oct 2019

Government of Alberta (2008) Alberta's 2008 climate change strategy: responsibility/leadership/action. Edmonton
Government of Alberta (2019) About natural gas. Available at: https://www.alberta.ca/about-natural-gas.aspx. Accessed 15 July 2019
Government of British Columbia (2017) BC Energy Step Code: A Best Practices Guide for local governments. Victoria. Available at: http://www2.gov.bc.ca/assets/gov/farming-natural-resources-and-industry/construction-industry/building-codes-and-standards/guides/bcenergystepcode_guide_v1.pdf
Government of British Columbia (2018) Climate Action in B.C.: 2018 progress to targets. Victoria
Government of Canada (2016) Pan-Canadian framework on clean growth and climate change: Canada's plan to address climate change and grow the economy. Ottawa: Government of Canada https://www.canada.ca/content/dam/themes/environment/documents/weather1/20170113-1-en.pdf
Government of Manitoba (2015) Manitoba's climate change and Green economy action plan. Winnipeg
Government of New Brunswick (2013) New Brunswick climate change action plan 2014–2020. Fredericton
Government of Newfoundland and Labrador (2015) Strengthening resilience to climate change. Available at: http://www.releases.gov.nl.ca/releases/2015/exec/0709n03.aspx. Accessed 13 May 2016
Government of Ontario (2016) Ontario's five year climate change action plan 2016–2020. Toronto
Government of Prince Edward Island (2019) Greenhouse gas emissions. Available at: https://www.princeedwardisland.ca/en/information/environment-water-and-climate-change/greenhouse-gas-emissions. Accessed 4 July 2019
Government of Quebec (2012) Quebec in action: greener by 2020. Gatineau
Government of Yukon (2015) Yukon government climate action plan. Whitehorse
Green, K. P. *et al.* (2016) Energy costs and Canadian households: how much are we spending? Vancouver
Gyberg P, Palm J (2009) Influencing households' energy behaviour-how is this done and on what premises? Energy Policy 37(7):2807–2813. https://doi.org/10.1016/j.enpol.2009.03.043
Hernández D, Bird S (2010) Energy burden and the need for integrated low-income housing and energy policy. Poverty Public Policy 2(4):668–688. https://doi.org/10.2202/1944-2858.1095
Höök M, Tang X (2013) Depletion of fossil fuels and anthropogenic climate change—a review. Energy Policy 52:797–809. https://doi.org/10.1016/j.enpol.2012.10.046
Höök M, Hirsch R, Aleklett K (2009) Giant oil field decline rates and their influence on world oil production. Energy Policy 37(6):2262–2272. https://doi.org/10.1016/j.enpol.2009.02.020
Huebner GM, Cooper J, Jones K (2013) Domestic energy consumption—what role do comfort, habit, and knowledge about the heating system play? Energy Build 66:626–636. https://doi.org/10.1016/j.enbuild.2013.07.043
Hughes D (2018) Canada's energy outlook: current realities and implications for a carbon-constrained future. Vancouver. Available at: www.policyalternatives.ca
International Energy Agency (2013) Tracking clean energy Progress 2013. Paris
International Energy Agency (2015) CO2 emissions from fuel combustion: highlights. Paris
International Energy Agency (2018) Energy security. Available at: https://www.iea.org/topics/energysecurity/. Accessed 23 Apr 2018
International Energy Agency (2019a) Global energy and CO2 status report. The latest trends in energy and emissions in 2018. Paris. Available at: https://www.iea.org/geco/
International Energy Agency (2019b) Global energy demand rose by 2.3% in 2018, its fastest pace in the last decade. Available at: https://www.iea.org/newsroom/news/2019/march/global-energy-demand-rose-by-23-in-2018-its-fastest-pace-in-the-last-decade.html. Accessed 13 July 2019
International Energy Agency (2020) Canada. Available at: https://www.iea.org/countries/Canada/. Accessed 3 Apr 2020

Joseph R (2018) Confused about carbon taxes and rebates? Here's what you need to know, Global News, 28 October. Available at: https://globalnews.ca/news/4586374/carbon-tax-rebate-what-you-need-to-know/

Karunathilake H, Hewage K, Sadiq R (2018) Opportunities and challenges in energy demand reduction for Canadian residential sector: a review. Renew Sustain Energy Rev 82(February):2005–2016. https://doi.org/10.1016/j.rser.2017.07.021

Karunathilake H et al (2019) Renewable energy selection for net-zero energy communities: life cycle based decision making under uncertainty. Renew Energy 130:558–573. https://doi.org/10.1016/j.renene.2018.06.086

Kelly G (2012) Sustainability at home: policy measures for energy-efficient appliances. Renew Sustain Energy Rev 16(9):6851–6860. https://doi.org/10.1016/j.rser.2012.08.003

Lewis, M. (2017) Canadian auto industry looking upward after explosive multi-year growth, Star Bus J. Available at: https://www.thestar.com/business/2017/05/19/canadian-auto-industry-looking-upward-after-explosive-multi-year-growth.html. Accessed 25 July 2019

Lindén AL, Carlsson-Kanyama A, Eriksson B (2006) Efficient and inefficient aspects of residential energy behaviour: what are the policy instruments for change? Energy Policy 34(14):1918–1927. https://doi.org/10.1016/j.enpol.2005.01.015

Mattauch L et al (2017) Making carbon pricing work, Munich Personal RePEc Archive. MPRA Paper No. 80943

McCarthy, S. (2019) Nearly two-thirds of Canadians oppose provincial governments spending taxpayers' dollars to battle federal carbon tax, poll says, The Globe and Mail Inc.. Available at: https://www.uottawa.ca/environnement/blogue/nearly-two-thirds-canadians-oppose-provincial-governments-spending-taxpayers-dollars-battle. Accessed 18 Nov 2019

Nahiduzzaman KM et al (2018) Households energy conservation in Saudi Arabia: lessons learnt from change-agents driven interventions program. J Clean Prod 185:998–1014. https://doi.org/10.1016/j.jclepro.2018.03.052

National Energy Board (2010) Industrial energy use in Canada emerging trends. Calgary

National Energy Board (2016) Canada's energy future 2016: energy supply and demand projections to 2040. Calgary

National Energy Board (2017) Canada's renewable power landscape: energy market analysis 2017. Ottawa. https://doi.org/10.1016/j.msea.2008.10.047

National Energy Board (2018) Canada's energy future 2018: energy supply and demand projections to 2040. Ottawa

National Energy Board (2019) Provincial and territorial energy profiles – Alberta. Available at: https://www.neb-one.gc.ca/nrg/ntgrtd/mrkt/nrgsstmprfls/ab-eng.html. Accessed 15 July 2019

National Research Council of Canada (2019) National Energy Code of Canada for Buildings 2017. Available at: https://nrc.canada.ca/en/certifications-evaluations-standards/codes-canada/codes-canada-publications/national-energy-code-canada-buildings-2017. Accessed 8 Aug 2019

National Round Table on the Environment and the Economy (2009) Achieving 2050: a carbon pricing policy for Canada. Ottawa

Natural Resources Canada (2007) Canadian Vehicle Survey: 2005 Summary Report. Ottawa. Available at: http://oee.nrcan.gc.ca/Publications/statistics/cvs05/pdf/cvs05.pdf

Natural Resources Canada (2011) Canadian vehicle survey: 2009 summary report. Ottawa. Available at: http://oee.nrcan.gc.ca/Publications/statistics/cvs05/pdf/cvs05.pdf

Natural Resources Canada (2014) Energy consumption of major household appliances shipped in Canada, trends for 1990–2011: summary report. Ottawa

Natural Resources Canada (2015) Energy-efficient products. Available at: https://www.nrcan.gc.ca/energy-efficiency/energy-efficiency-products/energy-efficient-product-information/12525. Accessed 14 Dec 2016

Natural Resources Canada (2016a) Energy efficiency trends in Canada: 1990 to 2013. Ottawa

Natural Resources Canada (2016b) Grants and financial incentives. Available at: http://www.nrcan.gc.ca/energy/funding/efficiency/49. Accessed 22 Oct 2016

Natural Resources Canada (2017) Market snapshot: some Canadian provinces have already met their 2030 GHG emissions targets. Available at: https://www.neb-one.gc.ca/nrg/ntgrtd/mrkt/snpsht/2017/07-01prvncsmssnstrgts-eng.html. Accessed 7 Aug 2019

Natural Resources Canada (2018a) Canada's national energy code. Available at: https://www.nrcan.gc.ca/buildings/canadas-national-energy-code/20675. Accessed 8 Aug 2019

Natural Resources Canada (2018b) Energy fact book 2018–2019. Ottawa

Natural Resources Canada (2019a) Data analysis software and modelling tools. Available at: https://www.nrcan.gc.ca/maps-tools-and-publications/tools/modelling-tools/7417. Accessed 8 Aug 2019

Natural Resources Canada (2019b) Energy and Greenhouse Gas Emissions (GHGs). Available at: https://www.nrcan.gc.ca/science-and-data/data-and-analysis/energy-data-and-analysis/energy-facts/energy-and-greenhouse-gas-emissions-ghgs/20063.. Accessed 26 July 2019

Natural Resources Canada (2019c) Energy and the economy. Available at: https://www.nrcan.gc.ca/science-and-data/data-and-analysis/energy-data-and-analysis/energy-facts/energy-and-economy/20062#L1. Accessed 15 July 2019

Natural Resources Canada (2019d) Energy use data handbook tables. Available at: http://oee.nrcan.gc.ca/corporate/statistics/neud/dpa/menus/trends/handbook/tables.cfm.. Accessed 23 July 2019

Natural Resources Canada (2019e) Reducing diesel energy in rural and remote communities. Available at: https://www.nrcan.gc.ca/climate-change/green-infrastructure-programs/reducing-diesel-energy-rural-and-remote-communities/20542.. Accessed 7 Aug 2019

Natural Resources Canada (2019f) Residential housing stock and floor space. Available at: http://oee.nrcan.gc.ca/corporate/statistics/neud/dpa/showTable.cfm?type=HB&sector=res&juris=00&rn=11&page=5.. Accessed 7 Aug 2019

Nejat P et al (2015) A global review of energy consumption, CO2 emissions and policy in the residential sector (with an overview of the top ten CO2 emitting countries). Renew Sustain Energy Rev 43:843–862. https://doi.org/10.1016/j.rser.2014.11.066

Nova Scotia Environment (2013) Amendments to greenhouse gas & air quality emissions regulations. Halifax

Price, L. *et al.* (2005) Tax and fiscal policies for promotion of industrial energy efficiency: a survey of international experience. Berkeley. Available at: http://escholarship.org/uc/item/4rh396ct.pdf

Public Works and Government Services Canada (2001) The environmentally responsible construction and renovation handbook, 2nd edn

Quinn, G. (2018) Ottawa eases carbon tax thresholds to help Canada's big industries compete, Financial Post. Available at: https://business.financialpost.com/commodities/energy/citing-competitiveness-pressures-feds-ease-carbon-tax-thresholds. Accessed 8 Aug 2019

Robins A (2017) Doing more with less: energy efficiency potential in Canada. Ottawa

Ruparathna R et al (2017) Climate conscious regional planning for fast-growing communities. J Clean Prod 165:81–92. https://doi.org/10.1016/j.jclepro.2017.07.092

Sorrell S (2015) Reducing energy demand: a review of issues, challenges and approaches. Renew Sustain Energy Rev 47:74–82. https://doi.org/10.1016/j.rser.2015.03.002

Stastna, K. (2015) How Canada's provinces are tackling greenhouse gas emissions, CBC News. Available at: https://www.cbc.ca/news/canada/how-canada-s-provinces-are-tackling-greenhouse-gas-emissions-1.3030535

Statistics Canada (2016) Statistical framework for energy in Canada. Available at: https://www150.statcan.gc.ca/n1/pub/57-602-g/2015001/chap2-eng.htm. Accessed 8 Aug 2019

Statistics Canada (2017) Housing in Canada: key results from the 2016 census. Ottawa

Statistics Canada (2018) Analysis: total population. Available at: https://www150.statean.gc.ca/n1/pub/91-215-x/2018001/sec1-eng.htm. Accessed 26 July 2019

Tax Policy Center (2019) What is a carbon tax? Available at: https://www.taxpolicycenter.org/briefing-book/what-carbon-tax. Accessed 8 Aug 2019

The Canadian Press (2019) People are ridiculing Ontario's anti-carbon tax ad for sending the opposite message on mute, CBC News. Available at: https://www.cbc.ca/news/canada/toronto/ontario-anti-carbon-tax-ad-backfires-1.5139384. Accessed 28 Nov 2019

The Climate Policy Info Hub (2019) Energy efficiency policy instruments in the European Union. Available at: https://climatepolicyinfohub.eu/energy-efficiency-policy-instruments-european-union. Accessed 1 Dec 2019

The Conference Board of Canada (2019) Energy intensity. Available at: https://www.conference-board.ca/hcp/provincial/environment/energy-intensity.aspx. Accessed 25 July 2019

The Globe and Mail (2019) Canada's carbon tax: a guide to who's affected, who pays what and who opposes it, The Globe and Mail Inc. Available at: https://www.theglobeandmail.com/canada/article-canadas-carbon-tax-a-guide/.. Accessed 23 Nov 2019

The Government of Northwest Territories (2011) A greenhouse gas strategy for the northwest territories. Yellowknife. Available at: https://www.enr.gov.nt.ca/en/services/changement-climatique/territorial-targets

The World Bank Group (2019) Report of the high level commission on carbon pricing and competitiveness, Washington, DC

Timmons D et al (2016) Decarbonizing residential building energy: a cost-effective approach. Energy Policy 92:382–392. https://doi.org/10.1016/j.enpol.2016.02.030

U.S. Energy Information Administration (2017) International energy outlook 2017. Washington, DC. www.eia.gov/forecasts/ieo/pdf/0484(2016).pdf

United States Environmental Protection Agency (2016) Climate change: basic information, climate change. Available at: https://19january2017snapshot.epa.gov/climatechange/climate-change-basic-information_.html. Accessed 7 June 2016

United States Environmental Protection Agency (USEPA) (2016) Climate impacts on energy, climate change impacts. Available at: https://19january2017snapshot.epa.gov/climate-impacts/climate-impacts-energy_.html.. Accessed 7 Aug 2019

Vanek FM, Albright LD (2008) Energy systems engineering: evaluation and implementation. McGraw-Hill, New York. https://doi.org/10.1036/0071495932

Worldwatch Institute (2019) Energy poverty remains a global challenge for the future. Available at: http://www.worldwatch.org/energy-poverty-remains-global-challenge-future.. Accessed 13 July 2019

Zimonjic, P. (2018) Nearly a third of Canadians don't believe humans, industry 'mostly' cause climate change: poll, CBC News. Available at: https://www.cbc.ca/news/politics/poll-abacus-carbon-tax-1.4603824.. Accessed 30 Sept 2019

Chapter 14
Climate Change and Energy Policies: European Union-Scale Approach to a Global Problem

Guller Sahin and Fatih Volkan Ayyildiz

Abstract The goal of this research was to examine the relationship between climate change and energy resources within the scope of the European Union by using static panel data method and Root Mean Square Error methodologies for the period of 1990–2018. As more specifically, the purpose was to evaluate the effects of the consumption in nonrenewable and ecological energy resources on CO_2 emissions. Under the related purpose, the variables of CO_2 emissions, which have the greatest impact on the climate change parameter among the greenhouse gases; oil, natural gas, coal, nuclear energy, hydroelectric energy, wind energy, geothermal energy, solar energy and biomass energy related to energy parameter were used. According to the findings of the unit effective fixed effects model; it was concluded that coal, natural gas and oil consumption increased CO_2 emissions while ecological energy consumption decreased CO_2 emissions. Oil consumption was the most influential variable on CO_2 emissions. Root Mean Square Error findings indicate that the variable which has the highest effect on CO_2 emissions is geothermal energy consumption; and the lowest effect variable is the consumption of oil energy.

Keywords European Union · Climate change · Energy sector · Energy policies · CO_2 emissions · GHG emissions · Panel data analysis · Root mean square error

G. Sahin
Department of Economics, Kutahya, Turkey

F. V. Ayyildiz (✉)
Department of Economics, Faculty of Economics and Administrative Sciences, Ardahan University, Ardahan, Turkey
e-mail: fatihvolkanayyildiz@ardahan.edu.tr

H. Qudrat-Ullah, M. Asif (eds.), *Dynamics of Energy, Environment and Economy*, Lecture Notes in Energy 77, https://doi.org/10.1007/978-3-030-43578-3_14

1 Introduction

Human effect on the climate system is highly apparent; anthropogenic emissions of greenhouse gases have reached the highest amount in the history of humanity in recent years. For the data of the Intergovernmental Panel on Climate Change (IPCC) in 2014, the period from 1983 to 2012 was the warmest 30 years of the last 1400 years in the Northern Hemisphere. With reference to the land and ocean surface temperature data around the world, the temperature increase in the period of 1880–2012 was 0.85 °C (Topçu 2018).

Anthropogenic greenhouse gas (GHG) emissions have increased from the pre-industrial period to the modern day because of the economic growth and population increase. This circumstance has caused the amount of atmospheric carbon dioxide (CO_2), methane (CH_4) and nitrous oxide (N_2O) concentrations that have not been observed at least in the last 800.000 years. In the course of events, it is revealed that especially the atmospheric CO_2 emissions have contributed to global warming as from the midst of the twentieth century (IPCC 2014).

Anthropogenic GHG emissions are directed by population growth, economic activities, energy consumption, land use, lifestyle dynamics, technology and climate in principle. Accordingly, future simulations are important based on these factors to determine the next projections.

Approximately, 40% of anthropogenic GHG emissions in atmosphere between the years of 1750 and 2011 stayed in the atmosphere; the rest of related emissions were conducted away the atmosphere. However, these emissions have been stored in plants, soils, and oceans. Oceans absorbed almost 30% of released anthropogenic CO_2 emissions; this situation has caused oceans to acidulate. Approximately half of the anthropogenic CO_2 emissions between 1750 and 2011 occurred in the last 40 years (See Fig. 14.1) (IPCC 2014).

Although the ever-mounting climate change decreasing policies, total anthropogenic GHG emissions continued to increase from 1970 to 2010 (by more absolute increases between 2000 and 2010). CO_2 emissions and industrial processes arising from fossil fuels combustion contributed to almost 78% of GHG emissions (See Fig. 14.2) (IPCC 2014).

According to the scenario of IPCC, global average surface temperature increases are projected to increase between 0.3 °C and 0.7 °C in the 2016–2035 in comparison to 1986–2005. It is expressed that global surface temperature will exceed 1.5 °C at the end of the twenty-first century (2081–2100) in comparison to the period of 1850–1900.

Figure 14.3, in 2017, GHG emissions in the EU decreased by 22% compared to 1990, shows a decrease of 1240 million tons (CO_2 equivalent). A general downward trend in emissions between 1990 and 1999 (except the peak in 1996) is seen. The course of GHG emissions in the period of 1999–2006 did not change in general in spite of the fact that it decreased till 2008 in small ratios. There was occurred a sharp drop in emissions as the result of industrial activities that decreased because of the 2009 global crisis. Emission amounts that increased again in 2010 have started to

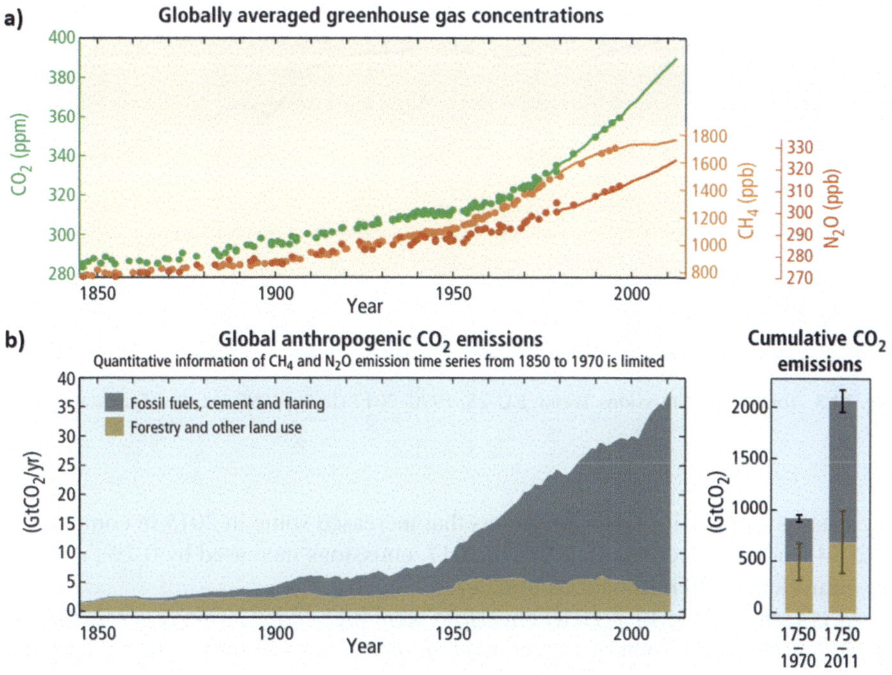

Fig. 14.1 Globally GHG Gas Concentrations and Global CO_2 Emissions, 1750–2011. (IPCC 2014)

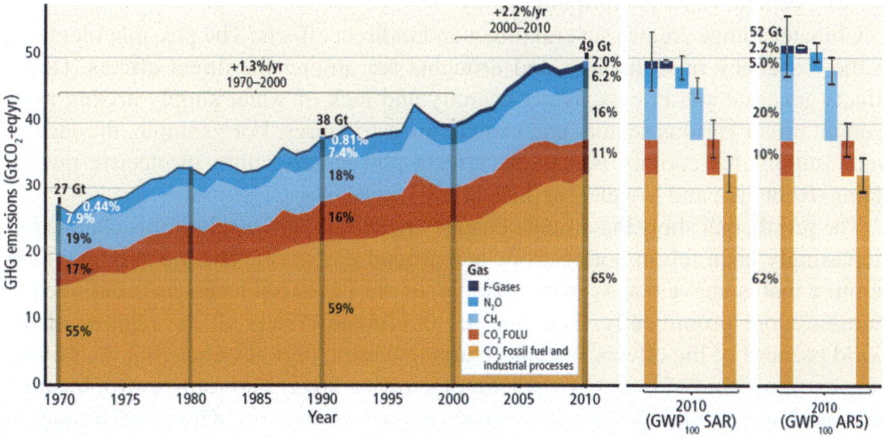

Fig. 14.2 Total Annual Anthropogenic GHG Emissions by Gases, 1970–2010. (IPCC 2014)

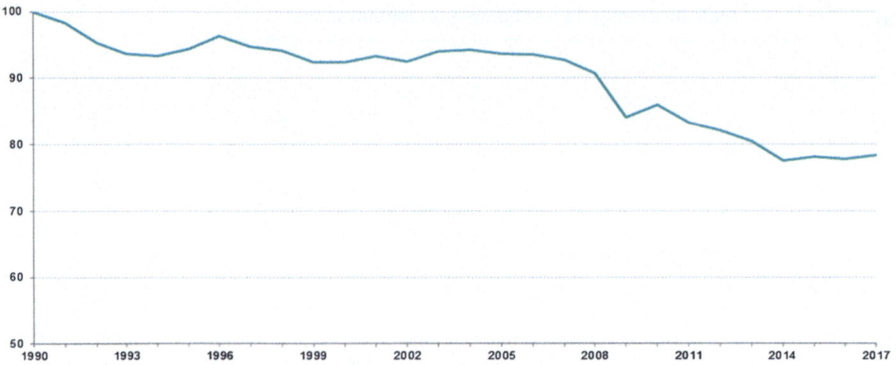

Fig. 14.3 Total GHG Emissions Trend, EU-28, 1990–2017 (Index 1990 = 100). (Eurostat 2019)

decrease as from 2011. GHG emissions that increased some in 2015 in comparison to 2014 tended to decrease in 2016. In 2017, emissions increased by 0.7%, namely, the equivalent of 30 million tons of CO_2.

As is seen in Table 14.1, GHG emissions were highest in Germany across the EU in 2017 (21% of the total of EU, equivalent of 936 million tons of CO_2). England and France followed Germany. The lowest emission amount in 2017 was seen in Malta (2.6%). The biggest declines in emission amounts in comparison to 1990 were recorded in Lithuania (57%), Latvia (56%), Romania (54%) and Estonia (48%). On the other hand, the biggest increases in emission were occurred in Cyprus (56%), Portugal (23%) and Spain (22%).

Climate change creates several direct and indirect effects. The possible increases in the frequency of heatwaves and droughts are among the direct effects. Direct effects generate sub-effects. Water scarcity and lack of water supply arising from drought affect various sectors and critical infrastructures. For example, the lack of water supply for cooling negatively affects power generation in electric power-plants (Rübbelke and Vögele 2010, 2011).

The proofs that show the climate change are certain; the result of this change has increasingly been felt in both Europe and around the world. The mean global temperature that is above 0.8 °C in comparison to pre-industrial levels has continued to increase more prominently. The changes in climate affects all the regions of the world because of the effects such as changing natural periods, changing the raining frequency, melting of icebergs, depleting the ozone layer, acid rains, eutrophication of coastal waters and elevation sea levels (Croce et al. 2018; Dovi and Battaglini 2015). There is a consensus among climate researchers about an assumption that serious natural disasters arising from excess weather events may occur as stronger and more frequent. In this sense, the importance of impact assessments of disaster is expected to increase for near future projections in climate change (Többen 2017). Thus, developments in economic disaster effect models form a basis to improve climate change adaptation strategies in different variations.

Table 14.1 Total GHG emissions, by Country, 1990–2017 (million tonnes of CO_2-equivalents). (Eurostat 2019)

	1990	1995	2000	2005	2010	2015	2017	Share in EU-28
EU-28	5 722.9	5 397.8	5 287.2	5 362.0	4 917.5	4 470.3	4 483.1	100.0%
Belgium	149.7	157.6	154.5	148.9	137.1	121.6	119.4	2.7%
Bulgaria	102.6	75.5	59.8	64.5	61.1	62.2	62.1	1.4%
Czechia	199.8	158.7	151.1	149.5	141.7	129.5	130.5	2.9%
Denmark	72.1	80.1	73.2	68.8	65.5	50.8	50.8	1.1%
Germany	1 263.2	1 138.1	1 064.7	1 016.5	967.0	931.8	936.0	20.9%
Estonia	40.5	20.3	17.4	19.3	21.3	18.3	21.1	0.5%
Ireland	56.5	60.3	70.3	72.0	63.4	61.7	63.8	1.4%
Greece	105.6	111.8	128.9	138.9	121.0	98.2	98.9	2.2%
Spain	293.3	335.3	397.1	452.6	370.1	351.8	357.3	8.0%
France	556.6	553.8	567.0	570.7	528.0	477.3	482.0	10.8%
Croatia	32.4	23.2	26.1	30.3	28.4	24.6	25.5	0.6%
Italy	522.1	538.3	562.1	589.2	514.7	443.7	439.0	9.8%
Cyprus	6.4	7.9	9.2	10.2	10.3	9.1	10.0	0.2%
Latvia	26.5	13.0	10.6	11.6	12.7	11.6	11.8	0.3%
Lithuania	48.6	22.5	19.6	23.0	20.9	20.5	20.7	0.5%
Luxembourg	13.1	10.7	10.6	14.3	13.4	11.6	11.9	0.3%
Hungary	94.2	75.9	73.9	76.2	65.7	61.3	64.5	1.4%
Malta	2.3	3.0	3.1	3.2	3.2	2.5	2.6	0.1%
Netherlands	226.4	239.3	229.8	225.8	224.1	207.5	205.8	4.6%
Austria	79.6	80.9	82.1	94.5	86.8	81.0	84.5	1.9%
Poland	475.0	445.7	396.3	404.3	413.1	392.3	416.3	9.3%
Portugal	60.8	70.8	84.3	88.1	71.7	71.1	74.6	1.7%
Romania	248.9	187.8	143.6	151.7	124.4	117.2	114.8	2.6%
Slovenia	18.7	18.8	19.1	20.6	19.7	16.9	17.5	0.4%
Slovakia	73.4	53.3	49.2	51.3	46.4	41.8	43.5	1.0%
Finland	72.3	72.8	71.3	71.2	77.4	57.2	57.5	1.3%
Sweden	72.7	74.7	70.4	68.6	66.4	55.7	55.5	1.2%
United Kingdom	809.9	767.6	741.9	726.2	642.1	541.5	505.4	11.3%

In analogy to the situation across the world, for IPCC (2014) data, the number of periods which are characterized the frequency and density of heatwaves and droughts in Europe will increase in the future because of the climate change. Current data on European scale points out that damages arising from the river and coastal flood; there is water scarcity; damages increase because of extreme temperatures and forest fires. In many countries, as well as in Europe, natural disasters are the main source of the economic effects which accelerate climate change. The long-standing public discussions in Europe focus on the economic costs of climate change mitigation policies like promoting the use of ecological energy resources. Natural disasters arising from the climate change are required to see the whole picture for a comprehensive evaluation towards cost-benefit accounting. Therefore, types of policies that Europe will follow in the energy sector against possible climate change threats which are expected to increase.

Data that belong to global climate change reveals the truth that effective policies regarding struggle with climate change need to be urgently determined and applied. The recent increase in the share of ecological energy sources in total energy production in Europe and the CO_2 emissions from energy consumption have increased by 1.0% in the world in 2007–2017 period, while the (European Union) EU 1.7% decrease in the EU necessitates the examination of energy policies implemented by the EU in the fight against climate change. The reason given is the starting point of this paper. In parallel with this purpose, it was aimed to scrutinize the struggle with climate change at the EU scale that is accepted as the leader of ecological energy policies.

The goal of this paper was to review the relationship between climate change and energy resources within the scope of European Union countries for the period of 1990–2018. More specifically, the purpose was to evaluate the effects of consumption on nonrenewable and ecological energy resources on CO_2 emissions. This study consists of 7 parts including an introduction. The second chapter shows the cyclical relationship between climate change and energy; the third part explains the developments in the energy sector in the European Union. Then, the energy policies of the European Union were analyzed within the historical process. The fifth chapter shows the material, method, and findings relating to the empirical analysis. In the sixth part, the energy policies implemented are evaluated together with solutions and suggestions. The research is ended by the conclusion text.

2 Cyclical Relation Between Climate Change and Energy

Anthropogenic effects on climate change go back to the roots of transformation in the energy sector as the result of the Industrial Revolution. Energy revolutions which started coal and have continued by natural gas in the twentieth century brought welfare and wealth. Energy transitions provided more flexible and noncentral production processes by transforming economic activities; it also encouraged demographic and social transformations and changed the political structures at the same time. This process placed the global economy on energy consumption orbit that may cause serious CO_2 emissions (Fouquet 2016). Global economic growth continues to be the most important driving force of increases in CO_2 emissions arising from fossil fuels combustion. Separating CO_2 emissions stemming from energy consumption and economic growth generates the seed of climate change discussions. A successful separation proves us the energy efficiency precautions can be economically sustainable (Moreau et al. 2019). Within this framework, being understood the interaction between climate and energy parameters is required to specify the climate change adaptation strategies which aim to minimize the negative effects of climate change and maximize the possible benefits of the same issue.

The most critical issues in energy discussions in the period between 1973 and 1979 oil shocks after the Second World War was to specify the level of pollution to provide the environmental sustainability and deplete the nonrenewable energy resources (Dovi and Battaglini 2015). Huge investments were made to nuclear energy technologies and local level solutions were delivered to reduce the level of pollution as the solution for increasing energy need. However, it was seen at the end of the 1970s that policies and approached applied remained incapable to a large extent. Need for reducing fossil fuel use to avoid the serious results of climate change has proved by the scientific pieces of evidence. Innovations have been adopted in the production, consumption and distribution of energy to reduce the use of fossil fuels; energy has become an integral part of climate change surveys. It has been worked on technology to increase the energy efficiency by providing better constructions and improved industrial design procedures; to encourage the energy conservation by developing low-energy products; to accelerate the transition to energy environment without carbon by generating more ecological energy resources. Moreover, scientists designed models to estimate the results of increasing CO_2 levels and other greenhouse gases. However, it is seen that science and technology could not provide a solution to avoid the negative results of global warming by oneself.

The relationship between climate change and energy is a cyclical process that reflects the feedback effect. Energy that is required to be generated the industrial and commercial wealth at economic growth stages of countries increases the social and cultural level of societies at the same time. However, energy generation and consumption processes create a big pressure on the environment in terms of several factors such as waste production, oil slicks, land use and emissions of greenhouse gases and other air pollutants. Related pressures negatively affect climate change and damage natural ecosystems and environment (EEA 2019).

Emissions arising from the use of nonrenewable energy resources in the feedback mechanism between climate change and energy increase the greenhouse gas density. Accordingly, climate change as the result of increased greenhouse gas density causes summers and winters to be warmer; it also changes the energy consumption and generation models.

Anthropogenic GHG emission is the key reason for climate change. Changes in temperature and availability of water are the remarkable channels which can affect the energy sector. These effects are analyzed under three titles as (i) the effect on energy demand, (ii) the effect on energy supply, (iii) other complementary effects (effects on energy infrastructures). Climate change differentiates the demand and supply in the energy sector. Heating and cooling demand patterns balance change on the demand side of the energy sector because of increasing rising temperatures. For example, while the increasing temperatures in winter decrease the heating demand, increasing temperatures in summer increase the cooling demand. However, temperature increases cause electricity consumption in housing and industry sector to increase; accordingly, higher cooling demand is expected. Positive and negative effects can be seen also on the supply side of the energy sector. Effects such as outage hours of technology can be seen because of the changes in density and

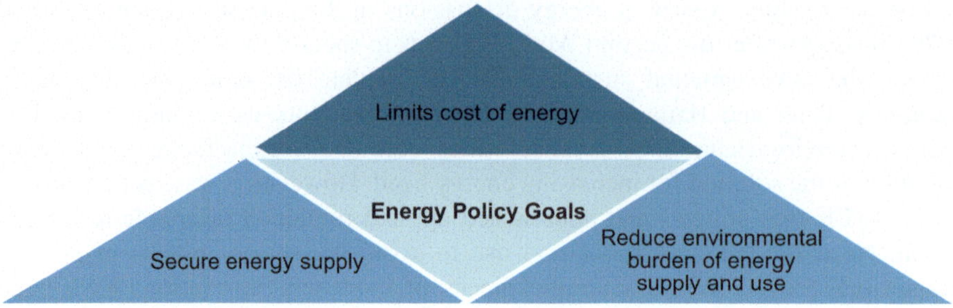

Fig. 14.4 Energy Policy Triangle or Trilemma. (Schmidt et al. 2019)

frequency of excess weather events; thermoelectric power plants and transmission lines productivity due to increasing temperature; costs and availability of fossil fuels; the presence of products for bioenergy resources; variability of wind, solar and hydroelectric resources. Again, for example, hydroelectric efficiency may increase in some of the regions due to the increasing rainfall; however, thermoelectric energy can create security gaps by the reasons of lower summer streams and higher water temperatures (Van Vliet et al. 2012; Ciscar and Dowling 2014; Auffhammer and Mansur 2014; Cronin et al. 2018; Popescu et al. 2018). Moreover, the effects that will be seen in the energy sector will reflect the whole economy because of being energy a key input during the production process.

Minimizing the climate change that is one of the main policy goals for the energy sector shows the importance that is given to the relationship between two parameters. Three main policy goals in the energy sector are as follows; limiting the energy resources, guaranteeing the energy supply and reducing the climate load including protection of the environment (See Fig. 14.4).

Four main issues should be considered while determining the energy policy targets to soften the effects of climate change (Akpan and Akpan 2012):

(a) Environmental Efficiency: Determining the level of policy to fulfill the environmental goals or actualize positive environmental results.
(b) Cost Efficiency: Level of policy goals to reach society by minimum cost.
(c) Issues on Distribution: Frequency or distribution results (justice, equality) of the policy.
(d) Institutional Feasibility: The legitimate view, acceptance, adoption and implementation of a policy instrument.

EU provides a long-term framework for its member states to overcome the cross-border effects of events that cannot be dealt with only at the national level. Within this scope, struggling with climate change has been accepted as a long-termed form factor that necessitates easy going EU activities in and out of unity. The commission suggested *Europe 2020 flagship* approach based on the use of efficient resources in Europe (EU 2011). A related approach is based on the thought that innovative solution offers are essential to activate energy, transportation, industry and technology investments; besides, there is a necessity to focus on energy efficiency policies more than ever. A series of long-dated policy plans in energy and climate fields have been determined in this context. Related connection explains the key factors that will shape the climate activities of EU for member countries to be a competitive low carbon economy till 2050.

Studies that are conducted about the impact channel of climate change for the energy sector are basically evaluated under two groups in the literature (Ciscar and Dowling 2014; Bonjean Stanton et al. 2016). The statistical relationship between climate change and energy parameters are analyzed in the first group. Those studies typically focus on a sector or a sub-sector of a system under the data limitation and remain on a regionally limited basis. In the second, the researchers, applied findings in empirical literature in larger modeling systems that are also called as integrated evaluation models.

Although the global effects of climate change on the energy sector have been researched often (Hitz and Smith 2004; Bouttes et al. 2006; Isaac and van Vuuren 2009; Ebinger and Vergara 2011; Bruckner et al. 2014), regional evaluations relating to the relationship between two parameters can be found less. Following studies in European sample cluster explain the interaction between climate change and energy; Pryor et al. (2005a, b), Tuck et al. (2006), Cleto et al. (2008), Olonscheck et al. (2011), Mima et al. (2011), Cosentino et al. (2012), Dowling (2013), Klein et al. (2013), Bardt et al. (2013), Hueging et al. (2013), Van Vliet et al. (2013), Gaetani et al. (2014), Tobin et al. (2014), Mima and Criqui (2015), Carvalho et al. (2017), Totschnig et al. (2017), Davy et al. (2018), Erdoğan (2018), Topçu (2018), and Tranberg et al. (2019).

3 Developments in Energy Sector in the European Union

Historical experiences of Europe emphasize how the economies develop and increase the energy consumption by decreasing the energy density (partially importing energy intensive goods from industrializing economies) and limiting the increases in energy (Marcotullio and Schulz 2007).

The EU has been conducting continuous research in the energy sector from the past to the present, involved in activities in issues like managing and using the available resources and creating resource diversity. Union's new alternative fund seeking of association has continued because of several reasons like nonrenewable and inadequate energy arising from fossil fuels; more energy import dependence; social security and labor costs stemming from energy generation; environmental pollution and climate change. These seekings gave result in time; ecological energy resources have become significant in the 1950s. However, uncertainties of projections in energy sector establish a remarkable challenge. The general run of developments in resources, demographic structure and technology cannot be clearly foreseen besides other events which shape the energy sector (AEO 2019; Adıgüzel 2018).

Increase in energy per capita in annual consumption volume is accepted as an indicator of civilization welfare in the modern world. Both oil and natural gas are the major energy resources that meet approximately 60% of the global demand. A person living in Europe uses an average of 27 (MWh) of energy annually, including all domestic, industrial and transport resources. This number greatly changes CO_2 emissions among countries, which are strongly linked to ecological and nuclear energy. Electricity and heat are the sectors which consume energy most in the EU in 2016 (See Fig. 14.5) (EEA 2019; Arshad 2018).

Northern European countries and EU's politically have technologically and ideologically lead ecological energies in the last 20 years (Uyanık 2018). In this direction, climate change concerns and energy security generally direct European Union to ecological energy resources.

For BP Statistical Review of World Energy (2019) data, global primary energy consumption significantly increased by the effect of natural gas and ecological energy resources in 2018. However, carbon emissions have also increased in the last decade. Primary energy consumption that decreased by 0.2% in the EU in 2018 decreased by 0.8% in the period of 2007–2017. In the same period, from non-renewable energy sources which are coal consumption was 3.3%; oil consumption 1.2%; natural gas consumption decreased by 0.8%. While nuclear energy use decreased by 0.3% in 2018, hydroelectric use increased by 15.7%. Again, while the consumption of ecological energy resources that have tended to increase as from 1989 increased by 4.8% in 2018, consumption of the same resources increased by 12.8% in the period of 2007–2017. The course of CO_2 emissions in member countries in 2018 and the period of 2007–2017 respectively are −2.0% and −1.7% (Graph 14.1).

European countries have consumed less energy in comparison to 10 years before because of the increase in energy efficiency. There is less need for fossil fuels because of energy saving and ecological energy generation quicker than expected. Some of the sectors and countries take the lead for clean energy. However, although the share of energy in the energy sector is continuously falling, fossil fuels still constitute the major weight of the energy used (Arkins 2017; Topçu 2018). With reference to 2019 calendar year estimations, energy need of Europe is met by non-renewable energy resources between 1st January and 4th September; ecological energy resources will met the same need as from 4th September (European Bioenergy Day 2019). The number of day's calculation shows that nonrenewable energy is used approximately two times more than ecological energy.

Energy dependence ratio that shows to what extent an economy hinges upon import to meet the energy demand was 54% in EU countries in 2016 (Eurostat 2018). Thus, net imports in energy equate to slightly more than half of gross domestic energy consumption. None of the European countries become self-sufficient relating to energy need; there is a dependence on external resources.

As is seen in Graph 14.2, energy resources generation increases most among energy resources both in the EU and across the world in 1990–2018. Nuclear energy among energy resources production is in the first place in the EU; about the world, oil is the first place. Natural gas energy has the lowest production rates in member countries; ecological energy has the lowest production rates in other world countries.

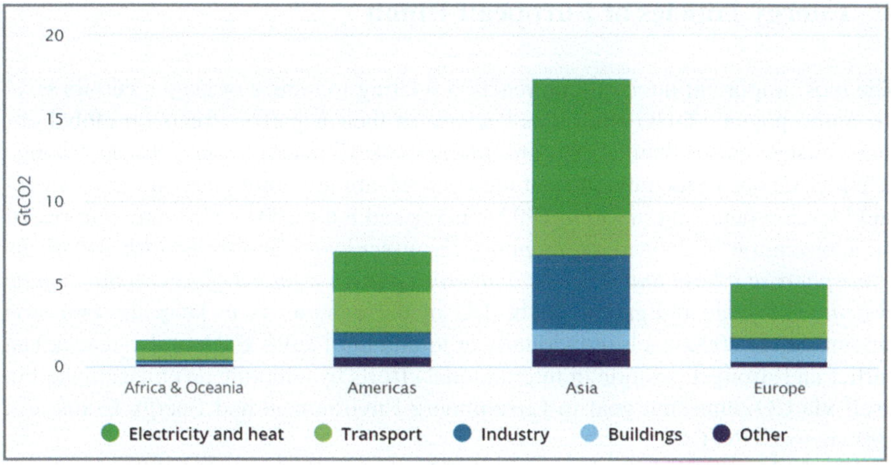

Fig. 14.5 CO_2 Emissions by Sector for Selected Regions, 2016. (IEA 2019)

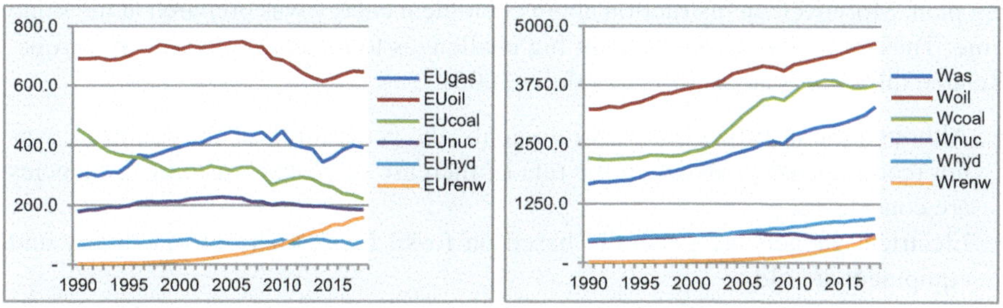

Graph 14.1 Tendency of Energy Resources Consumption in EU-28 and the World, 1990–2018. (Eurostat 2019)

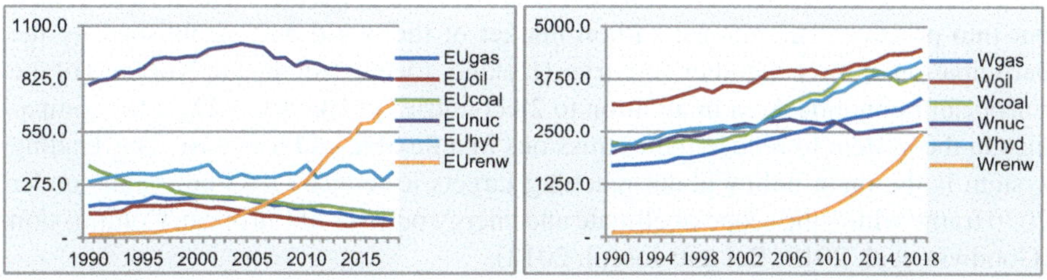

Graph 14.2 Tendency of Energy Resources Generation in EU-28 and the World, 1990–2018. (Eurostat 2019)

4 Energy Policies of European Union

The most important international reaction relating to reduce or keep a certain level the anthropogenic GHG emissions because of their negative effects on global climate change is the United Nations Framework Convention on Climate Change (UNFCCC) one of the outcomes of the United Nations Conference on Environment and Development held in Rio in 1992 (Türkeş and Kılıç 2004). The main purpose of the Convention is defined as, *stopping the greenhouse gas accumulations in the atmosphere at a level that prevents dangerous human-induced effects on the climate system*. The main obligation of the EU in the contract is to keep the emission amounts up to 2000 levels individually or jointly until 1990. EU has also reached an earlier and stronger position in international efforts by adopting an emission goal in itself via CO_2 emission goal in Luxembourg Environment and Energy Council in 1990 before UNFCCC.

Kyoto Protocol was accepted in 1997 to specify the actions after 2000 to strengthen the liabilities of Appendix 1 Parties according to thoughts of member countries about the insufficiency of UNFCCC about reducing GHG emissions (Türkeş and Kılıç 2004; Adıgüzel 2018). EU has adopted Kyoto goals to balance the greenhouse gas emissions and made the most developed emission buying and selling plan. Moreover, an instruction about ecological energy was prepared at the same time. These initiatives have brought big challenges to the energy sector of Europe. Related difficulties are as follows (Helm 2005):

- Although the insufficiency of Kyoto goals, the general run of the countries cannot reach related goals when the rate of increase of global warming tendencies are considered,
- Electrical gadgets are generally based on fossil fuels; many of machinery and equipment are old,
- Being more expensive the ecological technologies in comparison to fossil fuels increases the pressure on competitive capacity,
- Expiry of first generation nuclear power plants requires new investment areas in some countries.

EU leads for emission trade at the regional scale by focusing on the market-driven solution to meet the goals of the Kyoto Protocol by a minimum cost. For this purpose, the *EU Emissions Trading Scheme (EU ETS)* was established in 2005 and put into practice. The biggest carbon market of the world was established by the participation of three member countries (Iceland, Liechtenstein, and Norway) of the European Economic Area in addition to 28 countries of EU. EU ETS is for companies in the system to reduce their emissions in a flexible and low-cost. The Trading system is the main policy about reaching targets to reduce GHG emissions under 2030 frame within the scope of climate and energy policies of European Commission (Goodwin et al. 2018; Bel and Joseph 2018).

Issues of environment and energy are discussed together in the EU. There is an authority sharing between member countries and unity in related fields. European

Commission was entitled to conduct climate change negotiations within powers and limitations on behalf of the member countries by the *Lisbon Treaty* in 2007. International climate change contracts become a part of the judicial system of UE following signatures. EU institutions control the applications of member countries; this is one of the developments which contribute to the leadership of the EU about climate change. There is a European Commission that follows and reports the applications in member countries according to the declarations of member countries within EU system; there also is EU International Court of Justice that has the authority to rule against member countries (Erdoğan 2018).

EU made a commitment to decreasing emissions by 8% in comparison to the 1990 level within the scope of the Kyoto Protocol in the period of 2008–2012 (Erdoğan 2018). While it improved its position in international negotiations after protocol, there was adopted a series of a new policy initiative to consider. New policy initiatives include following items; longer-term carbon agreements; integration of road, air and railroad transportation into the emission trade plan; developing suggestions for other countries which are not member of the union as a part of negotiations after Kyoto; using clean development mechanisms and common applications on a large scale (Helm 2005).

Almost all the EU countries followed policies toward increasing ecological energy investments to develop ecological energy resources. These policies were organized within the frame of EU ecological energy instruction; there has generally been focused on wind energy. Scope of ecological energy resources was enlarged; it was also argued about how ecological energy liability and EU ETS can be banded together to provide carbon reduction results. Scale economies have been considered for some of the large-scaled technologies (coal, hydrogen, nuclear, etc.) toward collaboration in Research and Development (R&D) surveys. EU maintains its R&D activities in line with various directives to increase, diversify and use ecological energy resources in more sectors. Required investments have continuously been made for R&D activities; there is prepared a technological infrastructure at the same time (Helm 2005; EEA 2018; Adıgüzel 2018).

EU aims to proceed to an economy with lower carbon within the scope of climate change and clean and efficient energy use by the initiative of *Resource Efficient Europe*. It is decided in Union that framework conditions need to be improved to ideally use financial instruments (structural funds, community programs, etc.) and market-based instruments (emission trading, energy taxation, etc.) at Union level (Akbaş and Apar 2010).

Energy has become a policy priority of Europe in time. There are goals that are also called as 20-20-20 in struggling with climate change in the *Europe 2020* strategy that was adopted in 2010. Europe 2020 strategy aims member country economies to achieve an intelligent, sustainable, environmentalist and inclusive growth (Topçu 2018). The document emphasizes generalizing eco-friendly ecological energy resources besides minimizing GHG emissions. It is emphasized on achieving more effective use of energy resources due to new technologies (EEA 2018; Erdoğan 2018). Climate and energy goals for 2020 are as follows; reducing GHG emissions at least by 20% and 30% when the circumstances allow; increasing the share of ecological energy in energy consumption to 20%; providing energy efficiency by 20% (Topçu 2018). One of the five developmental areas among the targets of Europe 2020 strategy is the energy. Besides the specific goals of Europe 2020 energy strategy, extra interests have been discussed by the European Commission. Related policies include the issues below (EEA 2019; Adıgüzel 2018):

- Increasing energy supply security: EU energy supply security policy aims to protect from each kind of risks that can threaten both economy and social welfare order. It is aimed in parallel with this purpose to minimize the energy problem in the short terms and completely remove the related problem in the long term.
- Providing economic usability of energy and competitiveness of European economy: A competitive energy sector policy is essential in terms of increasing the social welfare level; increasing competitive capacity in production by reducing the energy costs; procuring energy by getting more efficiency with low cost.
- Encouraging to develop a competitive energy domestic market: Energy domestic market was founded by the European Coal and Steel Community Agreement. It was aimed for member countries to liberalize and enter into rivalry in some of the energy sectors by the energy directives that were published in the 1990s. It was decided to establish a single market especially in electricity and natural gas sectors. However, there were difficulties because of the problems of some of the companies during the adaptation process to energy prices. Besides, member countries have also problems in the phase of the implementation of energy prices.
- Determining minimum energy taxation levels: Especially the electricity and natural gas prices in the EU continuously increase in EU. There was seen about 4% increase in electricity prices for households in the period between 2008 and 2012. This increase reflected the economic units of the market in the supply and demand side as a tax boost. However, there has been incrementally shown decreases in electricity prices after 2012. Similarly, natural gas prices that tended to increase till 2015 have been in the tendency to decrease in the later process. The key reason for this circumstance is the increasing directional share of ecological energy resources within the total energy resources. Increasing the production of ecological energy resources causes prices to decrease; this issue decreases the effective rates of tax at the same time.

Another energy policy of the EU is to create a common market for the member countries which are geographically at different locations; expansion movement serves this purpose. Strategic activities that toward restricting energy import in the long term via creating an energy policy which is interdependent in it but independent from outside are among the targeted policies. In addition to all these, member countries follow policies to strengthen relationships by means of making exceptions for other countries. Energy policies are always built on sustainable growth targets. These targets are endeavored to be kept up-to-date by adapting to present conditions. By this means, it is tried to provide quality living standards for EU citizens in economic, social and cultural areas (Adıgüzel 2018; Topçu 2018).

Energy and environment indicators in EU are evaluated around six policy questions below (EEA 2019):

1. Has energy generation and consumption a decreasing effect on environmental pollution?
2. Does energy consumption decrease?
3. How fast is energy efficiency increasing?
4. Does less polluting fuel replace more harmful fuel?
5. How fast are ecological energy technologies applied?
6. How well are environmental costs included in the pricing system?

Related indicators play significant roles in reporting about greenhouse gas emissions and estimations in Europe according to the Kyoto protocol. This same significant role is also seen in preparing EU greenhouse inventory report for the United Nations Framework Convention on Climate Change.

Contracting countries of Warsaw Conference in 2013 were invited to submit Intended Nationally Determined Contribution; many countries informed their commitment. Within this framework, the EU promised a hard target of decreasing GHG emissions by at least 40% in comparison to the level in 1990 till 2030 (Topçu 2018).

Another remarkable step is the Paris Agreement in 2015. The chief goal of the agreement is to keep the global temperature rise under 2 °C in the twenty-first century and also offer resistance to climate change threat by making effort towards limiting temperature rise up to 1.5 °C (UNFCCC 2019).

Meaning of this target is a world in which countries dispense with oil and natural gas and start to use ecological energy while there is a use for fossil fuels by more than 80% by the moment (Karakaya 2016). Actualizing 2030 climate and energy goals of the Paris Agreement and EU so as to be in accord with Energy Association shows us energy supply solutions and fundamental changes in fossil fuels use that is the primary source of GHG emissions will be changeable (Goodwin et al. 2018).

Applications in the energy field within the borders of EU are performed by decision maker mechanisms at the level of conformity to common interests. Council of Europe makes decisions by determining policies toward supporting common interests by meeting at regular intervals.

Decisions in those meetings are expressed to member countries via directives. Developed policies provide energy need to be met; regional sectors that have strategically importance in terms of energy can remain on the market and compete with foreign countries due to these policies. Association head towards to make agreements for global competition. Protocols are brought into force to effectively use quality, sustainable and eco-friendly energy resources. Factors toward increasing energy supply security, energy generation, distribution, and consumption are considered in related agreements.

5 Empirical Analysis

The goal of this research was to analyze the relationship between climate change and energy resources within the scope of member countries of Europe for the period between 1990 and 2018.

5.1 *Materials and Method*

Relationships between variables belong to energy resource parameters and CO_2 emissions belong to climate change parameter were analyzed statics panel data method and Root Mean Square Error (RMSE) methodologies. Internal vectors belong to datasets that are established to determine explanatory parameters belong to the energy consumption of climate change are as follows:

$$CO_2 = \left(coal, gas, oil, renw\right) \tag{14.1}$$

CO_2 variable contained in Vector (1) shows carbon dioxide emission ($mtCO_2$); *coal* variable, coal consumption (mtoe); *gas* variable, natural gas consumption (mtoe); *oil* variable, oil consumption (mtoe); *renw* variable represents hydroelectric, wind, geothermal, solar, biomass and biofuel energy consumption (mtoe). Vector (1) is established for statics panel data model as follows:

$$CO2_{it} = \alpha_{it} + \beta_1 coal_{it} + \beta_2 gas_{it} + \beta_3 oil_{it} + \beta_4 renw_{it} + \mu_{it} \tag{14.2}$$

α_{it}, in model is constant parameter; β_1, $\beta_2 ve \beta_3$, shows slope parameter; μ_{it} is the error term. The sub-symbol *i*, means the number of units selected from 14 countries that are members of the European Union[1]; *t* is the time interval of 1990–2018. All the variables given were modeled in logarithmic form structure (Table 14.2).

$$CO_2 = \left(coal, gas, oil, nuc, hyd, wnd, geo, sol, bio\right) \tag{14.3}$$

CO_2 variable contained in Vector (3) explains carbon dioxide emissions ($mtCO_2$); *coal* variable, coal consumption (mtoe); *gas* variable, natural gas consumption (mtoe); *oil* variable, oil consumption; *nuc* variable, nuclear energy consumption; *hyd* variable, hydroelectric energy consumption (mtoe); *wnd* variable, wind energy consumption (mtoe); *geo* variable, geothermal energy consumption (mtoe); *sol* variable, solar power consumption; *bio* variable, biomass energy consumption. Vector (3) is established for time series model in 28 EU member countries integrated sample as follows:

[1] The panel data analysis was carried out in Belgium, Denmark, Finland, France, Germany, Greece, Hungary, Italy, Luxembourg, the Netherlands, Poland, Spain, Sweden and the UK sample set from EU member countries due to data limitation.

Table 14.2 Descriptive statistics belong to variables in vector (1)

Variable		Mean	Std. Dev.	Min	Max	Observations
lnCO$_2$	Overall	4.969674	1.161452	2.082592	6.910809	N = 406
	Between		1.197312	2.354203	6.725748	n = 14
	Within		.1205228	4.558136	5.359514	T = 29
lncoal	Overall	2.074261	1.573971	−3.180189	4.879233	N = 406
	Between		1.576451	−2.038161	4.449301	n = 14
	Within		.4049714	.5721997	4.208096	T = 29
lngas	Overall	2.250785	1.653274	−3.126442	4.473851	N = 406
	Between		1.637384	−.2540244	4.267764	n = 14
	Within		.4874864	−.8704506	3.678033	T = 29
lnoil	Overall	3.284635	1.111347	.4974198	4.960708	N = 406
	Between		1.143824	.9209949	4.851626	n = 14
	Within		.1311931	2.829916	3.67616	T = 29
lnrenw	Overall	.851439	1.934468	−4.070315	3.934934	N = 406
	Between		1.819971	−3.120948	2.863918	n = 14
	Within		.8117084	−1.156319	3.469181	T = 29

Table 14.3 Descriptive statistics belong to variables in vector (3)

Variable	Mean	Std. Dev.	Min	Max	Observations
CO$_2$	3991.205	279.3114	3458.022	4353.423	29
coal	320.4926	55.24047	222.3673	455.2439	29
gas	383.4858	46.93309	297.7198	448.2392	29
oil	696.0212	40.82710	616.2603	748.6718	29
nuc	205.0312	13.90354	179.8376	228.2815	29
hyd	75.51248	5.362622	65.60230	85.93772	29
wnd	24.37506	27.16986	0.176372	85.70727	29
geo	1.137798	0.253306	0.721139	1.502381	29
sol	6.408901	10.13096	0.002602	28.90683	29
bio	18.31528	14.10037	3.329364	43.55793	29

$$CO2_t = \alpha_t + \beta_1 coal_t + \beta_2 gas_t + \beta_3 oil_t + \beta_4 nuc_t + \beta_5 hyd_t + \beta_6 wnd_t + \beta_7 geo_t + \beta_8 sol_t + \beta_9 bio_t + \mu_t \tag{14.4}$$

α_t, in model shows constant parameter; β_1, β_2, β_3, β_4, β_5, β_6, β_7, β_8*ve*β_9 explains slope parameter; $\mu_{t,}$ represents error term; sub-symbol shows the time interval of 1990–2018. All the variables given were included in model by natural form structures (Table 14.3).

All the data in the analysis were reached by the world energy statistics reviews website. As is seen in Graph 14.3 that shows the change of time path of series, demand for ecological energy resources have increased in recent years.

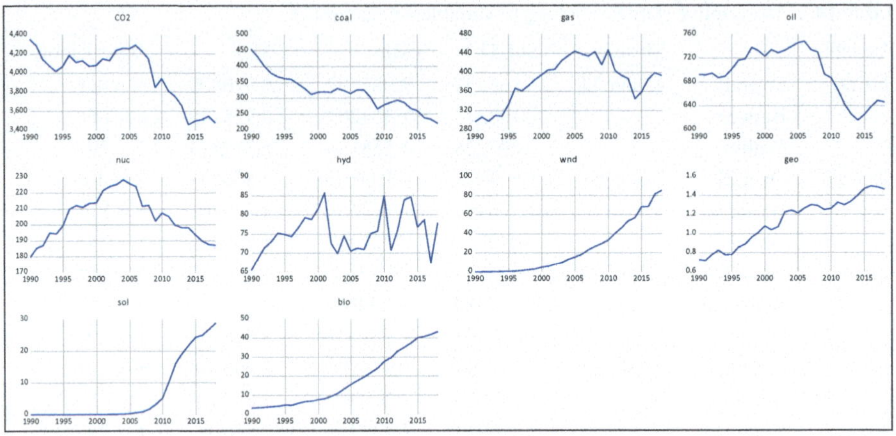

Graph 14.3 Time Path Change of Series

Table 14.4 Tests used to decide among estimators

Tests	Statistical values	
F_stat.	F(13, 388) = 95.46	Prob > F = 0.0000
LR test	$LR_{\text{\_unit effect}}$ x^2 (01) = 473.78	Prob > = x^2 = 0.000
	$LR_{\text{\_time effect}}$ x^2 (01) = 0.00	Prob > = x^2 = 1.000
	$LR_{\text{\_unit-time}}$ x^2 (2) = 474.70	Prob > = x^2 = 0.000
Score test	x^2 (01) = 473.78	Prob > = x^2 = 0.000
ALM test	$ALM_{\text{\_one sided}}$ = 35.99	Prob > *N* (0,1) = 0.000
	$ALM_{\text{\_two sided}}$ = 1295.03	Prob > x^2 (1) = 0.000
Wooldridge test	F (1, 13) = 32.666	Prob > F = 0.0001
Hausman test	x^2 (4) = 60.45	Prob > x^2 = 0.000

5.2 *Findings*

Preference procedures for panel data estimators are determined by the test findings based on the classical model, fixed-effects and random-effects models (Table 14.4).

F test findings in model based on fixed effects estimator and *F* statistics with (13, 388) degree of freedom and probability value which show that significance of panel unit effects is tested by composite hypothesis prove us there are unit effects. For *LR* test statistics, unit effects are valid in the model; thus, the classical model is not proper as of the moment. *LR* ratio test findings mean there is no time effect. Moreover, Score test that is derived from *LR* ratio test and suggested due to stronger small sample characteristics was performed. With reference to the findings, the classical model is not proper because of the presence of unit effects when *LR* test is supported. Results that are obtained by *LM* test do not reflect the truth in case of the model has autocorrelation problem. This is because *ALM* test that gives reliable results of autocorrelation was performed instead of *LM* test. Findings show us

Table 14.5 Test findings of assumptions belong to the fixed effects model with unit effect

Assumptions	Tests	Statistical values
Heteroscedasticity	Wald test	$x^2(14) = 5113.23$ $Prob > x^2 = 0.0000$
Autocorrelation	Bhargava et al. Baltagi-Wu	DW = .1991 LBI = .3347
Cross- sectional dependence	Frees' test	CD_stat. = 3.463 (alpha = 0.10: 0.0892; alpha = 0.05: 0.1160; alpha = 0.01: 0.1660)

one-way test statistics comply with standard normal distribution. It is seen when one-way and two-way test findings are reviewed together that unit effect is accepted; classical model is inappropriate. Wooldridge test surveys autocorrelation problem besides unit effects. For *F* statistics and probability value, there are first order autocorrelation and unit effects in model. Accordingly, it is expressed by denying H_0 hypothesis that classical model is inappropriate.

For findings of *F*, *LR*, *Score*, *ALM* and Wooldridge test findings for the classical model, the classical model is inappropriate for estimator preference that is used for analysis.

Findings belong to Hausman test statistics show that assumptions of random effects model are not met. The rejection of the H_0 hypothesis reveals that random effects estimator is not consistent and unbiased; estimation process needs to continue by fixed effects model.

Table 14.5 explains the test findings belong heteroscedasticity, autocorrelation and cross-sectional dependence that was conducted to determine the efficiency of one-way fixed effects model.

With reference to the findings, there are heteroscedasticity and first-order autocorrelation problems based on units in the model; remains between cross-section units are in correlation with each other. Table 14.6 shows the findings belong to Driscoll-Kraay estimator that is used to correct deviations of assumption to remove the problems.

Findings which reflect fixed effects model prove us *F* statistics from descriptive statistics is significant as a whole in 95% confidence interval. For determination coefficient intragroup R^2 value, independent variables in model explain almost 67% of variability in CO_2 emissions.

With reference to *t* statistics value and 0.01 significance levels, effects of coal, natural gas, oil and ecological energy resources consumption on CO_2 emissions are statistically significant.

For findings that were obtained under statistical significance for variables in Vector (2), below equation can be founded;

$$CO2_{it} = 0.1173 coal_{it} + 0.0600 gas_{it} + 0.5446 oil_{it} - 0.0443 renw_{it}$$

It is seen when the equation that is obtained by the coefficient tale that shows estimation results belong to the model that coal, natural gas, and oil consumption

Table 14.6 Correcting deviations by assumptions for fixed effects model with unit effect

lnCO₂	*Coefficients*	*Drisc/Kraay Std. Err.*	$t\,(\lvert P > t\rvert)$	[95% Conf.	Interval]
Lncoal	.1173096	.0169725	6.91 (0. 000)	.0806427	.1539766
Lngas	.0600236	.011523	5.21 (0.000)	.0351297	.0849175
Lnoil	.5445692	.0508253	10.71 (0.000)	.4347678	.6543707
Lnrenw	−.0442938	.0071047	−6.23 (0.000)	−.0596426	−.028945
C	2.840245	.1889616	15.03 (0.000)	2.432019	3.248472
id_2	−.0974457	.0387197	−2.52 (0.012)	−.1735724	−.0213189
id_3	−.0581547	.0364097	−1.60 (0.111)	−.1297397	.0134303
id_4	.4055302	.0420833	9.64 (0.000)	.3227904	.4882701
id_5	.7856545	.0609222	12.90 (0.000)	.6658756	.9054335
id_6	.0733027	.0291277	2.52 (0.012)	.0160348	.1305706
id_7	−.0960718	.0462003	−2.08 (0.038)	−.1869061	−.0052375
id_8	.5391457	.0399267	13.50 (0.000)	.460646	.6176454
id_9	−.6665905	.0919024	−7.25 (0.000)	−.8472794	−.4859015
id_10	.251085	.0218297	11.50 (0.000)	.2081657	.2940043
id_11	.8283758	.0323014	25.65 (0.000)	.7648681	.8918835
id_12	.37147	.0369635	10.05 (0.000)	.2987961	.4441439
id_13	−.0981414	.035929	−2.73 (0.007)	−.1687814	−.0275014
id_14	.6303309	.0428833	14.70 (0.000)	.5460181	.7146436
Diagnostic statistics					
$F\_stat\,(4, 13) = 40.90\ Prob > F = 0.000$					
$withinR^2 = 0.6717$					

increase CO_2 emissions; ecological energy consumption decreases CO_2 emissions. Hereunder, while coal, natural gas and oil consumption increase by 1%, CO2 emissions respectively increase by 0.12%, 0.06% ve 0.54%. A 1% increase in ecological energy consumption decreases CO_2 emissions by 0.04%.

Root Mean Square Error (RMSE) method is generally used in climatology, estimation and regression analyses to confirm results based on an experiment. Therefore, the RMSE method was applied to compare the results of statics panel data analysis.

RMSE method is the standard deviation of estimation remains. Related remains reveal the distance between data point or points and regression. RMSE gives circulation of related remains; it also shows how much data is concentrated around the most appropriate line (Barnston 1992):

$$RMSE = \sqrt{(f - o)^2} \tag{14.5}$$

f notation in formula shows estimation values; *o* notation explains values observed; the symbol on square differences describes the average $(f - o)^2$.

Below equation is seen when RMSE is expressed by a different formula:

$$RMSE_{fo} = \left(\sum_{N}^{i=1} \left(z_{fi} - z_{oi} \right)^2 / N \right)^{1/2} \tag{14.6}$$

$\sum$= total in equation explains $(z_{fi} - z_{oi})$ *sup* > 2=differences; squares and N= sample size.

Based on size of data,

$$RMSE = \sqrt{1 - r^s} SD_y \tag{14.7}$$

SD_y, is standard deviation of Y.

RMSE inputs have a linear relationship with correlation coefficient when standard observations and estimations. RMSE value is between0 *and*∞. The value that closes to 0 means those estimators has better performance.

Effects of nonrenewable and ecological energy resources on CO_2 emissions were reviewed in the analysis that was conducted for the time interval of 1990–2018 in EU-28 sample. Graph 14.4 show the findings.

Graph 14.4 shows that the lowest RMSE ratio for CO_2 emissions and input parameters belong to energy resources is in geothermal energy variable. The related finding explains to us geothermal energy has the maximum effect on CO_2 emissions; the variable that has the lowest effect is oil energy consumption.

For RMSE values in Table 14.7; the alignment from high too low for energy variables that affect CO2 emissions is geothermal, solar, biomass, wind, hydroelectric, nuclear, coal, natural gas, and oil energy resource consumption. Within the context, it is seen that ecological energy use on climate change delivers better performance on climate change. Thus, it can be deduced that there is a need to concentrate on ecological energy resources consumption in struggling with climate change.

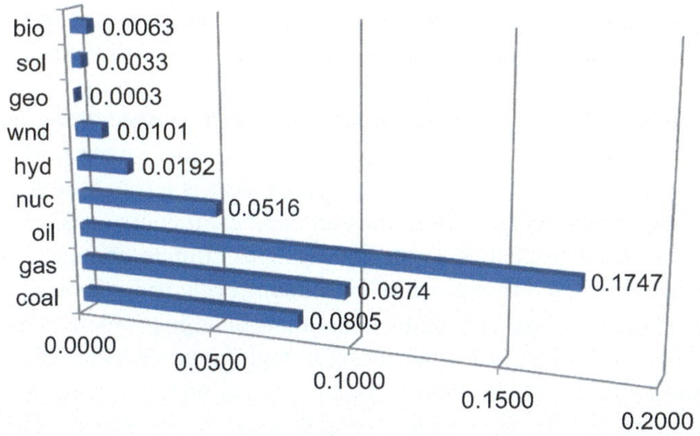

Graph 14.4 Effects of Energy Resources on CO_2 Emissions

Table 14.7 RMSE values belong to relationship between energy resources and CO_2 emissions

Output -CO_2	Input1-coal	Input2-gas	Input3-oil	Input4-nuc	Input5-hyd	Input6-wnd	Input7-geo	Input8-sol	Input9-bio
RMSE	0.0805	0.0974	0.1747	0.0516	0.0192	0.0101	0.0003	0.0033	0.0063

6 Solutions and Recommendations

Historical experiences represent those solutions toward climate policies are formulated based on short or long-termed perspectives. With reference to the available technological, institutional and behavioral patterns change in energy systems hang on. Thereby, climate strategies and policies are obliged to look at future centuries about the transformation of the energy sector (Fouquet 2016).

The truth of the modern day there is a revolution that transformed the global energy sector. This is because adopting a multidisciplinary and holistic approach with natural and social sciences besides economy and investment theories is important in terms of determining, developing and applying distribution strategies. This revolution is not only a technological process but also a social and political process in which technology is a determinant. Therefore, the needs and demands of society should be considered besides providing support of different stakeholders for transition stages to nonrenewable energy resources to ecological energy resources to be effective. In addition to international agreements, it is obligatory to consider small and large scale infrastructure projects which support the transition stage. A strong legal framework at both international and national scale should be established to guide investors and minimize the risks.

A more multidisciplinary approach is at the center of sustainable development activities. Policymakers ought consistently to follow parallel objectives of protecting the natural environment and minimizing or removing climate change. There is a need for extensive active participation at both the national and international level. Energy policies should focus on minimizing climate change that is one of the most urgent problems of the modern world; solve water scarcity; protecting food chains from deterioration; avoiding the collapse of political corporations.

Bioenergetics is the only energy resource that can provide three main energy resources as bio-warming/cooling, bio-power, and bio-fuel which both households and companies need. Increasing bioenergy generation is a remarkable point based on the thought that bioenergy is more than an ecological energy resource.

Many factors need to change in struggling with climate change. This circumstance provides to be used this point of investment cycle to make an investment in resources except for carbon. In conclusion, climate change problem needs to be met with big and new investments. Furthermore, it ought to be provided effects of modeling skills of several results including climate change on the European energy market to develop at EU level. Energy policies need to be modernized and kept up-to-date to get better results about passing to the European economy with lower carbon and minimizing supply security arising from external dependence; avoiding possible oil shocks.

Integrating energy infrastructure into other infrastructure services via long termed plans and also completing proper investment incentives is another policy proposal (Helm 2005). There is a need for reforms in regulatory rules and market conditions with competition policy to ease long termed contracts and encourage investments. Moreover, it should be endeavored to establish gas security and storage regimes at the regional level; deepen and enlarge EU ETS; consider long termed carbon contracts; ease information interchange relating to following capacity plans; focus on issues like coordination of supply security.

The 2030 year climate and energy goals of the Paris Agreement and EU necessitate fundamental changes in the use of fossil fuel that is the primary source of GHG emissions which cause climate change (Goodwin et al. 2018). This is because changeable energy supply solutions need to be sought. Big investments should be made in new energy infrastructures, systems and supply chains for laying foundations for a more sustainable future for the next generations. It is accepted as the best choice for the EU to maintain its leadership role by more prime targets.

7 Conclusion

Statics panel data analysis and Root Mean Square Error methodology were followed to monitor the effectiveness of energy policies that are applied in struggling with climate change by the EU for the period between 1990 and 2018. Variables of oil, natural gas, coal, nuclear power, hydro-electric power, wind energy, geothermal energy, biomass energy and CO_2 emissions which have the biggest effect on climate change were used in parallel with this purpose.

It is concluded in line with the fixed effects model with the unit effect that coal, natural gas, and oil consumption increase CO_2 emissions; ecological energy consumption decreases CO_2 emissions. The variable that has the biggest effect on CO_2 emissions is oil consumption. For RMSE findings, the variable that has the highest impact on CO_2 emissions is geothermal energy consumption; however, the variable that has the lowest impact on CO_2 emissions is oil energy consumption.

References

Adıgüzel GE (2018) Avrupa Birliği'nin yenilenebilir enerji dinamikleri ve Türkiye'nin uyumu. Tez, Başkent Üniversitesi

AEO Annual Energy Outlook (2019) With projections to 2050. U.S. Energy Information Administration Office of Energy Analysis U.S. Department of Energy, Washington, DC 20585. www.eia.gov/aeo. Accessed 5 Apr 2019

Akbaş G, Apar A (2010) Avrupa 2020 Stratejisi: Akıllı, sürdürülebilir ve kapsayıcı büyüme için Avrupa Stratejisi özet bilgi notu. T.C. Başbakanlık Avrupa Birliği Genel Sekreterliği Sosyal, Bölgesel ve Yenilikçi Politikalar Başkanlığı, 14 Eyl 2010, Ankara

Akpan UF, Akpan GE (2012) The contribution of energy consumption to climate change: a feasible policy direction. Int J Energy Econ Policy 2(1):21–33
Arkins K (2017) Energy in Europe-State of play. Environment & Me/EEA. https://www.eea.europa.eu/signals/signals-2017/articles/energy-in-europe-2014-state-1. Accessed 9 Apr 2019
Arshad M (2018) Cogeneration policy to contest climate changes and the future of sustainable energy. IOP Conf Ser Earth Environ Sci 200. https://doi.org/10.1088/1755-1315/200/1/012050
Auffhammer M, Mansur ET (2014) Measuring climatic impacts on energy expenditures: a review of the empirical literature. Energy Econ 46:522–530. http://tuck-fac-cen.dartmouth.edu/images/uploads/faculty/erin-mansur/Auffhammer_Mansur_climateadaptenergy.pdf
Bardt H, Biebeler H, Haas H (2013) Impact of climate change on the power supply in France, Germany, Norway and Poland: a study based on the IW climate risk indicator. CESifo Forum 14(4):33–41
Barnston AG (1992) Correspondence among the correlation, RMSE, and Heidke forecast verification measures; refinement of the Heidke score. Weather Forecast 7(4):699–709
Bel G, Joseph S (2018) Policy stringency under the European Union Emission trading system and its impact on technological change in the energy sector. Energy Policy 117:434–444
Bonjean Stanton MC, Dessai S, Paavola J (2016) A systematic review of the impacts of climate variability and change on electricity systems in Europe. Energy 109:1148–1159. https://doi.org/10.1016/j.energy.2016.05.015
Bouttes JP, Leban R, Trochet JM (2006) A low carbon electricity scenario – As a contribution to the energy policy and climate change debate. Working Paper:1–30. https://www.researchgate.net/publication/238796351
BP (2019) BP statistical review of world energy 2019: 68th edition. https://www.bp.com/content/dam/bp/business-sites/en/global/corporate/pdfs/energy-economics/statistical-review/bp-stats-review-2019-full-report.pdf. Accessed 5 Apr 2019
Bruckner T, Bashmakov IA, Mulugetta Y et al (2014) Climate change 2014: mitigation of climate change. In: Edenhofer O, Pichs-Madruga R, Sokona Y et al (eds) Energy systems, contribution of working group III to the fifth assessment report of the intergovernmental panel on climate change, Cambridge
Carvalho D, Rocha A, Gomez-Gesteira M et al (2017) Potential impacts of climate change on European wind energy resource under the CMIP5 future climate projections. Renew Energy 101:29–40. https://doi.org/10.1016/j.renene.2016.08.036
Ciscar JC, Dowling P (2014) Integrated assessment of climate impacts and adaptation in the energy sector. Energy Econ 46:531–538
Cleto J, Simoes S, Fortes P et al (2008) Renewable energy sources availability under climate change scenarios – impacts on the Portuguese energy system. In: 2008 5th international conference on the European electricity market. https://doi.org/10.1109/EEM.2008.4579074
Cosentino SL, Testa G, Scordia D et al (2012) Future yields assessment of bioenergy crops in relation to climate change and technological development in Europe. Ital J Agron 7(2):154–166
Croce P, Formichi P, Landi F et al (2018) The snow load in Europe and the climate change. Clim Risk Manag 20:138–154
Cronin J, Anandarajah G, Dessens O (2018) Climate change impacts on the energy system: a review of trends and gaps. Clim Chang 151:79–93. https://doi.org/10.1007/s10584-018-2265-4
Davy R, Gnatiuk N, Pettersson L et al (2018) Climate change impacts on wind energy potential in the European domain with a focus on the Black Sea. Renew Sust Energ Rev 81(2):1652–1659
Dovi V, Battaglini A (2015) Energy policy and climate change: a multidisciplinary approach to a global problem. Energies 8:13473–13480. https://doi.org/10.3390/en81212379
Dowling P (2013) The impact of climate change on the European energy system. Energy Policy 60:406–417
Ebinger JO, Vergara W (2011) Climate impacts on energy systems: key issues for energy sector adaptation. World Bank, Washington, DC. http://documents.worldbank.org/curated/en/580481468331850839/Climate-impacts-on-energy-systems-key-issues-for-energy-sector-adaptation

EEA European Environment Agency (2018) Trends and projections in Europe 2018 tracking progress towards Europe's climate and energy targets. EEA Report No: 16/2018. https://www.eea.europa.eu/publications/trends-and-projections-in-europe-2018-climate-and-energy

EEA European Environment Agency (2019). https://www.eea.europa.eu/themes/energy/intro. Accessed 5 Apr 2019

Erdoğan S (2018) İklim değişikliğine karşı verilen küresel mücadele ve Avrupa Birliği. MANAS Sosyal Araştırmalar Dergisi 7(4):704–718

EU European Union (2011) Communication from the Commıssıon to the European Parliament, the Council, the European Economic and Social Committee and the Committee of the Regions: a roadmap for moving to a competitive low carbon economy in 2050: 1–15. https://www.cbss.org/wp-content/uploads/2012/12/EU-Low-Carbon-Road-Map-2050.pdf

European Bioenergy Day (2019). http://www.europeanbioenergyday.eu/. Accessed 16 July 2019

Eurostat (2018) Key figures on Europe statistics illustrated 2018 edition. In: Strandell H, Wolf P (eds) Publications Office of the European Union, Luxembourg. https://ec.europa.eu/eurostat/documents/3217494/9309359/KS-EI-18-001-EN-N.pdf/0b8d8b94-541d-4d0c-b6a4-31a1f9939a75. ISBN: 978-92-79-77871-1

Eurostat (2019). https://ec.europa.eu/eurostat/statistics-explained/index.php/Greenhouse_gas_emission_statistics. Accessed 5 May 2019

Fouquet R (2016) Lessons from energy history for climate policy: technological change, demand and economic development. Energy Res Soc Sci 22:79–93

Gaetani M, Huld T, Vignati E et al (2014) The near future availability of photovoltaic energy in Europe and Africa in climate-aerosol modeling experiments. Renew Sust Energ Rev 38:706–716

Goodwin J, Young K, Tomescu M et al (2018) Integrating climate and energy projections in practice. Eionet Report – ETC/ACM 2018/5

Helm D (2005) European energy policy: securing supplies and meeting the challenge of climate change. Paper prepared for the UK Presidency of the EU, pp 1–10. http://www.offnews.info/downloads/european_energy.pdf

Hitz S, Smith J (2004) Estimating global impacts from climate change. Glob Environ Chang 14(3):201–218

Hueging H, Haas R, Born K et al (2013) Regional changes in wind energy potential over Europe using regional climate model ensemble projections. J Appl Meteorol Climatol 52(4):903–917

International Energy Agency (2019). https://www.iea.org/statistics/co2emissions/. Accessed 12 Apr 2019

IPCC Intergovernmental Panel on Climate Change (2014) Climate change 2014 synthesis report summary for policymakers. http://www.ipcc.ch/pdf/assessmentreport/ar5/syr/AR5_SYR_FINAL_SPM.pdf

Isaac M, van Vuuren DP (2009) Modeling global residential sector energy demand for heating and air conditioning in the context of climate change. Energy Policy 37(2):507–521

Karakaya E (2016) Paris anlaşması: içeriği ve Türkiye üzerine bir değerlendirme. Adnan Menderes Üniversitesi Sosyal Bilimler Enstitüsü Dergisi 3(1):1–12

Klein DR, Olonscheck M, Walther C et al (2013) Susceptibility of the European electricity sector to climate change. Energy 59:183–193. https://doi.org/10.1016/j.energy.2013.06.048

Marcotullio PJ, Schulz NB (2007) Comparison of energy transitions in the United States and developing and industrializing economies. World Dev 35(10):1650–1683

Mima S, Criqui P (2015) The costs of climate change for the European energy system, an assessment with the POLES model. Environ Model Assess 20(4):303–319

Mima S, Criqui P, Watkiss P (2011) The climate cost project (final report). In: Watkiss P (ed) The impacts and economic costs of climate change on energy in Europe. Summary of results from the EC RTD climate cost project, vol 1: Europe. The Stockholm Environment Institute, Sweden

Moreau V, Neves CA, De O et al (2019) Is decoupling a red herring? The role of structural effects and energy policies in Europe. Energy Policy 128:243–252

National Oil and Gas Services Association (2019). https://nangs.org/analytics/bp-statistical-review-of-world-energy. Accessed 6 July 2019

Olonscheck M, Holsten A, Kropp J (2011) Heating and cooling energy demand and related emissions of the German residential building stock under climate change. Energy Policy 39(9):4795–4806

Popescu GH, Mieila M, Nica E et al (2018) The emergence of the effects and determinants of the energy paradigm changes on European Union economy. Renew Sust Energ Rev 81(1):768–774

Pryor SC, Barthelmie RJ, Kjellström E (2005a) Potential climate change impact on wind energy resources in northern Europe: analyses using a regional climate model. Clim Dyn 25(7–8):815–835

Pryor SC, Schoof JT, Barthelmie RJ (2005b) Climate change impacts on wind speeds and wind energy density in Northern Europe: empirical downscaling of multiple AOGCMs. Clim Res 29(3):183–198

Rübbelke D, Vögele S (2010) Impacts of climate change on European critical infrastructures: the case of the power sector, BC3 working paper series: 2010-08. Basque Centre for Climate Change (BC3), Bilbao

Rübbelke D, Vögele S (2011) Impacts of climate change on European critical infrastructures: the case of the power sector. Environ Sci Pol 14(1):53–63

Schmidt TS, Schmid N, Sewerin S (2019) Policy goals, partisanship and paradigmatic change in energy policy – analyzing parliamentary discourse in Germany over 30 years. Clim Pol 19(6):771–786

Többen J (2017) Effects of energy-and climate policy in Germany: a multiregional analysis. Dissertation, University of Groningen

Tobin I, Vautard R, Balog I et al (2014) Assessing climate change impacts on European wind energy from ENSEMBLES high-resolution climate projections. Clim Chang 128(1–2):99–112

Topçu FH (2018) Düşük karbon ekonomisine geçme(me): İklim değişikliği ve enerji politikaları bağlamında bir bakış. Akdeniz İ.İ.B.F. Dergisi 18:115–154

Totschnig G, Hirner R, Müller A et al (2017) Climate change impact and resilience in the electricity sector: the example of Austria and Germany. Energy Policy 103:238–248

Tranberg B, Corradi O, Lajoie B et al (2019) Real-time carbon accounting method for the European electricity markets. Energ Strat Rev 26:100367. https://doi.org/10.1016/j.esr.2019.100367

Tuck G, Glendining MJ, Smith P et al (2006) The potential distribution of bioenergy crops in Europe under present and future climate. Biomass Bioenergy 30(3):183–197

Türkeş M, Kılıç G (2004) Avrupa Birliği'nin iklim değişikliği politikaları ve önlemleri. Çevre, Bilim ve Teknoloji, Teknik Dergi 2:35–52

UNFCCC United Nations Framework Convention on Climate Change (2019). https://unfccc.int/process-and-meetings/the-paris-agreement/the-paris-agreement. Accessed 1 July 2019

Uyanık S (2018) Uluslararası yankılarıyla enerji politikalarında bir sürdürülebilirlik deneyimi: Almanya ve yenilenebilir enerji. Electron J Soc Sci 17(68):1570–1584

Van Vliet MTH, Yearsley JR, Ludwig F et al (2012) Vulnerability of US and European electricity supply to climate change. Nat Clim Chang 2:676–681

Van Vliet MTH, Vögele S, Rübbelke D (2013) Water constraints on European power supply under climate change: impacts on electricity prices. Environ Res Lett 8:1–10. https://doi.org/10.1088/1748-9326/8/3/035010

Part V
Finally

Chapter 15
Conclusions and Future Research

Muhammad Asif

Abstract The energy and environmental landscapes of the world are experiencing rapid changes on multiple fronts. While a significant proportion of the global population continues to suffer when it comes to the broader aspects of energy security, climate change is also posing serious threats to large segments of population across the world. The primary sufferers from both the energy and environmental problems, however, are those who are at the lower end of the economic bracket. Energy, environment and economy are therefore interwoven areas – the rich and developed nations are typically well placed when it comes to energy prosperity and environmental wellbeing while it is the other way round for the poor and developing countries where many still lack access to refined forms of energy. Meaningful global efforts led by the developed and rich nations are needed to help address the energy and environmental problems in the world.

Keywords Energy · Environment · Economy · Climate change · Renewables · Sustainability

1 Introduction

Energy, environment and economy are three distinctive yet interwoven areas. The book has tried to advance the understanding on the dynamics of the energy-environment-economy triangle. The contributed chapters have been organized under three themes: (i) dynamics of energy and economy, (ii) dynamics of energy and environment, and (iii) dynamics of energy policy and climate change. Important conclusions that can be drawn from the chapters are reflected as followings.

M. Asif (✉)
School of Computing, Engineering and Built Environment, Glasgow Caledonian University, Glasgow, UK
e-mail: Mas1@gcu.ac.uk

H. Qudrat-Ullah, M. Asif (eds.), *Dynamics of Energy, Environment and Economy*, Lecture Notes in Energy 77, https://doi.org/10.1007/978-3-030-43578-3_15

2 Climate Change in Focus

The year 2020 dawns with climate abnormalities around the world. While Australia recorded all time high temperatures twice in December 2019, Russia experiences the mildest month since 1886 and the Indian subcontinent records the coldest day in over a century. Bushfires in Australia started in September 2019 continue to rage through February 2020, killing 33 people and burning 110,000 km^2, adding to the catalogue of record bushfires in other parts of the world. Extreme bushfires are now being widely tipped to be a new normal as experts predict more fires and higher degrees of devastation as each fire season comes (Horton 2020; Peischel 2020). According to the US National Aeronautics and Space Administration (NASA), 2014–2018 have been the hottest 5 years on record (NASA 2018). With the average atmospheric temperature in 2018 recorded to be 0.83 °C warmer than the 1951–1980 mean, almost 400 all-time high temperatures were reportedly set in the northern hemisphere over the 2019 summer (Stylianou and Guibourg 2019).

Global warming is being widely acknowledged as arguably the biggest challenge faced by mankind. The latest reports by the UN Intergovernmental Panel on Climate Change (IPCC) suggest that the human-induced warming of the atmosphere has reached approximately 1 °C above the pre-industrial levels in 2017, increasing at 0.2 °C per decade (IPCC 2019). Forecasts suggest that unless a serious collective effort is made at the global level, the average atmospheric temperature is expected to further rise by as much as 6 °C by the end of this century (Ghosh 2012). The rise in sea level due to the melting of glaciers is one of the most prominent implications of global warming. Estimates suggest that Antarctica is now annually losing around 160 billion tonnes of ice to the ocean, which is twice previous findings (McMillan et al. 2014). Climate change as a result of global warming is resulting into wide ranging problems such as seasonal disorder, a pattern of intense and more frequent weather related events such as floods, droughts, storms, heat waves and wild fires. It has been reported that since the advent of the twentieth century natural disasters such as floods, storms, earthquakes and bushfires have resulted in an estimated loss of nearly 8 million lives and over $7 trillion of economic loss (KIT 2016; Amos 2016). Future projections suggest that by the year 2060 more than 1 billion people around the world might be living in areas at risk of devastating flooding due to climate change (McMillan et al. 2014). Majority of the affected will be from developing countries with limited resources to mitigate the challenges and to rebuild their lives after extreme environmental events.

Through the Paris Agreement signed in 2016, 196 countries have adopted the first-ever universally legally binding global climate deal to avoid dangerous climate change by limiting global warming to well below 2 °C. A recent Intergovernmental Panel on Climate Change (IPCC) report, however, has warned that the world is seriously overshooting this target, heading instead towards a higher temperature rise. It concludes that there is need for major changes in the four big global systems: energy, land use, cities and industry. The report warns that to limit warming to 1.5 °C, world needs to invest around $2.4 trillion in renewable and energy efficiency initiatives every year through 2035 (IPCC 2018; Landberg 2018).

Climate change is truly a global issue in its very nature. It will affect all countries, though the poorest countries will suffer the earliest and the most (Stern 2007). Ironically, the poor and developing countries, despite having a marginal contribution towards the greenhouse gas emissions, are mainly at the suffering end when it comes to the implications of climate change. Low-lying island nations are particularly facing serious challenges form the rising sea level. The grave implications of global warming and climate change call for an urgent and paradigm shift in human activities and use of resources. The progress made in the world at national and international levels is falling short of what is needed to fight climate change. The major polluters and the developed industrialized countries need to share the due burden. On one hand they need to curb their emissions and on the other hand they need to support the developing countries that are suffering from climate change. The fight against climate is required to be embedded into the developmental policies across the world. There is need for greater seriousness and cooperation at the international level and more importantly words need to be matched by actions.

3 The New Energy Transition

Human dependence on energy is ever growing with all major sectors including industry, transport, household, agriculture, and trade and commerce becoming increasingly reliant on this vital commodity. Historically, biomass, mainly in the form of wood and agriculture waste, has served the mankind for thousands of years. Until the dawn of the industrial revolution in the eighteenth century, these biomass resources were sufficient to meet human energy needs, which were quite basic in nature. In the wake of the industrial revolution, as human creativity exceeded expectations, producing more effective and efficient energy technologies based on coal and then on oil were needed. Gradually, over the next two centuries, the world moved from wood and coal fuels to oil and then to natural gas, with the exception of nuclear energy being used in the developed countries. This energy transition from wood to coal to oil to natural gas has been the different phases of traditional fossil fuels.

With the advent of the twenty-first century the world faces a string of energy challenges including rapid growth in energy demand, depletion of fossil fuel reserves, volatile energy prices, geopolitical instability in and around the major fossil fuel producing countries and greenhouse gas emissions. These challenges have prompted an unprecedented energy transition, which is much more diverse and vibrant compared to the earlier one. This energy transition is manifested by four major drives: decarbonization, distributed generation, smart grid and metering, and energy conservation. These strands are being deemed central to energy policies and frameworks across the world to pursue sustainable development. European Union's 20-20-20 directive is an interesting example in this respect which aimed to increase energy efficiency by 20%, reduce CO_2 emissions by 20% and have renewables making up 20% of supplies by 2020. Smart metering and smart grids are also reported to have an effect of up to 20% increase in energy efficiency (EP 2008).

Decarbonization of the energy sector is being spearheaded by renewables. Renewable energy, with the exception of biomass, for its minimal Carbon dioxide (CO_2) emissions during the operational phase, is regarded as environmentally clean and has become an integral part of the energy developments in almost every corner of the world. E-mobility is another important dimension of the decrbonization initiative as electric cars are being regarded as the future of automobile industry with increasing number of car makers committing to cease production of traditional engines, and cities around the world targeting to ban petrol and diesel vehicles. Decentralized generation also termed as distributed generation or on-site generation is being deemed as the way forward to offer customized and local energy solutions, also to help curb transmission and distribution investments and the involved energy losses. Distributed generation can also contribute to grid reliability and load management, besides offering economic and environmental benefits. The distributed generation concept, which goes hand-in-hand with renewables, is becoming increasingly popular. USA, for example, has over 12 million distributed generation units accounting for almost one sixth of the country's centralized power generation capacity. Energy sector has been proactive in embracing digital technologies. The future of energy sector is projected to have digital technologies making energy systems intelligent, efficient, reliable and more connected. Finally, reduction in energy use through energy conservation and management is critical to address energy challenges. Energy conservation has become a foundational block of energy policies across the world as it is recognized to be the easiest and the most cost effective solution to growing energy needs. A unit of energy saved is far better than a unit generated.

4 Renewable Energy: Coming of Age

Renewable energy is widely being appreciated as a promising solution to help address the global energy and environmental problems. Renewable resources – i.e. solar energy, wind power, hydropower and biomass – are abundant and inexhaustible offering environmentally clean energy. Another major advantage renewables have in comparison to other types of energy resources is their vast distribution. Unlike fossil fuels, for example, renewable resources are widely distributed across the world. Almost every country has some form of renewable energy available within its own geographic borders thus offering a kind of energy freedom.

Over the last couple of decades renewable energy has made tremendous progress overtaking the supplies from fossil fuels and nuclear power both in terms of investment and capacity addition. The fast expansion of the renewable energy base has been mainly propelled by the success of solar energy and wind power. Factors like conducive policies, technological advancements and economy of scale have contributed towards the accomplishments of renewable energy. The global renewable power capacity at the end of 2018 stood at 2,378GW. For the fourth year in a row, increment in the installed capacity of renewables has outpaced the net capacity addition of fossil fuel and nuclear power combined. Solar photovoltaic (PV) is one of the most promising and fast growing renewable technologies, accounting for 55% of the total renewable capacity addition. With 100GW of new capacity addition in 2018, its total installed capacity stood at 505GW at the end of the year (REN 21 2019). China, Germany, Japan and the United states are leading the PV market in terms of installed capacity. In USA the distributed solar PV generation increased by 23%, while the total solar PV generation increased by 27% (USEER 2019).

The building sector, accounting for around 40% of world's total consumption is an important stakeholder in the global energy scenario. Potential renewable technologies for application in buildings include solar PV, solar water heating, wind turbines, hydroelectric systems, and ground and air source heat pumps. Some of these technologies especially wind turbines and hydroelectric systems are heavily site dependent and are typically not applicable in urban environment. Solar PV offers great diversity in terms of scale and type of application. One of the most successful applications of solar PV has been in buildings. Although the utility scale projects are constituting the larger share of the solar market in most part of the world, small scale applications are growing rapidly and are encouraged through incentives for the end users. For example, the share of installed rooftop PV is about 60% of the total PV market in Germany, with 35% installed on small to medium residential and commercial buildings (BSW-Solar 2013). Small scale solar market in India has grown dramatically with a 90% compounded annual growth rate in recent years. Australia's solar market has witnessed a significant increase in the number of small-scale solar projects as more than 1.51 million PV systems were installed by December, 2015. This represented about 16% of Australia's renewable power generation and 2.4% of the total electricity produced in the country (Dehwah et al. 2018). In the USA, 64.4% of the PV projects are utility-scale facilities while residential and non-residential systems have respective share of 20.5% and 19.1% (USEER 2019).

5 The Energy-Environment-Economy Conundrum

Energy has become an imperative commodity in our lives. The significance of energy is ever growing in the existence and advancement of modern societies. In the present age, energy is closely coupled with economy as the two do hand-in-hand – economic activities in a society require use of energy. For its vital role, availability of energy is an indicator of socio-economic prosperity in a society. The traditional use of energy is also intrinsically linked with environment – use of every unit of energy simply has an associated environmental load. Energy, environment and economy are therefore extremely interwoven factors and the energy-environment-economy triangle is quite complex and dynamic in its very architecture.

Despite all the realization about the importance of energy in the socio-economic uplift and advancement of societies, the global energy landscape continues to face numerous challenges, above all, acute energy security issues facing billions of people around the world. The nature of challenges for countries at different socio-economic and technological strata can hugely vary. While considerable segments of population in the developed countries continue to experience fuel poverty, electricity deprivation remains a challenge for many in parts of Africa and Asia. Although globally there has been a significant improvement in recent years in terms of grid penetration with the number of people lacking access to grid falling below 1 billion, the quality of grid remains an issue for many in the developing countries. Issues like poor grid quality, power outages and breakdowns, planned load shedding and brown shedding are a regular phenomenon even in major cities. Lack of access to reliable and refined energy resources can also be gauged from the fact that around 2.3 billion people cook on inefficient raw biomass. The situation is even more alarming when energy security is looked from the 4As perspective: availability, adequacy, affordability and acceptability.

Environmental spill-overs from traditional energy systems is an issue needing urgent and meaningful attention. Though there has been a global drive to shift away from coal and oil based power generation, energy sector still remains to be one of the biggest contributors to the GHG emissions. Renewable energy is a big step forward in addressing energy challenges and satisfying energy needs without damaging the environment. Different types of renewable technologies especially solar PV and wind turbines have made significant inroads in the energy mix at national and international levels. Large scale deployment of renewable energy comes with its own set of challenges – it is capital intensive as well as it requires significant changes in the existing energy markets, institutions, infrastructure, and political and cultural practices. Although the cost trends in energy markets are favoring renewables with solar and wind power already becoming economically competitive in many places. The price of PV projects for example has experienced a dramatic fall in recent years – power purchase agreements have been signed at prices as low as US Cents 2.34/kWh. Intermittency is, however, a major shortcoming of renewables like solar energy and wind power which can lead to grid integration issues with their increased penetration in the system.

6 Final Words

Different types of energy resources and their utilization have diverse range of associated environmental impacts. Both energy and environment on the other hand have strong interaction with economy. The energy-environment-economy triangle has a complex relationship between the three elements, dictated by such a wide range of technological, economic, political and socio-cultural factors. Given the significant scientific and technological advancements on the front of energy technologies, diverse socio-economic patterns, changes in regional and global geopolitical landscapes, and stark climate change threats, this triangle is fast evolving, presenting new challenges and opportunities.

Sustainable development requires energy and economic prosperity without having implications on environment. A balanced progress on all of these three facets is an extremely complicated task. There are further dimensions under these three areas too. In the case of energy, for example, there are major differences between different types of resources and technologies, such as coal, oil, gas, electricity, and renewables, in terms of policy and price influences. There is need for robust global governance on energy and climate change. Typically, governments have their political and economic comfort zones. While the energy sector lacks any universally accepted governance mechanism, the success of global governance towards climate change is also debatable as important major polluting nations have not meaningfully complied. The US disengagement from the Paris Agreement is another example in this respect. Given the dynamic and complex architecture of the energy-environment-economy triangle, any cost effective solution to the energy and environmental problems would require extensive trade-offs.

While answering some questions the book is expected to have also triggered certain queries necessitating the need for further scholarship on the intriguing energy-environment-economy triangle. Given the technological advancements and policy evolution surrounding the energy and environmental scenarios impacting the socio-economics around the world, new sets of problems as well as solutions will continue to unfold in the years ahead. The ongoing energy transition manifested by decarbonization, decentralization and digitalization will be offering interesting avenues to be explored.

References

Amos J (2016) Economic losses from natural disasters counted. BBC, 18 April 2016

BSW-Solar (2013) German Solar Industry Association (BSW-Solar), 2013, Result from the PV grid research in Germany. Available at: http://www.hollandsolar.nl. Accessed on 30 Dec 2019

Dehwah A, Asif M, Tauhidurrahman M (2018) Prospects of PV application in unregulated building rooftops in developing countries: a perspective from Saudi Arabia. Energ Buildings 171:76–87

EP (2008) Climate change and energy package 20 20 20 by 2020, European Parliament, 11 December 2008. http://www.europarl.europa.eu/sides/getDoc.do?type=IM-PRESS&reference=20081208BRI43933&secondRef=ITEM-002-EN&language=EN. Accessed on 30 Dec 2019

Ghosh P (2012) Science adviser warns climate target 'Out the window. BBC NEWS. 25 August 2012. Available online: http://www.bbc.co.uk/news/science-environment-19348194. Accessed on 30 Dec 2019

Horton O (2020) Australia Bushfires the New Norm, Radio France International, 7 January 2020. http://www.rfi.fr/en/international/20200107-australia-bushfires-new-norm-climate-change-smoke-environment-biodiversity. Accessed on 30 Dec 2019

IPCC (2018) Global warming of 1.5°C, Special report for policy makers. Intergovernmental Panel on Climate Change, Paris

IPCC (2019) Global warming of 1.5°C, IPCC special report. Intergovernmental Panel on Climate Change

KIT (2016) Natural disasters since 1900: over 8 million deaths and 7 trillion US dollars damage; Press release 058/2016. Karlsruhe, Karlsruhe Institute of Technology

Landberg R (2018) Climate crisis spurs UN call for $2.4 trillion fossil fuel shift, Bloomberg, 8 October 2018
McMillan M, Shepherd A, Sundal A, Briggs K, Muir A, Ridout A, Hogg A, Wingham D (2014) Increased ice losses from Antarctica detected by CryoSat-2. American Geophysical Union Geophys Res Lett 2014(41):3899–3905
NASA (2018) Fourth warmest year in continued warming trend, NASA, News, 6 February 2019. https://climate.nasa.gov/news/2841/2018-fourth-warmest-year-in-continued-warming-trend-according-to-nasa-noaa/. Accessed on 30 Dec 2019
Peischel W (2020) Australia's Wildfire Catastrophe Isn't the "New Normal." It's Much Worse Than That, 8 January, 2020. https://www.motherjones.com/environment/2020/01/michael-mann-australias-wildfires-podcast-climate-change/. Accessed on 30 Dec 2019
REN 21 (2019) International Renewable Energy Network, Annual Report 2019, Paris
Stylianou N, Guibourg C (2019) Hundreds of temperature records broken over summer, BBC, 9 October 2019
Stern N (2007) The Stern Review, The Economics of Climate Change, Cabinet Office, HM Treasury, UK
USEER (2019) The 2019 U.S. Energy & employment report. EFI & NASEO, Washington, DC

Index

H. Qudrat-Ullah, M. Asif (eds.), *Dynamics of Energy, Environment and Economy*, Lecture Notes in Energy 77, https://doi.org/10.1007/978-3-030-43578-3

The manufacturer's authorised representative in the EU is Springer Nature Customer Service Centre GmbH, Europaplatz 3, 69115 Heidelberg, Germany. If you have any concerns regarding our products, please contact ProductSafety@springernature.com

Printed and bound by CPI Group (UK) Ltd, Croydon, CR0 4YY
30/09/2026
02225697-0010